An Introduction to Fiber Optic Systems

**The Aksen
Associates
Series
in Electrical
and Computer
Engineering**

Principles of Applied Optics
Partha P. Banerjee / University of Alabama, Hunstville
Ting-Chung Poon / Virginia Polytechnic Institute and State University

**Object-Oriented Engineering
Building Engineering Systems
Using Smalltalk 80**
John R. Bourne / Vanderbilt University

**Discrete Event Systems:
Modeling and Performance Analysis**
Christos G. Cassandras / University of Masschusetts, Amherst

An Introduction to Fiber Optic Systems
John P. Powers / Naval Postgraduate School

Practical Software Engineering
Stephen R. Schach / Vanderbilt University

Software Engineering, second edition
Stephen R. Schach / Vanderbilt University

**Introduction to Applied Statistical
Signal Analysis**
Richard Shiavi / Vanderbilt University

High Speed Communications Networks
Pravin Varaiya and Jean Walrand / University of California, Berkeley

**Communication Networks:
A First Course**
Jean Walrand / University of California, Berkeley

Advisory Editors

Jacob A. Abraham / University of Texas at Austin
Leonard A. Gould / Massachusetts Institute of Technology
Frederic J. Mowle / Purdue University
James D. Plummer / Stanford University
Stuart C. Schwartz / Princeton University

An Introduction to Fiber Optic Systems

John P. Powers
Naval Postgraduate School

Aksen
Associates
Incorporated
Publishers

IRWIN

Homewood, IL 60430
Boston, MA 02116

Project Supervision: Business Media Resources
Electronic Compositor: Professional Book Center
Electronic Figure Rendering: Professional Book Center
Cover and Text Designer: Harold Pattek
Printer: R. R. Donnelley & Sons

Library of Congress Cataloging-in-Publication Data
Powers, John P.
    An introduction to fiber optic systems / John P. Powers
        p.   cm. — (The Aksen Associates series in electrical and
    computer engineering)
    ISBN 0-256-12996-7
    1. Fiber optics.   2. Telecommunications systems.   I. Title
II. Series.
TA1800.P69   1993
621.382'75—dc20   92-31650
                                                                   CIP

Printed in the United States of America

1 2 3 4 5 6 7 8 9 0 DOC 0 9 8 7 6 5 4 3

About
the Author

John P. Powers is a professor of electrical and computer engineering at the Naval Postgraduate School in Monterey, California. He received a BSEE from Tufts University, a MSEE from Stanford University, and a PhD from the University of California, Santa Barbara. His entire teaching career has been at the Naval Postgraduate School, beginning in 1970, where he served both as associate chair for research and as chairman of the department. His research has been in the areas of undersea fiber optic communications, acoustic imaging, scalar wave propagation, and acoustic–optic signal processing. In 1987, he was awarded the NPS Sigma XI Chapter award for outstanding research. He is a member of IEEE, the Optical Society of America, SPIE—The Optical Engineering Society, and the Acoustical Society of America. Dr. Powers has published extensively in professional journals and conference proceedings, contributed chapters in numerous books, and is the editor of *Acoustical Imaging, Volume II* (Pergamon Press, 1982).

On Bell's Photophone...

The ordinary man . . . will find a little difficulty in comprehending how sunbeams are to be used. Does Prof. Bell intend to connect Boston and Cambridge . . . with a line of sunbeams hung on telegraph posts, and, if so, what diameter are the sunbeams to be . . . ?. . . will it be necessary to insulate them against the weather . . . ?. . . . . . until (the public) sees a man going through the streets with a coil of No. 12 sunbeams on his shoulder, and suspending them from pole to pole, there will be a general feeling that there is something about Prof. Bell's photophone which places a tremendous strain on human credulity.

New York Times Editorial
30 August 1880

Useful Constants

Speed of light (vacuum)	$c = 3.00 \times 10^8$ m/s
Planck's constant	$h = 6.63 \times 10^{-34}$ joule $\cdot$ s
Boltzmann's constant	$k = 1.38 \times 10^{-23}$ joules $\cdot$ (K)$^{-1}$
Electron charge	$q = 1.60 \times 10^{-19}$ coulombs

Contents

Contents

Contents

Contents

Contents

Preface

This textbook is written for the beginning user of optical fibers in communications. My purpose is to introduce the terminology used in optical fibers, to describe the building blocks of an optical fiber system, to facilitate the initial design of optical links, and to gain entry to the research literature of optical fiber system components. As a result, the book is more pragmatic than most texts on fiber optics. Few derivations of formulas are given; the formulas are introduced to support design applications and to aid the reader's understanding of the physical phenomena being described. Detailed discussion of advanced topics currently being pursued is left for more advanced texts and the research literature.

The assumed prerequisites include

- an introductory knowledge of electromagnetic theory of waveguides (including the existence of modes, mode cutoff, the concepts of phase and group velocity, as well as reflection and refraction),

- an introduction to communications theory (amplitude modulation, pulse modulation, data rate, and bandwidth concepts), and

- fundamentals of electronics (amplifier frequency response, comparators, transistor circuits and logic gates).

A major goal of the text is to bring the reader to the point where he or she can intelligently read and use the information presented on the data sheets to incorporate the information into the system analysis. (It is suggested that the instructor provide the students with an up-to-date collection of representative manufacturer data sheets of typical fibers, sources, receivers, connectors, couplers, and other appropriate components to supplement and illustrate the material presented in the text.) Another goal of this text is provide the reader with the capability to evaluate the potential of a device for use in the synthesis of an optical link.

The book is divided into five parts. The first part is an introduction to the basic principles of fiber optics. Chapter 1 introduces the advantages and disadvantages of optical fibers and reviews their history; Chapter 2 presents an introduction to the structure of the optical fibers and their ability to guide light. Many of the terms used to describe fibers are introduced in this chapter. (It should be noted that the treatment of the electromagnetic modes in a fiber is different in this text than most other texts. A detailed mathematical description of the modes is not included——there are many lucid treatments referenced. The results of these electromagnetic analyses are described only in terms of how they impact the fiber performance in communication systems.) Chapter 3 presents a description of more properties of optical fibers, including the all-important concepts of fiber loss and fiber dispersion. In addition, a discussion of the strength of optical fibers is included, since many special-purpose applications such as tethers for remotely-operated vehicles (ROVs) depend critically on this property.

Part Two describes the other components used in fiber-optic communication links. Optical sources used in fiber optics (semiconductor lasers and light emitting diodes) are described in Chapter 4 with an emphasis on the devices' output characteristics and the factors that limit the performance as a communications device, such as modulation response times and reliability factors. Chapter 5 describes optical detectors and their performance limits and, then, goes on to a discussion of the performance of these detectors when combined with preamplifier and equalization amplifiers. Emphasis in this chapter is on the signal-to-noise performance of the receiver with the goal of estimating the power required at the detector to provide a required signal-to-noise ratio or to achieve a desired bit error rate. Chapter 6 is a discussion of the alignment requirements for connectors and splicing techniques that join fibers together and a survey of the techniques currently used to meet those objectives. This chapter also contains a discussion of optical couplers that are used for splitting and combining light in fibers.

Part Three brings the building blocks together in an optical link. Chapter 7 is the climactic chapter that integrates the building blocks introduced in earlier chapters into a method for the design of an optical link. Both power budgets and timing budgets are introduced to show that links can be either attenuation limited or dispersion limited. Dynamic range considerations, which are of special importance for shorter links, are also introduced. One of the revolutionary concepts being introduced to fiber links is the optical amplifier that eliminates the need for electronic repeaters. These in-line fiber amplifiers are also described in this chapter. Chapter 8 describes the application of optical fibers to modern networks, including the proposed standards of the Fiber Distributed Data Interface (FDDI) for computer network applications and the Synchronous Optical Network (SONET) for telecommunications applications.

Advanced concepts that are further out in the development cycle are described in Part Four. Techniques that use coherent detection for improved receiver sensitivity or increased link capacity are described in Chapter 9, along with the device requirements required to achieve these potential benefits. Chapter 10 describes the use of wavelength-division multiplexing to increase channel capacity, the technology required to implement this multiplexing, and introduces some network architectures to implement interconnections. A short description of the use of optical fibers to sense and measure field disturbances, such as acoustic or magnetic fields, is included in Chapter 11. (This use of fibers as sensors is, itself, the subject of other textbooks.)

Finally, Part Five contains a description of the fabrication of the optical fiber and the means of testing the fiber performance. The emphasis of the fabrication material in Chapter 12 is on vapor phase deposition techniques and preform technology with short descriptions of other techniques included. The intent of including this material is to provide background material to allow an appreciation of the fabrication processes and is not meant to provide detailed information. Chapter 13 introduces the techniques that are used to measure the optical performance parameters of optical fibers. Again, the intent is to provide a broad introduction to the techniques, rather than describing them in great detail. Measurement techniques are becoming standardized and the reader who is interested in more detail should consult the literature for more detailed descriptions of the measurement techniques.

Most of Chapters 1 through 7, along with selected topics from the later chapters, can be covered in a one-quarter course on optical fiber communications, offered at the senior year level. The addition of other subjects from the later chapters or topics of the instructor's choice (e.g., the electromagnetic mode analysis) leads to a one-semester course.

In all cases, references to the literature have been included for a more detailed or alternative treatment. The literature cited is heavy on tutorial papers on the subjects and offers a gateway into the research literature of fiber optics for those who seek more detailed information. Inevitably, omissions will have occurred and I apologize to those whose work has not been included.

I would like to take the opportunity to thank my thesis students who have sparked my interests in fiber optics, my research sponsors (DARPA, the Space and Naval Warfare Command, and the Naval Underwater Warfare Engineering Station) for providing the means to allow the interest to flourish, and to the students of my fiber optics course at the Naval Postgraduate School who have tested this manuscript, making wise and helpful comments. Their efforts are greatly appreciated.

I would like to thank the reviewers of the text who took the time and effort to provide complete, comprehensive reviews of the material. Their suggestions helped to improve the text. They were Dr. Monish Chatterjee of SUNY—Binghamton, Dr. Ahmed Safaai-Jazi of Virginia Polytech and State University, and Dr. Greg Sonek of the University of California, Irvine.

John P. Powers
Pacific Grove, California

Part One

Fiber Fundamentals

Chapter 1

Introduction

This text focuses on the application of optical fibers to the task of carrying information from point to point at high data rate–distance products and with high data integrity. It will omit imaging applications of fibers, even though they have historically antedated the communications application.

Figure 1.1 shows a representative fiber link. In the top line of the figure, an optical source, such as a semiconductor laser or LED, is modulated by a signal. (The modulator can be either external to the source, as shown, or the source can be modulated directly.) The modulated output light is introduced into the fiber link through a set of connectors or through a permanent fiber splice. If the link is long, the light intensity diminishes because of attenuation in the fiber, and the optical signal may need to be regenerated by a repeater, as shown in the second line of the figure. This device consists of an optical detector that converts the light into a voltage, some electronic circuitry to detect the signal, and an optical source that regenerates the detected signal. Thus strengthened, the optical signal proceeds on its way through more fiber. An alternative to repeating the signal is to optically amplify the light, as shown in the third line of the figure. The "WDM" box represents an optical wavelength-division multiplexer, which combines the signal beam and a pump laser beam (at two different wavelengths) in an optical-fiber amplifier. The fiber amplifier strengthens the signal beam using the power from the pump laser beam, without requiring optical/electronic and electronic/optical conversions. Finally, the optical signal arrives at its destination (as in the fourth line of the figure) and the information is recovered and converted back into its original format.

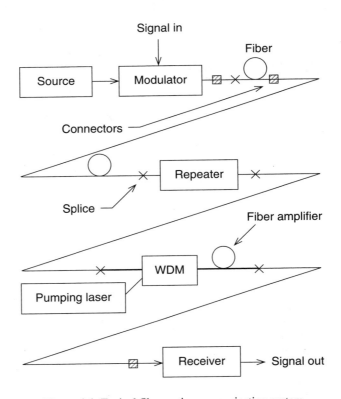

Figure 1.1 Typical fiber-optic communication system.

The design of a fiber-optic link similar to the one shown in Figure 1.1 is an interactive process. To illustrate some of the decisions that are required, we consider some of the choices that must be made. Selections in one area will impact the choices in the other area.

- **Signal to be transmitted**: Is the signal in digital or analog form? What is the signal's (analog) bandwidth or (digital) data rate? What is the dynamic range (i.e., what are the maximum and minimum signal strengths)? What is an acceptable (analog) signal-to-noise ratio at the output of the receiver? What is an acceptable (digital) bit-error rate?

- **Source:** Should one choose an LED or laser source? At what wavelength? What modulation format should be used—AM or FM for analog signals; pulse amplitude modulation (PAM) or pulse position modulation (PPM), etc. for a digital waveform? What is the source cost, reliability, output power level? How stable is the source in the face of temperature changes? A more powerful source will allow longer distances—how far can the signal go in the user's application?

■ **Detector:** Of what material should the detector be made to provide the maximum sensitivity at the wavelength of interest? At what cost? Does the user require enough sensitivity to justify the extra expense and complexity of an avalanche detector? How stable is the performance of the detector in the presence of changes of operating temperature?

■ **Fiber:** What attenuation is required to meet the design objectives? What bandwidth-distance product? Is single-mode fiber required or will multimode fiber suffice? What cabling of the fiber is required—strength members, power conductor, size, weight?

■ **Connectors and splices:** What splices and/or connectors will be required to meet the required attenuation limits? How easy is it to make a splice under operating conditions? Does the connector have to keep out water or gases?

These and other questions must be answered to accomplish a successful design that meets the link objectives. To the beginning worker in fiber-optics technology, the choices can be overwhelming. It is the purpose of this text to allow the user of fiber-optics technology to accomplish a first-order design of a link for an application, while recognizing the tradeoffs that are required to achieve the desired goal of performance.

1.1 Fiber Advantages

Among the potential advantages offered by optical-fiber communications are the following:

■ wide bandwidth,

■ light weight,

■ immunity to electromagnetic interference (EMI),

■ elimination of crosstalk,

■ elimination of sparking,

■ compatibility with modern solid-state devices,

■ lower costs than copper-based media, and

■ no radio-emission licenses required.

1.1.1 Bandwidth

Historically, bandwidth requirements for communications have shown an increasing trend. The traditional way of meeting this requirement has been to increase the carrier frequency, as the information bandwidth is constrained to be, at most, equal to the carrier (for baseband transmission) or some fraction of the carrier. Hence, to meet the increased bandwidth demands, information carriers have transitioned from HF to VHF to UHF to microwaves to millimeter waves and, finally, to light waves.

The bandwidth of a copper-based medium is tied to the losses in that medium. In coaxial-line technology, the cable losses (in decibels per length) increase linearly with the carrier frequency. Reduction of the losses at any particular frequency can be achieved by increasing the diameter of the cable, but eventually the cable becomes too large and bulky. Fiber-optic cables do not exhibit this linear increase in loss with frequency, and the losses can be made quite small by the proper fiber construction and the proper choice of operating wavelength.

In considering data rates, some benchmark data rates might be useful. The telephone industry has established *standard data rates* for various applications. Table 1.1 lists these data rates for the North American system, and Table 1.2 lists them for the Japanese and European systems.

Table 1.1 Standard United States telecommunications data rates.

Name	Data rate	Number of voice channels
DS − 0	64 kb/s	1
DS − 1 (T1)	1.544 Mb/s	24
DS − 2 (T2)	6.312 Mb/s	96
DS − 3 (T3)	44.736 Mb/s	672
DS − 4 (T4)	274.176 Mb/s	4,032

Table 1.2 Standard data rates in Japan and Europe.

Japan	Europe
1.544 Mb/s	2.048 Mb/s
6.312 Mb/s	8.448 Mb/s
32.064 Mb/s	34.364 Mb/s
97.728 Mb/s	139.264 Mb/s
396.20 Mb/s	565.148 Mb/s

Data Rate and Bandwidth

Let us consider the relationship between the *analog bandwidth* (BW) of a signal and the *data rate* (DR) required to transmit a digitized version of that signal.

Suppose that the signal extends from 0 Hz to an upper frequency of BW Hz. (For example, an audio signal extends from dc to 20,000 Hz; a video signal in the United States extends from 0 to 5 MHz.) Such a signal, before it is modulated onto a carrier wave, is called a *baseband signal*.

- *Sampling:* The *Nyquist criterion* tells us that, to faithfully reconstruct an analog signal, we must sample the wave at a rate equal to or greater than twice the highest frequency (i.e., we must sample at a rate greater than or equal to $2 \times BW$). Allowing for a sampling factor of S (where $S \geq 2$), the sampling frequency is given by $S \times BW$. Reasonable values of S range from 6 to 10.

- *Digitization:* Once sampled, each sample of the waveform must be digitized. The number of bits (N) per sample depends on the accuracy required. Eight bits allows the data to be divided into 256 (2^8) *quantization levels*. More accuracy requires more bits. Twelve bits allows the sample to be represented as one of 4,096 levels; sixteen bits would give 65,536 levels. Currently eight bits usually represents the low end of acceptable accuracy and sixteen bits represents the high end (except in cases calling for extreme accuracy).

So, we find that the data rate of the digitized signal (in units of bits per second) will be

$$DR = S \times N \times BW, \qquad (1.1)$$

where $S \times BW$ is seen to be the number of samples per second and N is the number of bits per sample. We can estimate the bandwidth B (in Hz) of a channel that carries a data rate of DR as

$$B = DR/2 . \qquad (1.2)$$

(It is important to separate in your mind the bandwidth [BW] of the information signal from the bandwidth [B] of the carrier required for the digitized version of the signal.)

Hence, we see that the data rate will always be a multiple of the bandwidth of the information signal. The size of the multiplier is $S \times N$, and the values can range from a typical low of $2 \times 8 = 16$ to a typical value of $10 \times 16 = 160$. From this multiplier factor we understand that increased accuracy in the data requires a significant increase in the data rate—the perfect justification for fiber optics.

1.1.2 Weight

Because of their small volume and lower density, optical-fiber cables enjoy considerable weight advantages over typical coaxial cables. (As a measure of the size of a reel of fiber-optic cable, a rule of thumb is that one can achieve "fifty miles per gallon," i.e., a spool containing fifty miles of fiber occupies the volume of a one-gallon gasoline can [approximately 30 cm × 15 cm × 18 cm].)

1.1.3 Immunity to Electromagnetic Interference

Since optical fibers are nonconducting, they will neither generate nor receive *electromagnetic interference* (EMI). This feature allows the use of fibers in regions of high electric fields, as with power electronics, radar feed horns and antennas, nuclear explosions, and other sources of intense electromagnetic fields. Indeed, one of the thriving applications of fiber optics is sending control signals into power stations, where switching transients could obliterate them. Telemetry links for bringing information out of a system exposed to high electromagnetic signals, such as EMP (electromagnetic pulse) or lightning-strike testing of military aircraft and missiles, also use fiber optics. Finally, optical fibers are used to telemeter information out of underground atomic-bomb test caverns, where the EMP from the blast would contaminate the data from the experiment.

1.1.4 Lack of Crosstalk

One form of EMI occurs when two conducting lines lie near enough to each other to allow the signal from one to leak into the other (called *crosstalk*) because of overlapping electromagnetic fields. Traditional solutions have included further separation of the cables or increased shielding (i.e., including a grounded wire mesh around the individual cables to short out the fields), thereby increasing the size, weight, and cost of the coaxial cable. However, the optical fields extending from an optical fiber are negligible, eliminating optical pickup between adjacent cables.

1.1.5 Lack of Sparking

For special purpose applications that require transmission of information through hazardous cargo areas (e.g., areas with explosive or flammable fuel), fibers offer the potential advantage of not sparking if there is a break in the transmission line. For example, circuitous routes are followed in electrically transmitting information from an aircraft fuselage to the wing stations, to avoid the fuel tanks in the wings. Use of fiber optics allows the line to be routed by the most direct path.

1.1.6 Compatibility with Solid-State Sources

The physical dimensions of the fiber-optic sources, detectors, and connectors, as well as the fiber itself, are compatible with modern miniaturized electronics. Most components are available in dual in-line packaging (DIP packs), making mounting on a printed circuit board extremely easy. This compact size of components is of prime importance in making this new technology acceptable to today's electronic designer.

1.1.7 Low Cost

Copper is a critical commodity on the world market; as such, it is subject to rapid upward and downward fluctuations in price. The primary ingredient of silica-based glass fibers, on the other hand, is widely available and is not a critical commodity. The price tradeoffs, comparing a fiber-optic link with alternative technology, are usually dependent on data rate, since low data rates can use more economical coaxial cable. All economic analyses show that, at some value of data rate, a fiber link becomes cost effective. The exact location of the price crossing point is sensitive to many variables, but the trend is clear—fiber-optic technology is cost effective only for wideband signals (with data rates typically in excess of tens of megabits per second). Of course, if one is using another of the special properties of fiber optics (e.g., immunity to EMI), then this benefit alone might justify their use (as frequently occurs in military applications). The requirement for wideband signals to justify an economic advantage for fiber optics works to a disadvantage in the American market, where regulation inhibits a telephone carrier from providing other services such as cable TV. While other nations combine their telephone, postal, and broadcasting services and are investigating the use of a single-fiber drop to provide broadband services, in the United States only cable television companies currently have the potential bandwidth demand to justify fiber-optic home hookups (unless new services, such as electronic-catalog shopping, are offered to telephone subscribers).

1.1.8 No Emission Licenses

Since fiber optics is a nonradiating means of information transfer, no government licenses are required to implement a link. This offers potential advantages to companies that desire a point-to-point communication link between two locations for temporary use and that wish to avoid the time-consuming licensing process. Although rights-of-way must be negotiated for the fiber-cable route, this is frequently preferable to the licensing process.

1.2 Fiber Disadvantages

Along with the advantages, of course, come some potential disadvantages. These include:

- lack of bandwidth demand,
- lack of standards, and
- radiation darkening.

1.2.1 Lack of Bandwidth Demand

Although fiber-optic systems have been demonstrated with gigabit data rates, few users have requirements for such data rates today. Since at lower data rates, fiber links cost more than conventional links, they are not yet suited for these applications. Many installed fiber links have excess capacity beyond their present usage. Historic trends predict, however, that society will require high data rates as we become more information dependent. Even at present, applications appear on the horizon that require the real-time transmission of high-resolution pictorial information, or the multiplexing of many channels into a data superhighway.

In addition, lower-cost components are being devised for low-data-rate links. The cost of low-loss fibers, plastic connectors, long-wavelength light-emitting diode sources, and other economic components is steadily reducing the cost of the lower-performance links.

Users of fiber optics find it advantageous, in some cases, to anticipate future demands for bandwidth, since installation of a fiber-optic link is typically a capital investment with an anticipated lifetime of over twenty years. It is possible in some links to upgrade performance by exchanging the sources and receivers, while keeping the same installed fiber.

1.2.2 Lack of Standards

Because fiber optics is a relatively new technology, standards are just evolving. Classically, the decision time to set standards has been difficult to establish, since one freezes technology in order to establish the standards. In a rapidly developing field such as fiber optics, committees are hesitant to freeze the technology prematurely, lest they stifle a major breakthrough. The penalty for this behavior is that the link components are in a constant state of flux, with each manufacturer touting product superiority. Until the marketplace makes a decision on the worth of the various products, the casual user is unable to make the necessary technical design decisions and consequently sits on the sidelines. Coupled with the capital investment required for point-to-point communications links, conservatism is highly rewarded.

Fiber-optic standards are being set by manufacturer trade groups, telecommunications groups (both national and international), and governmental agencies (e.g., military standards for defense systems). Currently, standards exist for fiber dimensions, some measurement techniques, and two networks (FDDI and SONET, discussed in the chapter on networks). Committees exist to standardize some connector designs and electro-optic components, but few decisions have been made.

The consequence of this lack of standards is seen, for example, in the incompatibility of connectors. Connectors of varying design are available from a variety of manufacturers, and most are incompatible with each other. Consequently, a major portion of the design process is an evaluation of connector designs for the link under consideration.

Standardization will come eventually as the technology matures and a consensus is reached regarding the best techniques.

1.2.3 Radiation Darkening

Optical glass darkens under the exposure to nuclear radiation (Sigel, 1980; Greenwell, 1991). Although the specifics of the interaction depend on the dose rate and time history of the dose, as well as the type of radiation and the material and dopants of the glass (Friebele et al., 1988; Friebele et al., 1990; Taylor et al. 1990), optical fibers are generally susceptible to interruption by nuclear radiation. While research continues to study the interaction and to identify glasses that minimize the effect (Kakuta et al., 1986; Iino and Tamura, 1988; Greenwell, 1991), most optical fibers cannot be used in a nuclear environment, such as reactor spaces or platforms exposed to nuclear effects.

1.3 History

The principle of total internal reflection was studied by Tyndall in the 1850s. Development of optical fibers occurred in the 1950s, driven by imaging applications in the medical and nondestructive testing fields. Plastic fibers were also used for lighting effects. During the late 1960s (Kao and Hockham, 1966; Werts, 1966), the communications aspects of fiber optics came into prominence when various schemes for atmospheric transmission or elaborate guided-wave schemes proved to be impractical for the transmission of light. The telecommunications industry had a rule of thumb that a guided-wave–transmission system would have to achieve transmission losses of 20 dB/km or less to be economically competitive with then-current repeater spacings. Fibers at that time were highly lossy, but it became evident that purification techniques similar to those developed for silicon in the semiconductor industry could be used to reduce those losses (which were determined by impurities in the glass). Developments in the late 1960s and early 1970s brought losses steadily lower, until current loss-levels of a few tenths of a dB/km were achieved.

Now approaching maturity, the technology has been through three distinct stages. The first stage used visible and near-IR (600–920 nm) light sources combined with fiber bundles. The fibers carried redundant optical signals to the receiver. The second stage used single, multimode fibers as the channel, with the sources still in the visible and near IR (called *short-wavelength* sources). Many current short-distance commercial installations are of this type. The third stage, widely used in long-haul communications, uses so-called *long-wavelength* sources, operating between 1300 and 1600 mm in concert with fibers of smaller diameters (called *single-mode* fibers). The data rate and distance between repeaters of this latter combination are superior to the second-stage systems, but the lower cost of the second-stage systems will ensure their continued use in applications requiring modest bandwidths and/or short transmission distances. Current (1992) efforts in the development of fiber-optic–communication systems are proceeding in two directions. The first is high–data-rate systems for long-distance or high throughput applications using long-wavelength sources and single-mode fibers. The second thrust is toward high-density, short-distance, moderate–data-rate applications, such as home data services (e.g., video on demand or database applications such as airline-ticket booking) and computer local-area networks (LANs). Such applications are focusing on reducing the cost of the components used, defining services and providers, and standardizing the technology to allow a multivendor environment, where the components and systems are treated as commodities rather than pieces of custom-designed systems.

1.4 Textbooks and Other Sources

The following lists textbooks on fiber-optic systems (listed alphabetically by author). The reader is encouraged to consult this material for alternative or more complete viewpoints.

Allard, F. C., ed., *Fiber Optics Handbook for Engineers and Scientists*. New York: McGraw-Hill, 1990.

Bendow, B. and S. Mitra, *Fiber Optics Advances in Research and Development*. New York: Plenum Press, 1979.

Chaffee, C. D., *The Rewiring of America: The Fiber Optics Revolution*. New York: Academic Press, 1988.

Cherin, A., *Introduction to Optical Fibers*. New York: McGraw-Hill, 1983.

CSELT, Staff of, *Optical Fibre Communications*. New York: McGraw-Hill, 1981.

Gagliardi, R. and S. Karp, *Optical Communications*. New York: Wiley, 1976.

Gallawa, R., *A User's Manual for Optical Waveguide Communications*. Pub. No. OTR76–83: US Dept. of Commerce, 1976.

Gloge, D., ed., *Optical Fiber Technology*. New York: IEEE Press, 1976.

Howes, M. and D. Morgan, *Optical Fibre Communications*. New York: Wiley, 1980.

Jones, W. B., Jr., ed., *Introduction to Optical Fiber Communications Systems*. New York: Holt, Rinehart and Winston, 1988.

Kao, C., ed., *Optical Fiber Technology*. New York: IEEE Press, 1981.

Kao, C., *Optical Fiber Systems*. New York: McGraw-Hill, 1982.

Keiser, G., *Optical Fiber Communications*. New York: McGraw-Hill, 1983.

Keiser, G., *Optical Fiber Communications, 2nd Edition*. New York: McGraw-Hill, 1991.

Kressel, H., ed., *Semiconductor Devices for Optical Communications*. New York: Springer, 1980.

Lin, C., ed., *Review of Optoelectronic Technology and Lightwave Communication System*. New York: Van Nostrand Reinhold, 1989.

Marcuse, D., *Principles of Optical Fiber Measurement*. New York: Academic Press, 1981.

Midwinter, J., *Optical Fibers for Transmission*. New York: Wiley, 1979.

14

Miller, S. E. and Chynoweth, A. G., eds., *Optical Fiber Telecommunications*. New York: Academic Press, 1979.

Miller, S. E. and Kaminow, I. P., eds., *Optical Fiber Telecommunications II*. New York: Academic Press, 1988.

Palais, J. C., *Fiber Optic Communications, Third Edition*. Englewood Cliffs, New Jersey: Prentice Hall, 1992.

Personick, S., *Optical Fiber Transmission Systems*. New York: Plenum Publishing, 1981.

Personick, S., "Protocols for fiber-optic local area networks," *J. Lightwave Technology*, vol. LT-3, no. 3, pp. 426–431, 1988.

Sandbank, C., ed., *Optical Fibre Communication Systems*. New York: Wiley, 1980.

Senior, J., *Optical Fiber Communications: Principles and Practice*. Englewood Cliffs, NJ: Prentice Hall, 1985.

Research Journals: Through 1983, fiber-optic research results in the United States were published primarily in the *IEEE Journal on Quantum Electronics*, *Applied Optics*, the *Bell System Technical Journal*, and other journals. Since then, US research results are typically published in the *Journal of Lightwave Technology*, published jointly by the IEEE and the Optical Society of America since 1983. Device research results are presented in *IEEE Photonics Technology Letters* and in *Electronic Letters*.

1.5 Problems

1. (a) How many US standard voice channels could a 20 Gb/s channel accommodate?

   (b) What is the bit period (T_b) of each bit in a 20 Gb/s signal?

2. (a) Calculate the data rate required to transmit a 20,000 Hz audio signal at a sampling rate of four times the Nyquist rate with a digitization of 8 bits per sample. Calculate the channel bandwidth.

   (b) ... for a 5 MHz signal?

References

Friebele, E., E. Taylor, G. Turquet, J. Wall, and C. Barnes, "Interlaboratory comparison of radiation-induced attenuation in optical fibers. Part I: Steady-state exposures," *J. Lightwave Technology*, vol. 6, no. 2, pp. 165–171, 1988.

Friebele, E., P. Lyons, J. Blackburn, A. J. H. Henschel, J. Krinsky, A. Robinson, W. Schneider, D. Smith, E. Taylor, G. Y. Turquet de Beauregard, R. West, and P. Zagarino, "Interlaboratory comparison of radiation-induced attenuation in optical fibers. Part III: Transient exposures," *J. Lightwave Technology*, vol. 8, no. 6, pp. 977–989, 1990.

Greenwell, R. A., "Reliable fiber optics for the adverse nuclear environment," *Optical Engineering*, vol. 30, no. 6, pp. 802–807, 1991.

Iino, A. and J. Tamura, "Radiation resistivity in silica optical fibers," *J. Lightwave Technology*, vol. 6, no. 2, pp. 145–149, 1988.

Kakuta, T., N. Wakayama, K. Sanada, O. Fukuda, K. Inada, T. Suematsu, and M. Yatsuhashi, "Radiation resistance characteristics of optical fibers," *J. Lightwave Technology*, vol. LT-4, no. 8, pp. 1139–1143, 1986.

Kao, K. and G. Hockham, "Dielectric-fiber surface waveguides for optical frequencies," *Proc. IEE*, vol. 133, pp. 1151–1158, 1966.

Sigel, G., Jr., "Fiber transmission losses in high radiation fields," *Proc. IEEE*, vol. 68, no. 10, pp. 1236–1240, 1980.

Taylor, E., E. J. Friebele, H. Henschel, R. H. West, J. Krinsky, and C. Barnes, "Interlaboratory comparison of radiation-induced attenuation in optical fibers. Part II: Steady-state exposures," *J. Lightwave Technology*, vol. 8, no. 6, pp. 967–976, 1990.

Werts, A., "Propagation de la lumiere coherente dans les fibres optique," *L'Onde Electrique*, vol. 46, pp. 967–980, 1966.

Chapter 2

The Optical Fiber

2.1 Introduction

In this chapter we will describe the geometry of the optical fiber and define some of the fiber parameters that are used to characterize its physical shape. We begin by describing the confinement process that makes optical fibers useful, defining several parameters, and introducing several terms that are associated with optical fibers. We will see that fibers can propagate various electromagnetic modes. Some fibers are designed to propagate a single mode; others will propagate multiple modes. Each type of fiber has its own advantages.

Gloge (1971a, 1971b, 1979), Felsen (1974), Marcuse (1974), Conradi et al. (1978), Marcuse et al. (1979), Olshansky (1979), Ramsey and Hockham (1980), Keck (1981), and Payne et al. (1982) represent earlier writings, and other descriptions can be found in most fiber-optics texts.

2.2 Optical Confinement

Optical fibers work by confining the light within a long strand of glass. The confinement process that traps the light inside the fiber and allows it to propagate down the length of the fiber is based on the principle of *total internal reflection* at the interface of two dielectric media. Consider an optical–plane-wave incident on an infinite planar interface between two dielectric media, as shown in Figure 2.1. The

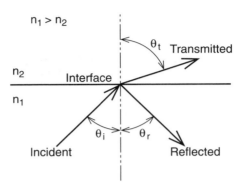

Figure 2.1 Planar interface geometry.

propagation direction of the plane wave makes an *angle of incidence* θ_i with the normal to the interface as shown in the figure. The *index of refraction* (or *refractive index*), n, of a medium is given by the equation

$$n = \frac{c}{v} , \tag{2.1}$$

where c is the velocity of light in a vacuum (3×10^8 m/sec) and v is the velocity of light in the medium. (Note that n is always greater than 1; for glass, n is between 1.4 and 1.5.) For total internal reflection to occur, the index of refraction of the medium containing the incident plane wave is required to be larger than the index of refraction of the other medium (i.e., the wave is travelling from the higher-index medium into the lower-index medium).

Snell's law governs the transmission of the plane wave through the interface and is given by

$$n_1 \sin \theta_i = n_2 \sin \theta_t , \tag{2.2}$$

where θ_i is the angle of incidence and θ_t is the angle of transmission. From this relation, one can see that θ_t reaches 90 degrees when θ_i reaches the value of

$$\theta_i = \theta_c \equiv \sin^{-1}\left(\frac{n_2}{n_1}\right), \tag{2.3}$$

where θ_c is called the *critical angle of incidence* and is given by the latter part of the equation. For angles of incidence equal to or exceeding the critical angle, the energy of the incident wave is totally reflected back into medium 1. It is this total internal reflection that allows the light to propagate with no loss. Light that is incident at an angle below the critical angle is partially transmitted and partially reflected, losing a significant fraction of the power into the transmitted beam.

Example:

(a) Assuming that n_2 is 1% smaller than n_1, find n_2 if $n_1 = 1.45$.

Solution: Since n_2 is 1% smaller than n_1, it is 99% of n_1, so we have

$$n_2 = 0.99 n_1 \qquad (2.4)$$
$$= (0.99)\,(1.45) \qquad (2.5)$$
$$= 1.435\,. \qquad (2.6)$$

(b) Find the value of the critical angle.

Solution: The critical angle is found from

$$\theta_c = \sin^{-1}\left(\frac{n_2}{n_1}\right) \qquad (2.7)$$

$$= \sin^{-1}\left(\frac{0.99 n_1}{n_1}\right) \qquad (2.8)$$

$$= \sin^{-1}(0.99) \qquad (2.9)$$

$$= 81.9°\,. \qquad (2.10)$$

A typical optical fiber is shown in Figure 2.2. The central portion, called the *core*, is cylindrically shaped and is surrounded by an annular shaped outer region called the *cladding*. The index of refraction radial variation, $n(r)$, for a *step-index fiber* is plotted in Figure 2.3. The lower index of refraction of the cladding is related to the index of the core by

$$n_2 \approx n_1\,(1 - \Delta)\,, \qquad (2.11)$$

where Δ is the *fractional change in the index of refraction*, given by

$$\Delta = \frac{n_1^2 - n_2^2}{2 n_1^2} \qquad (2.12)$$

$$\approx \frac{n_1 - n_2}{n_1}\,. \qquad (2.13)$$

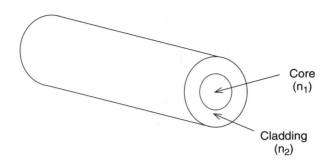

Figure 2.2 Typical step-index fiber showing the core and cladding.

Table 2.1 Representative parameters for standard fibers

Type	Core Diam. μm	Cladding Diam. μm	Δ	Application
8/125 single mode	8	125	0.1% to 0.2%	Long distance, high date rate
50/125 multimode	50	125	1% to 2%	Short distance, moderate data rate
62.5/125 multimode	62.5	125	1% to 2%	Local area networks
100/140 multimode	100	140	1% to 2%	Local area networks, short distance

(We have used the approximation for Δ [good for Δ ≪ 1] in Equation 2.11.) Typical values of Δ range between 0.001 and 0.02 (i.e., 0.1% to 2%) with a nominal core refractive index of 1.47. Typically the cladding is also surrounded by a plastic protective *jacket* for handling purposes and to protect the fiber surface from forming flaws that will weaken it.

Table 2.1 summarizes some of the parameters associated with standard fibers. (While the sizes are standardized [except for the core diameter of single-mode fiber], the other values are only representative.)

2.3 Waveguide Modes in Step-Index Fibers

The fibers just described are called *step-index* fibers due to the step discontinuity in the index of refraction at the core-cladding interface, as seen in Figure 2.3.

Such cylindrically symmetric dielectric waveguides can be modeled and solved for the electromagnetic fields that propagate in the waveguide. The solutions involve Bessel functions and will not be displayed here. (A lucid description of the electromagnetic modes in optical fibers has been published by Yeh (1987), as well as in many fiber-optic texts.)

Like other waveguides, this waveguide has certain characteristic electromagnetic modes that can propagate. The simplest modes are TE (transverse electric) and TM (transverse magnetic) modes with radial symmetry, while hybrid modes also exist and are called HE_{mn} and EH_{mn} modes. Some modes are linearly polarized and are called LP modes. Each mode not only has its own geometric electric and magnetic fields, but also will have a unique propagation constant. Much of the electromagnetic analyses of optical fibers is the description of these modes and their propagation constants. We will selectively extract the results of these analyses for our use.

A key parameter that describes the mode structure is the *V-parameter*, *V*, defined by

$$V = \frac{2\pi a}{\lambda} \sqrt{n_1^2 - n_2^2} \qquad (2.14)$$

$$\approx \frac{2\pi a}{\lambda} n_1 \sqrt{2\Delta} , \qquad (2.15)$$

where *a* is the radius of the core of the fiber, λ is the nominal free-space wavelength of the light, and n_1 and n_2 are the indices of the core and cladding, respectively. The V-parameter is important because it determines the number of electromagnetic modes in the fiber. Many other properties of the fiber (e.g., the ability to couple light into the fiber or the losses in a fiber splice) rely, in turn, on these modes.

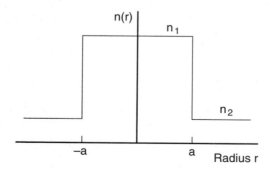

Figure 2.3 Refractive index profile of a step-index fiber.

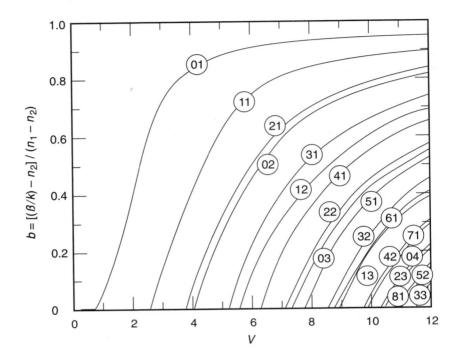

Figure 2.4 Relative modal propagation index $b = [(\beta/k) - n_2]/(n_1 - n_2)$ as function of V-parameter for a step-index fiber. (From D. Gloge, *Applied Optics*, vol. 10, no. 10, pp. 2252–2258, 1971.)

Figure 2.4 shows the relative modal index for the linearly polarized modes of step-index fibers as a function of V (Gloge, 1971b). (Similar plots can be obtained for fibers with different refractive-index profiles.) Each mode shown has two polarizations associated with it and is described as *doubly degenerate*. The *relative modal index* (shown on the vertical axis) is defined as

$$b = \frac{(\beta/k)^2 - n_2^2}{n_1^2 - n_2^2} \tag{2.16}$$

$$\approx \frac{(\beta/k) - n_2}{n_1 - n_2}, \tag{2.17}$$

where β is the mode's propagation constant and $k = 2\pi/\lambda$.

We note from Figure 2.4 that, for any given value of V, there are, in general, several modes that can propagate in a step-index fiber. The number of modes present in a fiber is dependent on the value of V (i.e., the number of modes present increases as V increases). The value of the propagation constant of any particular mode can vary from a minimum of $n_2 k$ (i.e., the propagation constant of the cladding) to a maximum of $n_1 k$ (i.e., the propagation constant of the core). The actual value of the propagation constant (and, hence, the mode's velocity) is somewhere between these extremes; each mode travels at its own velocity.

2.4 Multimode Step-Index Fibers

2.4.1 Number of Modes

For the step-index profile curve of Figure 2.4, we observe that, for any value of V larger than 2.405, more than one mode will exist. For large values of V, many electromagnetic modes can be supported by the fiber waveguide structure. An estimate of the *total number of modes*, N, supported at any given large value of V ($V \gg 2.405$) is

$$N \approx \frac{V^2}{2} \quad (\text{for } V \gg 2.405) \tag{2.18}$$

$$\approx (kan_1)^2 \, \Delta \tag{2.19}$$

$$\approx \left(\frac{2\pi an_1}{\lambda}\right)^2 \Delta \tag{2.20}$$

(Cherin, 1983). Such a fiber, operated in this region of many modes, is called a *multimode* fiber.

In a fiber with a large number of modes, we can use ray theory to describe the modes (otherwise, an electromagnetic field approach must be used, as is done with single-mode fibers). As seen in Figure 2.5, the highest-order modes correspond to the rays that are at the steepest angles (i.e., closest to the critical angle); the lowest-order modes correspond to those rays that strike the interface at low, grazing angles. This simple picture is useful for envisioning the behavior of the modes and in performing simple modeling of connectors, splices, and coupling of light from sources into fiber. (For single-mode fibers, we emphasize that the ray model of the fiber-optic modes is *not* valid. The electromagnetic wave is no longer confined primarily to the core and can have appreciable energy in the cladding. Hence, the propagation properties of the wave are a combination of the core propagation properties

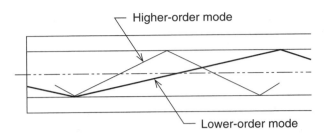

Figure 2.5 Ray diagram illustrating high-order modes and low-order modes as rays in a step-index multimode fiber.

and the cladding propagation properties. This leads us to use the mode field diameter description of the single-mode fiber, described in a later section.)

Deviations from the ideal waveguide (because of geometry fluctuations, impurities, bends, etc.) cause energy to be coupled out of the lower-order modes into the higher-order modes and from the higher-order modes into the cladding modes, where the energy is lost. (The process of transferring energy between modes is called *mode mixing*.) Using the ray approach, a localized bend in the core-cladding interface, for example, will change the angle of incidence of the ray that would have been at the critical angle in an unperturbed fiber. Because of this new angle of incidence, some of the light will be transmitted into the cladding and lost. The total amount lost depends on the severity and frequency of such perturbations.

The excitation of the modes is a function of the beam pattern of the exciting source at the input end of the fiber, the geometry of the fiber, the optical properties of the fiber, and the alignment of the source and fiber. Phenomena involving multimode fibers are usually modeled assuming that the modes are all equally excited. Measurements of multimode fiber parameters (as described in the later chapter on fiber-optic measurements) must incorporate features to ensure excitation of all of the modes in the fiber.

2.4.2 Power Distribution Between Core and Cladding

The electromagnetic fields in the fiber must meet the boundary conditions across the core-cladding interface. Unlike the metallic waveguide problem, where the fields are zero inside the conductor, the field in the cladding is not zero. Fields exist in both the core and the cladding. Therefore, some of the power of the mode is carried in the core and some in the cladding. It is instructive to consider the power distribution in the core and the cladding for the different modes in a step-index multimode fiber, as shown in Figure 2.6 (Gloge, 1971a). Using the left axis, we can estimate, for a given mode, the fraction of the total power in the cladding; with the right axis, we can find the fraction of the power in the core. We also note that, as the V parameter approaches cutoff for any particular mode, more of the power of the mode is in the cladding. At cutoff, all of the power transfers to the cladding, the mode becomes radiative, and it ceases to exist as a propagating mode. For large values of V, the modes that are close to cutoff can be ignored compared to the propagating modes, and the fraction of the total power that is found in the cladding can be estimated by

$$\frac{P_{\text{cladding}}}{P_{\text{core}}} \approx \frac{4}{3\sqrt{N}} \quad (V \text{ large}) \tag{2.21}$$

(Gloge, 1971a), where N is the total number of modes in the fiber, as given by Equation 2.18. Efforts to reduce the number of modes by reducing V usually cause more

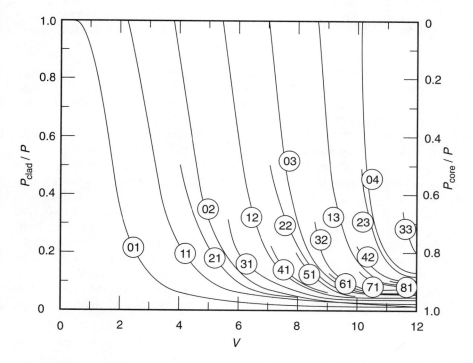

Figure 2.6 Power distribution in cladding (left axis) or core (right axis) vs. V parameter for a step-index fiber. (From D. Gloge, *Applied Optics*, vol. 10, no. 10, pp. 2252–2258, 1971.)

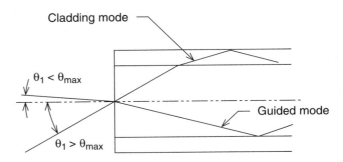

Figure 2.7 Entry conditions for a step-index fiber showing maximum acceptance angle.

of the total power to be carried in the cladding, where it is susceptible to effects from the outside environment, which can lead to undesirably high losses.

2.5 Numerical Aperture

One of the primary parameters of a fiber is its *numerical aperture*. This parameter is used in equations describing the coupling losses of the source light entering the fiber, the mode excitation, connector and splice losses, and other system performance equations. We will now use our ray theory that is applicable to multi-mode fibers to derive an expression for the numerical aperture.

Figure 2.7 shows a step-index fiber. Consider the ray, internal to the fiber, that is incident on the core-cladding interface at exactly the critical angle. If we extend that ray back through the fiber-air interface at the input by applying Snell's law, it will have an angle of incidence given by θ_{max}. Some thought should convince you that all incident rays with angles *less than* θ_{max} will have a core-cladding angle of incidence greater than the critical angle and will be guided down the (ideal) fiber. All input rays with an angle *greater than* θ_{max} will intercept the core-cladding interface at an angle less than the critical angle, will be partially transmitted through the interface, and, hence, will suffer some loss at each reflection. They will soon be attenuated into insignificance if the fiber is relatively long. (Low-loss fibers of short length [i.e., a few meters] can carry significant power to the receiver in these cladding modes.) It can be shown (see the problems at the end of chapter) that the expression for θ_{max} is given by

$$\theta_{max} = \sin^{-1}\left(\sqrt{n_1^2 - n_2^2}\right) \tag{2.22}$$

$$= \sin^{-1}\left(n_1\sqrt{2\Delta}\right). \tag{2.23}$$

Hence, this angle depends only on the indices of refraction of the core and cladding; it is independent of the diameter of the fiber. The *numerical aperture* (NA) of the fiber is the sine of this maximum input angle and is given by

$$NA = \sin \theta_{max} \tag{2.24}$$

$$= \sqrt{n_1^2 - n_2^2} \tag{2.25}$$

$$= n_1 \sqrt{2\Delta} \ . \tag{2.26}$$

(It worth noting that the numerical aperture of a fiber is also independent of the fiber-core radius.)

Example: Consider a 50/125 fiber with a core index of 1.49 and $\Delta = 1.5\%$.

(a) Calculate the maximum acceptance angle.

Solution: Using Equation 2.23,

$$\theta_{max} = \sin^{-1}(n_1 \sqrt{2\Delta}) \tag{2.27}$$

$$= \sin^{-1}(1.49\sqrt{2(0.015)}) \tag{2.28}$$

$$= 14.96° \ . \tag{2.29}$$

(b) Find the numerical aperture of the fiber.

Solution: Using Equation 2.26,

$$NA = n_1 \sqrt{2\Delta} \tag{2.30}$$

$$= 1.49\sqrt{2(0.015)} \tag{2.31}$$

$$= 0.258 \ . \tag{2.32}$$

(c) Calculate the maximum acceptance angle and NA for the same fiber, except with a 62.5/125 diameter.

Solution: Since θ_{max} and NA are independent of the fiber size, the same answers apply.

$$\theta_{max} = 14.96° \tag{2.33}$$

$$NA = 0.258 \ . \tag{2.34}$$

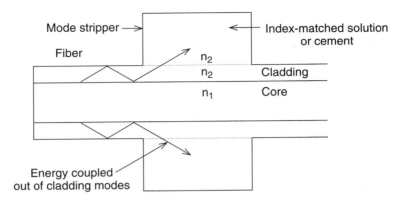

Figure 2.8 Removal of cladding power by a mode stripper.

2.5.1 Cladding Modes and Leaky Modes

As noted in Figure 2.7, certain rays in the incident radiation are not captured by the core of the fiber, but pass through the core-cladding interface into the cladding region. Because of the finite radius of curvature of the outer cladding surface, some of this light at this boundary will be reflected back into the cladding, where it can be trapped and propagated. This light forms the *cladding modes* of the fiber, and appreciable coupling can occur with the higher-order modes of the core, resulting in increased loss of the core power. Cladding modes are suppressed by placing a high-loss material outside of the cladding surface that will absorb the light as it strikes the interface, by increasing the scattering at the cladding interface to extract the cladding modes, or by surrounding a portion of the fiber with a material whose index of refraction matches that of the cladding, causing the cladding light to transmit into the index-matching material (see Figure 2.8). This latter technique is called *mode stripping*.

A third type of mode is the *leaky mode*, which is a nonpropagating mode with significant power shared between the core and the cladding. They are predicted by theory and occur near the cutoff conditions for propagating modes. These modes are attenuated in long fibers but can carry significant power in short fibers. They also play a role at connectors and splices, occurring in fibers where the connector or splice can cause conversion of energy from a propagating mode to a leaky mode.

Having considered the properties of multimode fibers, we now turn our attention to fibers that support a single electromagnetic mode.

2.6 Step-Index Single-Mode Fibers

Single-mode (or *monomode*) fibers are characterized as having only one propagating mode. In the next chapter, we will find that these fibers offer superior performance to multimode fibers for carrying high–data–rate signals.

2.6.1 Cutoff Wavelength

From the mode diagram of Figure 2.4, we observe that, for small values of V ($V < 2.405$), only one mode, called the LP_{01} mode, will propagate. Although this mode is doubly degenerate (i.e., two orthogonal polarizations of this mode can simultaneously exist), this region of operation is called *single-mode operation*. We note that, generally, one wants a small core radius (typically several wavelengths) and a small index difference (typically less than 1%) between the core and cladding to achieve this relatively low value of V. For a given fiber geometry, the value of λ that makes V equal to 2.405 is the (theoretical) *cutoff wavelength* of the fiber. (The actual cutoff wavelength is very susceptible to variations in the fiber parameters and, so, is usually a measured fiber parameter. The chapter on fiber measurements describes this measurement.)

From the power distribution curve (Figure 2.6), we note that, at $V = 2.405$, approximately 84% of the mode's power is in the core; while at $V = 1$, only 30% is in the core. At $V = 2.0$, about 75% of the power is in the core. The remaining power in the cladding is at risk to being removed by jacket losses, splices, and other loss mechanisms. For this reason, single-mode fibers keep V in the range $2.0 < V < 2.405$. Some compromise is required here. If V is made too close to 2.0, the cutoff wavelength of the fiber will be too close to the designed operating wavelength; if V is made too small, significant cladding power will be removed, raising the optical losses too much.

Example:

(a) Calculate the required Δ if a fiber with an 8 µm core and a 125 µm cladding is to be single-mode at 1300 nm. Assume that the core index is 1.46.

Solution: For single-mode operation we want V to be in the range $2.0 < V < 2.405$ as described in the text. We will choose $V = 2.1$ as an arbitrary choice that falls within the desired range.

$$V = \frac{2\pi a}{\lambda} n_1 \sqrt{2\Delta} \qquad (2.35)$$

$$2.1 = \frac{2\pi(4 \times 10^{-6})}{(1300 \times 10^{-9})}(1.46)\sqrt{2\Delta} \qquad (2.36)$$

$$\sqrt{2\Delta} \approx \frac{2.1(1300 \times 10^{-9})}{2\pi(4 \times 10^{-6})\,(1.46)} \qquad (2.37)$$

$$\sqrt{2\Delta} \approx 7.44 \times 10^{-3} \qquad (2.38)$$

$$\Delta \approx 0.277\% . \qquad (2.39)$$

Note: From this result we find that, to make single-mode fibers, tight control on the difference in refractive index is required, as well as the ability to make fibers with a small core.

(b) Calculate the cutoff wavelength for this single-mode fiber.

Solution: The cutoff wavelength λ_c is the value that will make $V = 2.405$.

$$V \approx \frac{2\pi a}{\lambda} n_1 \sqrt{2\Delta} \qquad (2.40)$$

$$2.405 \approx \left(\frac{2\pi a}{\lambda_c}\right) n_1 \sqrt{2\Delta} \qquad (2.41)$$

$$\lambda_c \approx \frac{2\pi a}{2.405} n_1 \sqrt{2\Delta} \qquad (2.42)$$

$$\approx \frac{2\pi(4 \times 10^{-6})}{2.405}(1.46)\sqrt{2(2.77 \times 10^{-3})} \qquad (2.43)$$

$$\lambda_c \approx 1136 \text{ nm} . \qquad (2.44)$$

This fiber, therefore, is no longer single-mode for wavelengths below 1136 nm.

We now want to turn our discussion to the parameters used to describe the behavior of single-mode fibers. Since the ray theory portrayal of the electromagnetic modes does not apply to single-mode fiber, we need an alternative representation. Also, since the size of the core a and the fractional index difference Δ are smaller than multimode fibers, some of the measurement techniques applied to single-mode are more difficult, due to reduced tolerances. (The techniques used to measure the general fiber parameters are described in a later chapter on optical-fiber measurements.)

2.6.2 Mode Field Diameter

In single-mode fibers we have seen that an appreciable part of the wave is contained in the cladding outside of the core. For such waves, the use of the core diameter to express the "width" of the wave is no longer possible, since the wave "width" is now larger than the core. We still want to know the "width" of the wave because this quantity is important in predicting the cabling losses, the losses due to bends, and the joining losses when the cables are connected or spliced together. (These losses are discussed in detail in the chapter on fiber properties and the chapter on splices and connectors.)

For single-mode fibers a useful measure of the field "width" has been the *mode field diameter* (usually abbreviated "MFD"). This measure has, unfortunately, been mathematically defined in several different ways, as we shall describe. If the fiber produces a Gaussian-shaped field, then the definitions all reduce to the same value; if the field is non-Gaussian (as occurs in varying degrees in dispersion-adjusted fibers) or the fiber is non-symmetric (as might intentionally be the case in a polarization-maintaining fiber), then the definitions diverge slightly. Originally, simple step-index single-mode fiber designs produced Gaussian-shaped beams; recent advances in designing dispersion-shifted fibers and dispersion-flattened fibers have produced non-Gaussian beams, resulting in revised definitions of the MFD.

Usually, circular fiber geometries are assumed. (New definitions are also required to accommodate noncircular geometries [e.g., elliptical core polarization-maintaining fibers]. Artiglia et al. (1989) suggest methods that can be applied to these nonsymmetric fibers.) We can measure or observe the near-field optical amplitude distribution (Kapron, 1990), $e(r)$ (where r is the radial coordinate), and the far-field amplitude distribution (Kapron, 1990), $E(\rho)$ where ρ is an angular variable defined below). These distributions are related by

$$E(R,\rho) = \frac{k \cos \theta}{iR} \exp (ikR) \int_0^\infty e(r)J_0(r\rho)r \, dr, \qquad (2.45)$$

where R is the observation distance from the end of the fiber and

$$\rho = k \sin \theta . \qquad (2.46)$$

Here r, φ, and θ are the spherical coordinates with their origin located at the center of the fiber end. To ensure that the far-field observation is truly in the far-field requires that the observation distance R from the end of the fiber obey $R \gg r_{max}^2/\lambda$, where r_{max} is the maximum extent of the near-field (typically estimated as being a little larger than the radius of the fiber core). Usually the $\cos \theta$ term is neglected for the small-valued range of θ expected. Then

$$E(\rho) = \int_0^\infty e(r)J_0(r\rho)r\,dr \qquad (2.47)$$

$$= \frac{1}{\sqrt{2\pi}}\,\mathcal{H}\{e(r)\}\,, \qquad (2.48)$$

where $\mathcal{H}$ is the Hankel-transform operator. This gives the far-field angular distribution $E(\rho)$ as the zero-order Hankel transform of the near-field distribution.

The light *intensity* distributions are given by $|e(r)|^2$ and $|E(\rho)|^2$. Since $e(r)$ and $E(\rho)$ are real functions (Artiglia et al., 1989), we can use $e^2(r)$ and $E^2(\rho)$ to represent the intensity distributions. The "width" of the wave, or MFD, can be defined from the measurement of the widths of these intensity patterns, either in the fiber near-field or far-field.

Several definitions of the mode field diameter have been made (Artiglia et al., 1989), depending on how the extent of the field is measured.

- MFD I: Near-field mode field diameter—The definition for the *near-field mode field diameter* (d_n) is

$$d_n = 2\sqrt{2}\,\sqrt{\int_0^\infty e^2(r)\,r^3\,dr\Big/\int_0^\infty e^2(r)\,r\,dr}\,. \qquad (2.49)$$

  This definition uses the square root of the ratio of the third moment of the near-field intensity pattern to the first moment of that pattern in its definition. This MFD is also called the *Petermann I MFD* (Petermann, 1985).

- MFD II: Far-field mode field diameter—Since the angular width of the far-field pattern (to first-order approximation of the diffraction phenomenon) is inversely proportional to the width of the near-field intensity pattern, we can also consider the *inverse rms (root means square) width* w_{ff} *of the far-field intensity pattern* (with units of inverse length) as

$$w_{ff} = \sqrt{\int_0^\infty E^2(\rho)\rho^3\,d\rho\Big/\int_0^\infty E^2(\rho)\rho\,d\rho}\,. \qquad (2.50)$$

  Note that this definition is the square root of the ratio of the third moment of the far-field intensity pattern to the first moment of that pattern.

A more convenient definition with units of length is

$$d_f = \frac{2\sqrt{2}}{w_{ff}} \tag{2.51}$$

$$= 2\sqrt{2} \ \sqrt{\int_0^\infty E^2(\rho)\rho \ d\rho \Big/ \int_0^\infty E^2(\rho)\rho^3 \ d\rho} \quad . \tag{2.52}$$

This far-field MFD is called the *far-field MFD* or the *Petermann II MFD* (Petermann, 1985). This definition of the mode field diameter is the one that has been adopted by various international standards–setting organizations.

A few observations (Artiglia et al., 1989) are in order about the near-field and far-field MFDs.

☐ The upper limit on the ρ integrals in Equation 2.52 should actually be $\rho = k$ (corresponding to $\theta = \pi/2$); we use ∞ to be able to use Hankel transforms, with the understanding that $E^2(\rho)$ is zero for $\rho \geq k$. For small values of θ, e and E are, then, a Hankel transform pair. If so, then d_n can be expressed in terms of $E(\rho)$, and d_f can be found from $e(r)$ as

$$d_n = 2\sqrt{2} \ \sqrt{\int_0^\infty |E'(\rho)|^2 \ \rho \ d\rho \Big/ \int_0^\infty E^2(\rho) \ \rho \ d\rho} \tag{2.53}$$

and

$$d_f = 2\sqrt{2} \ \sqrt{\int_0^\infty e^2(r) \ r \ dr \Big/ \int_0^\infty |e'(r)|^2 \ r \ dr} \ , \tag{2.54}$$

where the prime indicates differentiation. Hence, we are able to find the near-field MFD from a far-field measurement and the far-field MFD from a near-field measurement, if we so desire.

☐ It can be shown (Artiglia et al., 1989) that, as defined,

$$d_f \leq d_n \ . \tag{2.55}$$

☐ The near-field and far-field MFD definitions are equal if and only if $re(r)$ and $e'(r)$ are proportional to each other (Artiglia et al., 1989). This is true for a Gaussian-shaped wave. If

$$e(r) = e_g(r) = A \exp\left(-\frac{r^2}{2w^2} \right), \qquad (2.56)$$

where w is the Gaussian beam radius, then

$$d_f = d_n = 2\sqrt{2}\, w . \qquad (2.57)$$

(From this diameter, we can define $\sqrt{2}\, w$ as the *field beam radius*.)

How good is an assumption that the field in a single-mode fiber is Gaussian-shaped? A comparison of the measured near-field MFD and the far-field MFD would tell us. Artiglia et al. (1989) demonstrate the variation of d_n and d_f with wavelength for two single-mode fibers, a step-index fiber and a special fiber called a "dispersion-flattened" fiber. The two measured MFDs were found to be within 5% of each other for the conventional step-index single-mode fiber, indicating that the Gaussian assumption is pretty good for this fiber. However, the MFDs were found to vary significantly (20% to 45%) for the dispersion-shifted fiber, showing that a Gaussian assumption should not be made for this fiber.

■ MFD III: Transverse-offset mode field diameter—yet another definition of MFD is related to the transverse offset technique (Kapron, 1990) that can be used to measure the MFD (as described in more detail in the chapter on fiber parameter measurements). In this measurement, two identical single-mode fibers are butt-coupled (i.e., aligned with a tiny longitudinal separation) and laterally aligned for maximum transmission. One fiber is then laterally displaced with a micrometer and the transmitted power is recorded. As the translation increases, less of the power is collected by the receiving fiber. The MFD d_a is defined with this technique as the $1/e$ width of the transmission curve.

It can be shown (Artiglia et al., 1989) that all three definitions of mode field diameter are the same *if* the field is Gaussian (as in step-index fibers); otherwise, the alternative definition (d_a) has little utility (other than being fairly easy to measure) and has grown out of favor with the increased use of dispersion-shifted and dispersion-flattened fibers.

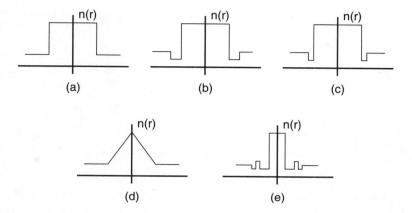

Figure 2.9 Variations of the step-index profile. (a) Regular profile. (b) Depressed cladding profile. (c) W profile. (d) Triangular profile. (e) Segmented profile.

2.6.3 Multi-Step Single-Mode Fibers

The expression for V that we have been using is valid for the simple step-index profile. The reader should be aware that other profiles are being increasingly used in single-mode fibers in an effort to improve the information-carrying capacity of the fiber. Some of these profiles are illustrated in Figure 2.9. Figure 2.9(a) is the simple step-index fiber as described earlier. Other profiles shown include the *depressed cladding*, the *W profile*, the *triangular profile*, and the *segmented profile*. Each of these profiles has its own electromagnetic mode distributions, and the value of V at cutoff will be different. These profiles are under research investigation by various manufacturers, and it is not yet obvious which ones offer superior properties over the others. These more elaborate profiles are discussed further in the section on dispersion shifting in Chapter 3.

2.7 Graded-Index Multimode Fibers

So far, we have described fibers with a step shape to their index profile $n(r)$. A second type of fiber profile does not have the abrupt jump in the index of refraction. It has a parabolic (or almost-parabolic) variation in the index, as shown in Figure 2.10. A fiber with this index profile is a *graded-index fiber*. Notice that the

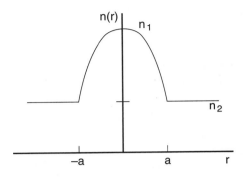

Figure 2.10 Index of refraction vs. radial position for a graded-index fiber.

index of refraction is a maximum at the center (with value n_1) and tapers off to a minimum value (n_2) at the edge of the core. The confinement process in such a fiber is a consequence of the solution of the electromagnetic propagation problem in a dielectric medium with such a refractive index profile. These waves follow sinusoidal paths, as shown in Figure 2.11. One popular mathematical model for the radial refractive index profile is the *power-law profile*, given by

$$
n(r) = \begin{cases} n_1\sqrt{1 - 2\Delta\,(r/a)^g} & \text{for } r < a \\[2mm] n_1\sqrt{1 - 2\Delta} \approx n_1(1 - \Delta) = n_2 & \text{for } r > a, \end{cases}
\tag{2.58}
$$

where n_1 is the index of refraction at the center, Δ is the fractional change in the index of refraction from the center to the edge of the core, r is the radial distance,

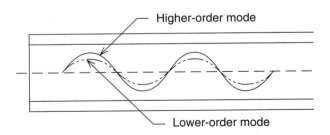

Figure 2.11 Ray path of modes in a graded-index fiber.

and a is the radius of the core. The quantity g is the *profile parameter* or the *gradient* of the graded-index fiber. Note that $g = 2$ is a parabolic profile, $g = 1$ is a triangular profile, and $g = \infty$ is a step-index profile. In these fibers the cladding ($r > a$) plays the role of a mechanical support and buffer from the outside world; it has no role in guiding the light. While a parabolic gradient is close to the optimum profile, other profiles also have been analyzed and evaluated in various laboratories.

The approximation in the above relation follows from the definition of Δ for the graded-index fiber as

$$\Delta = \frac{n_1^2 - n_2^2}{2n_1^2} \tag{2.59}$$

$$= \frac{(n_1 + n_2)\,(n_1 - n_2)}{2n_1^2} \tag{2.60}$$

$$\approx \frac{n_1 - n_2}{n_1}, \tag{2.61}$$

which is the same expression as in the step-index fiber.

A typical graded-index multimode fiber has an outside diameter of 125 or 140 μm with a fractional change of the index of refraction of 1–2% induced by varying the doping level of impurities radially.

2.7.1 Number of Modes

The V-parameter for graded-index fibers (for typically small values of Δ) is defined (Cherin, 1983) in the same way as step-index fibers,

$$V \approx n_1 k a \sqrt{2\Delta} \ . \tag{2.62}$$

The *number of modes* (N) in a multi-mode graded-index fiber can be estimated (Cherin, 1983) by the approximation

$$N \approx \left(\frac{4a^2 \pi^2 n_1^2 \Delta}{\lambda^2} \right) \left(\frac{g}{g + 2} \right) \tag{2.63}$$

$$\approx \left(\frac{g}{g + 2} \right) \left(\frac{V^2}{2} \right). \tag{2.64}$$

Example: Consider a 50/125 graded-index fiber with $g = 2$. If $n_1 = 1.48$ and $n_2 = 1.46$,

(a) Calculate Δ from the exact expression of Equation 2.59.

Solution:

$$\Delta = \frac{n_1^2 - n_2^2}{2n_1^2}$$

(2.65)

$$= \frac{(1.48)^2 - (1.46)^2}{2(1.48)^2}$$

(2.66)

$$= 1.342\% .$$

(2.67)

(b) Calculate Δ from the approximation of Equation 2.61.

Solution:

$$\Delta \approx \frac{n_1 - n_2}{n_1}$$

(2.68)

$$\approx \frac{1.48 - 1.46}{1.48}$$

(2.69)

$$\approx 1.351\% .$$

(2.70)

Hence we see that the approximation is quite good.

(c) Calculate the number of modes in the graded-index fiber at an operating wavelength of 850 nm.

Solution: Using Equation 2.63,

$$N \approx \frac{4a^2 \Delta \pi^2 n_1^2}{\lambda^2} \frac{g}{g+2}$$

(2.71)

$$\approx \left(\frac{4(25 \times 10^{-6})^2 (0.0135)\pi^2 (1.48)^2}{(850 \times 10^{-9})^2} \right)\left(\frac{2}{2+2} \right)$$

(2.72)

$$\approx 504 \text{ modes} .$$

(2.73)

2.7.2 Numerical Aperture

The *numerical aperture of a graded-index fiber* is more difficult to define than a step-index fiber, since, unlike the step-index fiber, the maximum acceptance angle θ_{max} of a ray is a function of the radial position of the entry location. A local NA can be defined by

$$NA(r) = \begin{cases} NA(0)\sqrt{1 - (r/a)^g} & r < a \\ 0 & r > a , \end{cases} \qquad (2.74)$$

where $NA(r)$ is the *local numerical aperture* and $NA(0)$ is the numerical aperture at the center of the fiber core (as computed by Equation 2.25 or 2.26). This equation predicts that the local NA decreases from the on-axis value as one moves away from the center of the fiber.

2.7.3 Single-Mode Graded-Index Fibers

It is possible to make single-mode graded-index fibers. The cutoff value of V for single-mode operation in a fiber with a graded refractive-index profile that follows Equation 2.58 is different from the 2.405 value used for a step-index fiber. For a parabolic-index fiber ($g = 2$), the cutoff value is 3.53; for a triangular-index profile ($g = 1$), it is 4.38 (Ainslie and Day, 1986). An estimate of the cutoff value of V for single-mode propagation in a graded-index fiber can be shown to be (Marcuse et al., 1979)

$$V \approx 2.405 \sqrt{1 + (2/g)} . \qquad (2.75)$$

If other parameters are the same, the core diameter of the graded-index single-mode fiber can be a factor of $\sqrt{1 + (2/g)}$ larger than the equivalent step-index fiber. This provides an increased ease of coupling light, an increased ease in splicing, and a reduction in the susceptibility of the fiber to microbend-induced losses (discussed in Chapter 3). A parabolic-index fiber ($g = 2$) provides an improvement by a factor of $\sqrt{2}$; the triangular-profile fiber ($g = 1$) provides a $\sqrt{3}$ improvement.

2.8 Summary

In this chapter we have introduced the fiber and some of the parameters used to describe the fiber's performance. The basic optical fiber consists of a core and a cladding; more advanced designs can add more cladding layers. The refractive index profile is generally modeled as either a step profile or a power-law profile described by Equation 2.58. We have seen that a fiber can be designed to be either single-mode or multimode. In a multimode fiber, the fiber V-parameter, diameter $2a$, and the fiber numerical aperture are three of the prime fiber parameters. In a single-mode fiber, the mode field diameter and cutoff wavelength describe its characteristics.

Multimode fibers (especially graded-index fibers) are used for applications calling for moderate distances or data rates or both. The larger core size makes coupling light into them relatively easy. Their primary disadvantage is a lack of bandwidth capacity (compared to single-mode fibers). Also, because of the mode mixing effects described, it is difficult to model and predict the loss mechanisms in the fiber and at connections and splices.

Single-mode fibers are currently the fiber of choice for applications combining long-distance and high data-rate. Their advantages in high data-capacity and low attenuation have overcome the disadvantages of more difficult fabrication tolerances and more difficult coupling of light from the source. Currently, single-mode fibers are price-competitive with most multimode fibers. Single-mode fibers are more susceptible to increases in losses (excess losses) due to bends in the fiber as it turns corners and to losses induced by core-diameter fluctuations because of spooling and handling.

Both multimode and single-mode fibers are readily available commercially. The choice of fiber type is dictated by the system design tradeoffs described in Chapter 7.

2.9 Problems

1. If $n = 1.5$, calculate the velocity of light in glass.
2. (a) Find n_2 if $\Delta = 1\%$ and $n_1 = 1.48$.
   (b) ... if $\Delta = 2\%$?
   (c) ... if $\Delta = 0.1\%$?
3. (a) Light traveling in air strikes a glass plate with an angle of incidence of 57 degrees. If the reflected and refracted beams make an angle of 90 degrees with each other, calculate the refractive index of the glass.

(b) What is the critical angle for this material if the light travels from glass into air?

4. A point source of light is located 1 m below a water-air interface. Find the radius of the light circle seen by an observer positioned over the source outside of the water. The refractive index of water is 1.333.

5. (a) Consider a fiber with a 100 μm core diameter and a 140 μm cladding diameter. If $n_1 = 1.48$ and $\Delta = 1\%$, calculate the V-parameter if the operating wavelength is 850 nm.

   (b) ... if the wavelength is 1300 nm?

   (c) Find the value of V at a wavelength of 850 nm if the diameter of the core is 50 μm?

6. (a) Calculate the number of modes for each case described in the previous problem.

   (b) Calculate the percentage of the optical power that is carried in the cladding for each case described in the prior problem.

7. (a) Calculate the numerical aperture of a step-index fiber having a core index of 1.47 and a cladding index of 1.45.

   (b) Find the value of the largest angle made by a ray that is accepted by the fiber if the outer medium is air.

   (c) ... if the outer medium is water? ($n = 1.33$)

8. (a) Determine the mode parameter V at 820 nm for a step-index fiber with a 50 μm core diameter, $n_1 = 1.47$, and $n_2 = 1.45$.

   (b) How many modes will propagate in this fiber at 820 nm?

   (c) ... at 1300 mm?

   (d) What percentage of the optical power is flowing in the cladding for each operating wavelength?

9. (a) Find the core diameter required to ensure single-mode operation of a step-index fiber with $n_1 = 1.485$ and $n_2 = 1.480$ at a wavelength of 820 nm. At 1300 nm?

   (b) What are the NA and θ_{max} for this fiber?

10. (a) Design a single-mode step-index fiber for operation at 1300 mm with a fused silica core ($n_1 = 1.458$). Find n_2 and the diameter of the core.

    (b) Is the fiber still single-mode at 820 nm? If not, how many modes are there?

    (c) Calculate the cutoff wavelength λ_c.

11. (a) Consider a step-index fiber with a core diameter of 50 μm, a core index of 1.450, and a fractional index difference of 1.3%. Find the values of the V parameter and NA of the fiber if the operating wavelength is 820 nm.

(b) Does the number of modes in the fiber increase or decrease if n_1 increases?

(c) ... if λ increases?

12. A step-index fiber has $n_{core} = 1.450$, $n_{cladding} = 1.440$, and will operate at 820 nm.

(a) Find the diameter of the core that will ensure single-mode operation.

(b) ... if the core diameter is 50 μm, how many modes will the fiber have?

(c) Calculate the numerical aperture of the fibers in parts (a) and (b).

13. Calculate V and the numerical aperture of a step-index multimode fiber if $n_1 = 1.450$, $\Delta = 1.3\%$, $\lambda_0 = 0.82$ μm, and $a = 25$ μm.

14. Design a single-mode fiber (with $V = 2.3$) for operation at 1300 mm with a fused silica core ($n_1 = 1.458$). The numerical aperture of the fiber is to be 0.10.

(a) Find the cladding index n_2 *and* the radius a of the fiber.

(b) Calculate the approximate number of modes in your fiber for operation at 820 nm.

15. Using a computer, plot the power-law refractive index profile from n_1 to n_2 vs. radial position for $g = 1,2,4,8$ and ∞. Assume a core diameter of 50 μm, $n_1 = 1.480$, and $\Delta = 1.00\%$.

16. (a) Calculate the number of modes in a 50/125 graded-index fiber having a parabolic index (i.e., $g = 2.0$), $n_1 = 1.485$. and $n_2 = 1.460$ at an operating wavelength of 820 nm.

(b) ... at a wavelength of 1300 mm?

(c) Calculate the number of modes in an equivalent step-index fiber at both wavelengths.

17. Prove Equation 2.22 by applying Snell's law at the fiber input face for the ray that meets the critical angle condition at the core-cladding interface. For generality, assume that the medium outside the fiber has an index n_0 and then check the equation for $n_0 = 1$.

References

Ainslie, B. J. and C. R. Day, "A review of single-mode fibers with modified dispersion characteristics," *J. Lightwave Technology*, vol. LT-4, no. 8, pp. 967–979, 1986.

Artiglia, M., G. Coppa, P. DiVita, M. Potenza, and A. Sharma, "Mode field diameter measurements in single-mode optical fibers," *J. Lightwave Technology*, vol. 7, no. 8, pp. 1139–1152, 1989.

Cherin, A., *Introduction to Optical Fibers*. New York: McGraw-Hill, 1983.

Conradi, J., F. Kapron, and J. Dyment, "Fiber optical transmission between 0.8 and 1.4 micrometers," *IEEE Trans. on Electron Devices*, vol. ED-25, pp. 180–193, 1978.

Felsen, L., "Rays and modes in optical fibers," *Electronics Letters*, vol. 10, pp. 95–96, 1974.

Gloge, D., "Weakly guiding fibers," *Applied Optics*, vol. 10, no. 10, pp. 2252–2258, 1971a.

Gloge, D., "Dispersion in weakly guiding fibers," *Applied Optics*, vol. 10, no. 9, pp. 2442–2445, 1971b.

Gloge, D., "The optical fiber as a transmission medium," *Rep. Prog. Phys.*, vol. 42, pp. 1777–1824, 1979.

Kapron, F. P., "Fiber-optic test methods," in *Fiber Optics Handbook for Engineers and Scientists*, (F. C. Allard, ed.), pp. 4.1–4.54, New York: McGraw-Hill, 1990.

Keck, D. B., "Optical fiber waveguides," in *Fundamentals of Optical Fiber Communications*, (M. F. Barnoski, ed.), pp. 1–107, New York: Academic Press, 1981.

Marcuse, D., "Theory of dielectric optical waveguides," in *Theory of Optical Waveguides*, New York: Academic Press, 1974.

Marcuse, D., D. Gloge, and E. A. Marcatili, "Guiding properties of fibers," in *Optical Fiber Telecommunications*, (S. E. Miller and A. G. Chynoweth, eds.), pp. 37–100, New York: Academic Press, 1979.

Olshansky, R., "Propagation in glass optical waveguides," *Review of Modern Physics*, vol. 51, pp. 341–367, 1979.

44

Payne, D., A. Barlow, and J. R. Hansen, "Development of low and high birefringence optical fibres," *IEEE J. on Quantum Electronics*, vol. QE-18, no. 4, pp. 477–487, 1982.

Petermann, K., "Constraints for fundamental spot size for single-mode fibers," *Electronics Letters*, vol. 20, no. 3, pp. 628–634, 1985.

Ramsay, M. and G. Hockham, "Propagation in optical fibre waveguides," in *Optical Fibre Communications System*, (C. Sandbank, ed.), pp. 25–41, New York: Wiley, 1980.

Yeh, C., "Guided-wave modes in cylindrical optical fibers," *IEEE Trans. on Education*, vol. E-30, no. 1, pp. 43–51, 1987.

Chapter 3

Fiber Properties

3.1 Introduction

In this chapter, we continue our discussion of the important optical fiber properties by considering optical attenuation, fiber dispersion, and fiber strength.

As the signal propagates through the fiber, the optical power loss will eventually attenuate the signal until it is lost in the noise at the receiver. We want to understand the mechanisms that cause this loss and to show that there is a minimum loss in silica optical fibers at an operating wavelength of 1550 nm.

A narrow pulse that originates at the optical source will spread out as it propagates down the fiber. This spread in the pulse width (called pulse dispersion) can limit the data rate, since we want to avoid overlapping pulses at the receiver. Multimode fibers are especially susceptible to this pulse spread; single-mode fibers have inherent advantages over multimode fibers in reducing the dispersion. In fact, we will find that a dispersion-free wavelength that eliminates the total dispersion in the fiber can be found near 1300 nm.

Optical fibers are not as strong as theory predicts. The strength of a fiber depends on flaws that are incorporated on the surface of the fiber during manufacture and on flaws that are induced in the surface by coating defects which contact the fiber surface. A statistical description of fiber strength is given, along with the proof-test method of screening optical fibers for those applications where fiber strength is especially important.

3.2 Fiber Losses

The power in the optical signal in a fiber decreases exponentially with distance. We can write

$$P(z) = P(0)e^{-\alpha_p z},$$ (3.1)

where $P(z)$ is the power at a position z from the origin, $P(0)$ is the power in the fiber at the origin, and α_p is the fiber *attenuation coefficient* (in units of m^{-1}). The attenuation coefficient α_p is a function of several variables to be described in this section.

To avoid having to compute exponentials repeatedly, the optical losses of a fiber are given by an *attenuation factor* α, usually expressed in *decibels per kilometer (dB/km)*. The expression for α is

$$\alpha = -\frac{10}{z} \log\left(\frac{P(z)}{P(0)}\right).$$ (3.2)

For a given fiber, these losses are wavelength-dependent, and either a spectral distribution should be given (as in Figure 3.1) or the losses must be specified at the operating wavelength. The value of the attenuation factor depends greatly on the fiber material and the manufacturing tolerances, but the general spectral shape of the loss curve shown in Figure 3.1 is representative. In particular, it is important to note that there is an optimum operating wavelength (1550 mm for silica fibers) that reduces fiber loss to a minimum.

Fiber losses are due to several effects; among the more important are

- material absorptions,
- impurity absorptions (particularly metallic ions and the hydroxyl ion from water vapor),
- scattering effects,
- interface inhomogeneities (both impurities and geometry imperfections), and
- radiation from bends.

Example: An optical fiber has losses of 0.6 dB/km at 1300 nm. If 100 μW of power is injected into the fiber at the transmitter, how much will the power be at a distance of 22 km down the fiber?

Solution: We will use a dB method of calculating the output power. We begin by choosing an arbitrary power reference; useful references in fiber

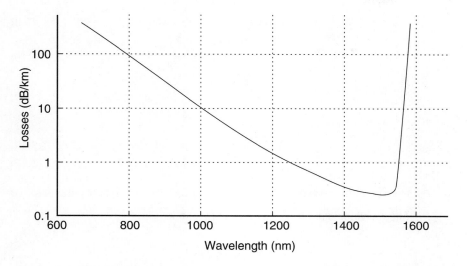

Figure 3.1 Spectral distribution of losses for a typical multimode silica fiber.

optics are 1 mW and 1 μW. We calculate the input power relative to 1 mW; the result will be expressed as *dBm* (i.e., in dB relative to 1 mW).

$$P(\text{dBm}) = 10 \log\left(\frac{P_{\text{in}}(\text{watts})}{1 \times 10^{-3}}\right). \tag{3.3}$$

(An alternative is to calculate the power in dB relative to 1 μW; the result is expressed as dBμ. The two results can be shown to related by the expression, $P(\text{dBμ}) = P(\text{dBm}) + 30$.)

$$P(\text{dBm}) = 10 \log\left(\frac{P_{\text{in}}(\text{watts})}{1 \times 10^{-3}}\right) \tag{3.4}$$

$$= 10 \log\left(\frac{100 \times 10^{-6}}{1 \times 10^{-3}}\right) \tag{3.5}$$

$$= 10 \log\left(10^{-1}\right) \tag{3.6}$$

$$= -10.0 \text{ dBm}. \tag{3.7}$$

The output power is reduced by 0.6 dB/km times the distance of 22 km (= 0.6 × 22 = 13.2 dB). Subtracting the losses, we have

$$P_{out}(dBm) = P_{in}(dBm) - losses(dB) \qquad (3.8)$$
$$= -10 - 13.2 \qquad (3.9)$$
$$= -23.2 \text{ dBm} . \qquad (3.10)$$

Finding the power, we compute

$$P_{out} = 10^{-\frac{23.2}{10}} \qquad (3.11)$$
$$= 4.78 \times 10^{-3} \text{ mW} \qquad (3.12)$$
$$= 4.78 \text{ } \mu W . \qquad (3.13)$$

Note that the result of a conversion from dBm has the units of milliwatts.

Alternative solution I: Working this problem in dBμ,

$$P_{in}(dB\mu) = 10 \log \left(\frac{P_{in}(watts)}{1 \times 10^{-6}} \right) \qquad (3.14)$$

$$= 10 \log \left(\frac{100 \times 10^{-6}}{1 \times 10^{-6}} \right) \qquad (3.15)$$
$$= 20 \text{ dB}\mu . \qquad (3.16)$$

$$P_{out}(dB\mu) = P_{in}(dB\mu) - losses(dB) \qquad (3.17)$$
$$= +20 - 13.2 \qquad (3.18)$$
$$= 6.8 \text{ dB}\mu . \qquad (3.19)$$

Finding the power we have

$$P_{out} = 10^{\frac{6.8}{10}} \qquad (3.20)$$
$$= 4.78 \text{ } \mu W . \qquad (3.21)$$

Note that the result of a conversion from dBμ has the units of microwatts. Also you should note that we subtract the same number of dB for the losses in both cases, regardless of our choice of reference power.

Alternative solution II: The −13.2 dB loss in the fiber can be expressed as an equivalent transmission factor of

$$T = 10^{-\frac{13.2}{10}} \qquad (3.22)$$
$$= 47.9 \times 10^{-3} . \qquad (3.23)$$

The output power, then, is

$$P_{out}(W) = T P_{in}(W) \qquad (3.24)$$
$$= (47.9 \times 10^{-3}) (100 \times 10^{-6}) \qquad (3.25)$$
$$= 47.9 \times 10^{-7} \text{ W} \qquad (3.26)$$
$$= 4.79 \text{ } \mu\text{W} . \qquad (3.27)$$

3.2.1 Material Absorptions

The *material absorptions* are those due to the molecules of the basic fiber material, either glass or plastic. These losses represent a minimum attainable loss and can be overcome only by changing the fiber material. Indeed, the search continues for materials with ultra-low losses with several candidates (especially halide glasses and metal halides) being identified with extremely low losses in the middle-infrared region from 2 μm to 4 μm.

The primary sources of *material impurities* in a glass fiber are metallic ions (primarily iron, cobalt, copper, and chromium) and the OH^- ion from water. Losses due to the metallic ions can be reduced to contributions below 1 dB/km by refining the glass mixture to an impurity level below 1 part per billion. The effects of water vapor in the glass are primarily located at wavelengths of 2.7 μm (a fundamental absorption) and of 950 nm and 725 nm (characteristic of second and third overtones of the fundamental). Peaks of attenuation located at these wavelengths are sometimes observed in the loss spectral distribution curve, similar to Figure 3.1. Minimization of this loss also requires hydroxyl-ion concentrations of 1 part per billion (which can be achieved by drying the glass in chlorine gas to leach out the water vapor).

Hydrogen Effects on Loss

An increased loss can occur when the fiber is exposed to hydrogen gas (which can be produced by corrosion of steel-cable strength members or by certain bacteria (Schick et al., 1991)). The hydrogen either can interact with the glass to produce hydroxyl ions and their losses or it can infiltrate the fiber and produce its own losses near 1.2 μm to 1.6 μm (Lemaire, 1991). The solution is to eliminate the hydrogen-producing sources or to add a coating to the fiber that is impermeable to hydrogen.

Ultra-Lowloss Fiber

Research is being conducted to fabricate fibers from materials with lower intrinsic losses than the silica that is used in current fibers (Tran et al., 1984; Bendow et al., 1985; Sakaguchi and Takahashi, 1987; Nagel, 1988). Most of these predicted loss minima occur in the mid-IR portion (3–6 μm) of the optical spectrum. Of course, low loss does not, by itself, mean that a material is suitable for a fiber; other properties, such as mechanical strength, ability to fabricate the fiber, and the index of refraction and its control, help to determine candidates for ultra-lowloss fibers.

Lowloss IR materials have already been studied, in general, for possible use in military systems and in mid-IR laser systems. They can be used to make optical windows, IR missile domes, and IR lenses, as well as optical fibers. Present candidates for materials suitable for ultra-lowloss fibers include the following:

- ■ *Glasses*
    - □ *Halide glasses* are available in single-component glasses (e.g., $BeFl_2$ and $ZnCl_2$) or multicomponent halide glasses. These materials offer a potential minimum loss on the order of 0.001 dB/km at 3.44 μm (compared with silica's minimum loss of 0.16 dB/km at 1.6 μm). The $BeFl_2$ glass has been available for over 50 years (Tran et al., 1984) and is as easy to work with as glass. Unfortunately, beryllium is a toxic material and the glass is hygroscopic, limiting its use. Much of the current work (Tran et al., 1984; Bendow et al., 1985; Sakaguchi and Takahashi, 1987) is being done on multicomponent heavy-metal fluoride glasses. A typical mixture (Tran et al., 1984) in the core of such a fiber might be ZrF_4 (60 mole %), BaF_2 (19 mole %), LaF_3 (6 mole %), and NaF (15 mole %). The cladding might be ZrF_4 (57 mole %), BaF_2 (12 mole %), LaF_3 (6 mole %), and NaF (25 mole %). Many other mixtures are also possible. Halide glasses have shown the most promise in the development of fiber prototypes (Sakaguchi and Takahashi, 1987).
    - □ *Chalcogenide glasses* are also available in single-component glasses (e.g., As_2S_3, GeS_2) or multicomponent glasses (with mixtures of As, Ge, P, S, Se, and Te). These glasses have a high index of refraction (between 2 and 2.5) and have predicted minimum losses of 0.01 dB at 4.54 μm. However, the halide glasses offer more promise, so not much work is being done on these materials as optical fibers.
    - □ *Heavy-metal oxide glasses,* such as GeO_2, Sb_2O_3, TeO_2, and La_2O_3, have a theoretical minimum loss of about 0.02 dB/km. Only GeO_2 has been studied as a fiber material; these materials have poor mechanical properties and a high index of refraction.

■ *Polycrystalline materials* can produce polycrystalline fibers. However, it has been difficult to process long lengths of fiber in polycrystalline form, with only extrusion showing any promise (Bendow et al., 1985). The losses of the short fibers that have been made were two to three orders of magnitude above the predicted minimum, and the fibers are mechanically brittle and easy to break. Among the kinds of materials are the following:

☐ *alkali halides,* such as AgCl, AgBr, TlBrI, TlBr, and an IR material known as KRS-5,

☐ *alkaline earth fluorides*, and

☐ some *oxides*, such as MgO and Al_2O_3.

■ *Single-crystal materials* such as AgBr, CsI, and KCl offer a theoretical minimum loss of 0.001 dB/km, but have proved hard to fabricate into fiber form (i.e., it is difficult to grow a crystal that is over 1 km long!).

Whether any of these materials are amenable to the growing of fibers and have the mechanical properties of glass fibers remains to be seen, but research is progressing rapidly.

3.2.2 Scattering Losses

Scattering losses occur when a wave interacts with a particle in a way that removes energy from the directional propagating wave and transfers it to other directions.

Linear Scattering

For *linear scattering*, the amount of light power that is transferred from a wave is proportional to the power in the wave. It is characterized by having no change in frequency in the scattered wave (unlike nonlinear scattering, where the frequency is changed upon scattering).

■ *Rayleigh scattering* results from light interacting with inhomogeneities in the medium that are much smaller than the wavelength of the light. Such inhomogeneities can be minute changes in the refractive index of the glass at some locations, caused by changes in the composition of the glass (controllable with improving quality control) or by fluctuations in the glass density (fundamental and not improvable). This scattering strength is proportional to $1/\lambda^4$.

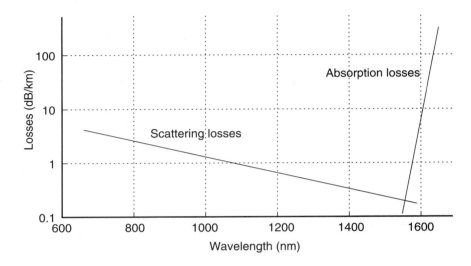

Figure 3.2 Attenuation valley between scattering and absorption in silica fibers.

Because the fundamental limits for a silica-based glass are the Rayleigh scattering at short wavelengths and the material absorption properties of silica at long wavelengths, a theoretical *attenuation minimum* for silica fibers can be predicted at a wavelength of 1550 mm, where the two curves cross, as shown in Figure 3.2. This predicted minimum of attenuation is one reason for the implementation of *long-wavelength* sources and receivers that work in this portion of the spectrum. Attenuations below 0.2 dB/km have been measured at these wavelengths.

■ *Mie scattering* occurs at inhomogeneities that are comparable in size to a wavelength. They can be core-cladding refractive index variations over the length of the fiber, impurities at the core-cladding interface, strains or bubbles in the fiber, or diameter fluctuations. This scattering can have a large angular dependence (especially when the size of the scatterer is bigger than $\lambda/10$). This scattering can be reduced by carefully removing imperfections from the glass material, carefully controlling the quality and cleanliness of the manufacturing process, and keeping Δ relatively large.

Nonlinear Scattering

High values of electric field within the fiber (associated with fairly modest amounts of optical power) leads to the presence of nonlinear scattering interactions

(Stolen, 1979; 1980; Agrawal, 1989; Chraplyvy, 1990). Such scattering causes significant power to be scattered in the forward, backward, or sideward directions, depending on the nature of the interaction. The scattering is accompanied by a frequency shift of the scattered light.

The major nonlinear scattering mechanisms when only one optical wave is present are Brillouin scattering and Raman scattering. At normal power densities (i.e., the power per unit area), the interactions are negligible; at high power densities, such as might be encountered with more powerful diode lasers or smaller fiber cores, significant power can be removed from the light wave.

- *Brillouin scattering* is modeled as a modulation of the light by the thermal energy in the material. The incident photon of light undergoes the nonlinear interaction to produce vibrational energy (or "phonons") in the glass as well as the scattered light (as "photons"). The scattered light is found to be frequency modulated by the thermal energy, and both upward and downward frequency displacement is observed. The size of the frequency shift and the strength of the scattering vary as functions of the scattering angle, with the maximum occurring in the backward direction and the minimum of zero being observed in the forward direction. Hence, Brillouin scattering is mainly in the backward direction, which directs the power toward the source and away from the receiver, thus reducing the power at the receiver. The optical power level at which Brillouin scattering becomes significant in a single-mode fiber is given by the empirical formula (Stolen, 1979; Stolen, 1980),

$$P_B = (17.6 \times 10^{-3})\, a^2 \lambda^2 \alpha\, \Delta v \,. \tag{3.28}$$

  where
  - $\square$ P_B is the power level (in watts) required for the onset of Brillouin scattering,
  - $\square$ a is the radius of the fiber (in µm),
  - $\square$ λ is the wavelength of the source (in µm),
  - $\square$ α is the fiber loss in dB/km, and
  - $\square$ Δv is the linewidth of the source (in GHz).

  Power levels above this threshold will lose significant amounts of the signal power to the scattering.

- In *Raman scattering*, the nonlinear interaction produces a high-frequency phonon instead of the low-frequency phonon and the scattered photon of Brillouin scattering. The scattering is predominately in the *forward* direction, hence the power is not lost to the receiver. The power level threshold for significant Raman scattering to occur is given by (Stolen, 1979; Stolen, 1980)

$$P_R = (23.6 \times 10^{-2})a^2\lambda\alpha \quad \text{(in watts)}, \tag{3.29}$$

where a, λ, and α are the quantities defined above, with the same units noted.

Example: Consider an 8/125 single-mode fiber operating at 1300 nm with a loss of 0.8 dB/km. The linewidth of the source is 0.013 nm.

(a) Calculate the threshold power level for Brillouin scattering in the fiber.

Solution: We begin by finding the linewidth in Hertz.

$$\Delta v = \frac{c}{\lambda^2}\Delta\lambda \tag{3.30}$$

$$= \frac{3 \times 10^8}{(1300 \times 10^{-9})^2}(0.013 \times 10^{-9}) \tag{3.31}$$

$$= 2.31 \times 10^9 \text{ Hz} \tag{3.32}$$

$$= 2.31 \text{ GHz}. \tag{3.33}$$

The threshold power for Brillouin scattering is then found as

$$P_B = (17.6 \times 10^{-3})\, a^2\lambda^2\alpha\,\Delta v \tag{3.34}$$

$$= (17.6 \times 10^{-3})\,(4)^2(1.3)^2(0.8)\,(2.31) \tag{3.35}$$

$$= 0.879 \text{ watts}. \tag{3.36}$$

(b) Calculate the threshold power for the beginning of Raman scattering in the same fiber.

Solution: The threshold power for Raman scattering is given by

$$P_R = (23.6 \times 10^{-2})\, a^2\lambda\alpha \tag{3.37}$$

$$= (23.6 \times 10^{-2})\,(4)^2(1.3)\,(0.8) \tag{3.38}$$

$$= 3.93 \text{ watts}. \tag{3.39}$$

(c) Find the ratio of the Brillouin scattering threshold to the Raman scattering threshold.

Solution: The ratio of the threshold power levels is given by

$$\frac{P_B}{P_R} = \frac{17.6 \times 10^{-3} \, a^2\lambda^2\alpha \, \Delta v}{23.6 \times 10^{-2} a^2\lambda\alpha} \tag{3.40}$$

$$= 0.0746\lambda\Delta v \tag{3.41}$$

$$= (0.0746)(1.3)(2.31) \tag{3.42}$$

$$= 0.224 \, . \tag{3.43}$$

So, we find that the threshold for Brillouin scattering is about 22% of that for Raman scattering, indicating that Brillouin scattering will be the limiting factor in nonlinear scattering.

When we have more than one signal present in the fiber, more complicated interactions can occur because of nonlinear interactions between the signals. Chraplyvy (1990) describes both single-signal effects and multiple-signal effects that occur when several discrete optical wavelengths are simultaneously present. He finds that Brillouin scattering will affect single-signal links but will not affect multiple-signal channels. Four-photon mixing, which requires at least two signals to occur, will have the greatest impact on multi-signal links.

3.2.3 Interface Inhomogeneities

Interface inhomogeneities have the effect of converting high-order modes into lossy modes extending into the cladding, where they are removed by the jacket losses. Such inhomogeneities may be due to impurities trapped at the core-cladding interface or impurities in the fiber buffering. The inhomogeneities may also be geometric changes in the shape and size of the core due to manufacturing tolerance allowances. Generally, single-mode fibers are more susceptible to losses from geometric irregularities or defects in the jacket material, because a defect of a given size will represent a larger fractional portion of the single-mode fiber diameter than a multimode fiber of larger diameter.

These losses are minimized by reducing the source of the problem. Manufacturing improvements have reduced the losses due to geometric variation in core diameter and coating imperfections.

3.2.4 Macrobending and Microbending Losses

Fibers show increased losses due to bending effects (Petermann and Kuhne, 1986; Artiglia et al., 1989; Kapron, 1990; Miyamoto et al., 1990). Large bends of the cable and fiber are *macrobends*; small-scale bends in the core-cladding interface are *microbends*. These latter localized bends can develop during deployment of the fiber, or can be caused by local mechanical stresses placed on the fiber (e.g., cabling the fiber or wrapping the fiber on a spool or bobbin). These latter losses are called the *cabling loss* and *spooling loss*, respectively. Typical additional losses caused by microbends added during cabling can be 1 to 2 dB/km.

Bend Losses: Multimode Fibers

Radiation losses occur at bends in the fiber path. At a bend the geometry at the core-cladding interface changes and some of the guided light in the core is transmitted from the core into the cladding. Most power is lost from the higher-order modes; less power is lost from the lower-order modes. The radius of curvature of a fiber bend is critical to the amount of power lost.

The loss coefficient associated with a fiber bend is given by (Senior, 1985)

$$\frac{P_{out}}{P_{in}} = e^{-\alpha_{bends} z} \tag{3.44}$$

and the attenuation coefficient (α_{bends}) is given by (Senior, 1985)

$$\alpha_{bends} = c_1 e^{-c_2 r} , \tag{3.45}$$

where r is the radius of curvature of the fiber bend and c_1 and c_2 are constants. The losses are negligible until the radius reaches a critical size, given by (Senior, 1985)

$$r_{critical} \approx \frac{3 n_2 \lambda}{4\pi (NA)^3} . \tag{3.46}$$

From this relation we see that, to minimize these losses, we want a fiber with a large NA and we want to operate at a short wavelength.

Fortunately, macrobending does not cause appreciable losses until the radius of curvature of the bend is below (approximately) 1 cm. This requirement does not present much of a problem in the practical utilization of fiber cables, but does present a minimum curvature to the fiber. Frequently the fiber jacket is stiffened to prevent an attempt to loop the fiber into too small a curvature.

Example:

(a) Calculate the critical radius of curvature for a multimode 50/125 fiber with an NA of 0.2 operating at 850 nm.

Solution: Lacking information about the value of n_2, we have to assume a reasonable value. (We note that the critical radius is linearly proportional to the index n_2, so any error in n_2 will be directly reflected in $r_{critical}$.) We will assume a value of $n_2 = 1.48$.

The critical radius of curvature is

$$r_{critical} \approx \frac{3n_2\lambda}{4\pi(NA)^3} \tag{3.47}$$

$$\approx \frac{3(1.48)\,(850 \times 10^{-9})}{4\pi(0.2)^3} \tag{3.48}$$

$$\approx 37.5 \text{ µm} . \tag{3.49}$$

(b) For a 9/125 single-mode fiber with an NA of 0.08 operating at 1300 nm?

Solution: The critical radius is found from

$$r_{critical} \approx \frac{3n_2\lambda}{4\pi NA^3} \tag{3.50}$$

$$\approx \frac{3(1.48)\,(1300 \times 10^{-9})}{4\pi(0.08)^3} \tag{3.51}$$

$$\approx 897 \text{ µm} . \tag{3.52}$$

We note that most fiber bends (typically greater than a few centimeters) will well exceed these values. We note, however, that the calculated values differed by more than an order of magnitude. We also note that some single-mode fiber designs carry more power in the cladding, making them more susceptible to this loss mechanism than designs that carry more power within the core. This observation helps to explain the increased sensitivity of single-mode long-wavelength fibers to microbends.

Bend Losses: Single-Mode Fiber

Bend losses are particularly important in single-mode fibers. In these fibers, the bend losses show a dramatic increase above a critical wavelength when the fiber is bent or perturbed. In particular, it has been observed that the bend losses can be appreciably high at 1550 nm in fibers designed for operation at 1300 nm (Kaiser and Keck, 1988). The susceptibility of a fiber to these losses depends on the mode-field diameter and the cutoff wavelength (Artiglia et al., 1986, 1987, 1989; Gartside et al., 1988). The analysis is beyond the level of this text, but, in general, the worst-case condition is in a fiber with a large mode-field diameter and a low cut-off wavelength. Bending losses are minimized in single-mode fibers by avoiding this combination of features.

Losses due to mechanical stresses can be reduced by the fiber design (e.g., choosing a small ratio of core diameter to fiber diameter and having a large Δ value) or by jacketing the fiber with a compressible material that distributes external mechanical stresses more uniformly along a portion of the fiber, or both. Further resistance to microbending losses can be achieved by encasing the fiber and its compliant jacket in a highly rigid sheath that resists bending.

3.3 Dispersion

As a pulse of light proceeds through the fiber, it widens in the time domain (see Figure 3.3). This spreading is caused by *dispersion*. The amount of pulse spreading determines how close (in time) two adjacent output pulses are. For any given receiver, there is a minimum spacing required between output pulses, since the receiver must be able to resolve the two separate pulses. Hence, the amount of pulse spreading in the fiber limits the maximum rate at which data can be sent (or, the spreading determines the maximum length of the fiber, if the data rate is fixed). Low dispersion means a high data rate, since pulses can be transmitted closer together with less overlap at the output. There are three primary sources of dispersion in fibers:

- material dispersion,
- waveguide dispersion, and
- modal delay (or group delay).

In actuality, material dispersion and waveguide dispersion are caused by the same physical effect (the dependence of the index of refraction of glass on wavelength), but they are accurately modeled as separate effects with additive results (Keiser, 1991).

We now want to describe material dispersion and waveguide dispersion and to consider their effects on the performance of a fiber.

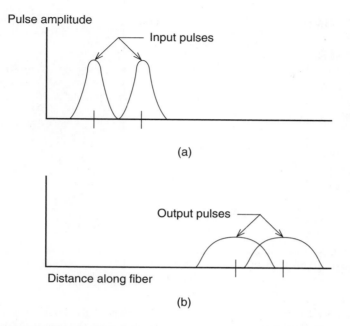

Figure 3.3 Effects of pulse spreading on data rate. (a) Well-resolved pulses at input. (b) Unresolved (overlapping) pulses at output.

3.3.1 Material Dispersion

Material dispersion is caused by the velocity of light (or, equivalently, the index of refraction) being a function of wavelength, as shown in Figure 3.4. In single-mode fibers operating at shorter wavelengths (i.e., 800–900 nm), this will be the dominant source of dispersion. All sources of light have some degree of spectral width to their output. (A laser source has a considerably narrower spectral width than an LED.) This nonzero spectral width implies that, even with single-mode propagation, the longer wavelengths with their faster velocities will arrive at the receiver before the shorter wavelengths, thereby stretching the pulse.

As shown in the following example, the pulse spreading due to material dispersion is given by

$$\Delta\tau_{mat} = -\frac{L}{c}\frac{\Delta\lambda}{\lambda}\left(\lambda^2\frac{d^2 n_1}{d\lambda^2}\right), \tag{3.53}$$

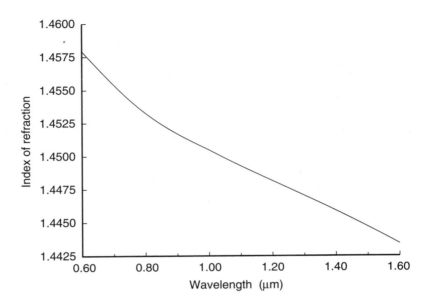

Figure 3.4 Index of refraction vs. wavelength for typical glass. (From Cherin, A.H., *An Introduction to Optical Fibers*, p. 102, New York: McGraw-Hill, 1983.)

where $\Delta\lambda$ is the spectral width of the source, λ is the nominal wavelength of the source, and $d^2n_1/d\lambda^2$ is the second derivative of the core index of refraction with respect to wavelength. The product of λ^2 and the second derivative characterizes the material dispersion of the fiber and is so grouped in writing the equation. A plot of $\lambda^2(d^2n/d\lambda^2)$ for the data of Figure 3.4 is shown in Figure 3.5. The interesting thing about the material dispersion of silica is that it changes sign at a wavelength of 1.27 μm (see Figure 3.5) and, hence, has a zero value at that wavelength, offering the possibility of a zero-dispersion fiber. (We shall see that the other sources of dispersion move the wavelength for zero total dispersion up to 1300 nm.)

Example: Derive the expression for the material dispersion in a fiber.

Solution: The arrival time τ of light after traversing a length L of fiber is

$$\tau = \frac{L}{v_g},$$

(3.54)

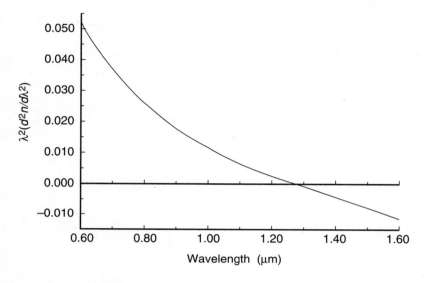

Figure 3.5 Plot of $\lambda^2(d^2n/d\lambda^2)$ vs. wavelength for typical glass. (From A.H. Cherin, *An Introduction to Optical Fibers*, p. 102, New York: McGraw-Hill, 1983.)

where v_g is the group velocity of the fiber, given by

$$v_g = \frac{1}{\dfrac{d\beta}{d\omega}} \cdot \tag{3.55}$$

We have, then,

$$\tau = L\frac{d\beta}{d\omega} \tag{3.56}$$

$$= L\frac{d\beta}{d\lambda}\frac{d\lambda}{d\omega}\cdot \tag{3.57}$$

Since $\lambda = c/v = 2\pi c/\omega$, we find

$$\frac{d\lambda}{d\omega} = -\frac{2\pi c}{\omega^2} \tag{3.58}$$

$$= -\frac{1}{\omega}\frac{2\pi c}{\omega} \tag{3.59}$$

$$= -\frac{\lambda}{\omega}\cdot \tag{3.60}$$

Substituting Equation 3.60 into Equation 3.57, we obtain

$$\tau = L\frac{d\beta}{d\lambda}\left(-\frac{\lambda}{\omega}\right) \tag{3.61}$$

$$= -\frac{L\lambda}{\omega}\frac{d\beta}{d\lambda} \tag{3.62}$$

$$= -\frac{L\lambda^2}{2\pi c}\frac{d\beta}{d\lambda}. \tag{3.63}$$

We know that

$$\beta = \frac{2\pi n(\lambda)}{\lambda}, \tag{3.64}$$

so

$$\tau = -\frac{L\lambda^2}{2\pi c}\frac{d\beta}{d\lambda} \tag{3.65}$$

$$= -\frac{L\lambda^2}{2\pi c}\left[-\frac{2\pi n}{\lambda^2} + \frac{2\pi n'}{\lambda}\right] \tag{3.66}$$

$$= -\frac{L}{c}[-n + \lambda n'] \tag{3.67}$$

$$= +\frac{L}{c}\left[n(\lambda) - \lambda\frac{dn(\lambda)}{d\lambda}\right]. \tag{3.68}$$

The pulse spread $\Delta\tau$ due to a source linewidth of $\Delta\lambda$ is

$$\frac{\Delta\tau}{\Delta\lambda} = \frac{d\tau}{d\lambda} \tag{3.69}$$

$$= \frac{L}{c}\left[\frac{dn(\lambda)}{d\lambda} - \lambda\frac{d^2n}{d\lambda^2} - \frac{dn}{d\lambda}\right] \tag{3.70}$$

$$= -\frac{L\lambda}{c}\frac{d^2n}{d\lambda^2}. \tag{3.71}$$

Multiplying by $\Delta\lambda$, we find

$$\Delta\tau = -\frac{L\lambda\,\Delta\lambda}{c}\frac{d^2n}{d\lambda^2} \tag{3.72}$$

$$= -\frac{L}{c}\frac{\Delta\lambda}{\lambda}\left(\lambda^2\frac{d^2n}{d\lambda^2}\right). \tag{3.73}$$

This is the desired expression for the material dispersion.

Example: Consider the material dispersion in a 62.5/125 fiber with $n_1 = 1.48$ and $\Delta = 1.5\%$.

(a) Calculate the material dispersion in normalized units of $ps \cdot km^{-1} \cdot nm^{-1}$ at 850 nm.

Solution: The pulse spreading is

$$\Delta\tau_{mat} = -\frac{L}{c}\frac{\Delta\lambda}{\lambda}\left(\lambda^2\frac{d^2n_1}{d\lambda^2}\right). \tag{3.74}$$

The normalized delay is

$$\frac{\Delta\tau_{mat}}{L\,\Delta\lambda} = -\frac{1}{c\lambda}\left(\lambda^2\frac{d^2n}{d\lambda^2}\right). \tag{3.75}$$

From Figure 3.5, we see that $\lambda^2 d^2 n_1/d\lambda^2$ is approximately 0.022 at $\lambda = 850$ nm, hence,

$$\frac{\Delta\tau_{mat}}{L\,\Delta\lambda} = -\frac{1}{c\lambda}\left(\lambda^2\frac{d^2n}{d\lambda^2}\right) \tag{3.76}$$

$$= -\frac{1}{(3.0\times10^8)\,(850\times10^{-9})}\,(0.022) \tag{3.77}$$

$$= -8.63\times10^{-5}\ s\cdot m^{-1}\cdot m^{-1} \tag{3.78}$$

$$= -86.3\ ps\cdot km^{-1}\cdot nm^{-1}. \tag{3.79}$$

(b) ... at 1500 nm?

Solution: From Figure 3.5 we estimate that $\lambda^2 d^2 n_1/d\lambda^2 \approx -0.016$, so

$$\frac{\Delta\tau_{mat}}{L\,\Delta\lambda} = -\frac{1}{c\lambda}\left(\lambda^2\frac{d^2n}{d\lambda^2}\right) \tag{3.80}$$

$$= -\frac{1}{(3.0\times10^8)\,(1500\times10^{-9})}\,(-0.016) \tag{3.81}$$

$$= +3.56\times10^{-5}\ s\cdot m^{-1}\cdot m^{-1} \tag{3.82}$$

$$= +35.6\ ps\cdot km^{-1}\cdot nm^{-1}. \tag{3.83}$$

3.3.2 Waveguide Dispersion

For the low material-dispersion region near 1.27 µm, a third type of dispersion, *waveguide dispersion*, becomes important. Usually this source of dispersion is negligible in multimode fibers and in single-mode fibers operated at wavelengths below 1 µm, but it is *not* negligible in single-mode fibers operated in the vicinity of 1.27 µm. This waveguide dispersion results from the propagation constant of a mode (and, hence, its velocity) being a function of a/λ. In particular, the net delay due to waveguide dispersion is expressed by (Keiser, 1991)

$$\tau_{wg} = \frac{L}{c}\frac{d\beta}{dk} \, .$$
(3.84)

We again define the normalized propagation constant b as

$$b = \frac{(\beta^2/k^2) - n_2^2}{n_1^2 - n_2^2} \, .$$
(3.85)

An approximation for b is

$$b \approx \frac{(\beta/k) - n_2}{n_1 - n_2} \, ,$$
(3.86)

thereby giving

$$\beta \approx n_2 k \, (b\Delta + 1) \, .$$
(3.87)

Here b is a function of V (and of k). Substitution of Equation 3.87 into Equation 3.84 gives

$$\tau_{wg} \approx \frac{L}{c}\left(n_2 + n_2\Delta\frac{d(kb)}{dk} \right).$$
(3.88)

Using the approximation,

$$V \approx kan_2\sqrt{2\Delta} \, ,$$
(3.89)

we can write

$$\tau_{wg} \approx \frac{L}{c}\left(n_2 + n_2\Delta\frac{d(Vb)}{dV} \right).$$
(3.90)

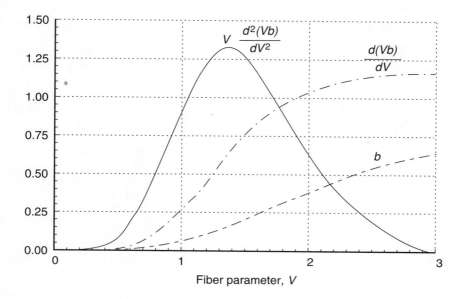

Figure 3.6 Plot of b, $d(Vb)/dV$, and $V\,d^2(Vb)/dV^2$ vs. V for the lowest-order fiber mode.

Neglecting the constant term in the delay, we have a wavelength-dependent time delay of

$$\tau_{wg}(\lambda) \approx \frac{n_2 \Delta\, L}{c}\, \frac{d(Vb)}{dV}\, . \qquad (3.91)$$

Figure 3.6 shows a plot of $d(Vb)/dV$ versus V for the lowest-order mode in the fiber. Similar curves (Gloge, 1971a) can be obtained for higher-order modes, if desired. (For large values of V [i.e., multimode fibers], however, the value of the delay due to waveguide effects is small and negligible compared to the delay due to modal or material dispersion.) In single-mode fibers, however, modal dispersion is not present and, in the region near 1270 nm, the waveguide dispersion is comparable with the material dispersion. Therefore, we want to concentrate the waveguide dispersion in that region of the spectrum.

Continuing the development, we now want to include the effects of source linewidth. We know that the difference in propagation time $\Delta\tau_{wg}$ is

$$\Delta\tau_{wg} = \Delta\lambda \frac{d\,\tau_{wg}}{d\,\lambda} \qquad (3.92)$$

$$= \Delta\lambda \frac{dV}{d\lambda} \frac{d\,\tau_{wg}}{dV} \, . \tag{3.93}$$

Since

$$V = \frac{2\pi a n_1 \sqrt{2\Delta}}{\lambda} \, , \tag{3.94}$$

we can show that

$$\frac{dV}{d\lambda} = -\frac{V}{\lambda} \, . \tag{3.95}$$

We can use Equation 3.91 to find $d\,\tau_{wg}/dV$ and, hence, can show that

$$\Delta\tau_{wg} = -\frac{V}{\lambda}\,\Delta\lambda \frac{d\,\tau_{wg}}{dV} \tag{3.96}$$

$$\approx -\frac{n_2 L \Delta}{c} \frac{\Delta\lambda}{\lambda} \left(V \frac{d^2(Vb)}{dV^2} \right) . \tag{3.97}$$

This is our desired expression for the waveguide dispersion.

Gloge (1971b) defines the *normalized propagation constant b* as

$$b(V) = 1 - \frac{u^2}{V^2} \tag{3.98}$$

$$= \frac{\left(\frac{\beta^2}{k^2}\right) - n_2^2}{n_1^2 - n_2^2} \tag{3.99}$$

$$\beta = k\sqrt{n_2^2 + (n_1^2 - n_2^2)b} \, . \tag{3.100}$$

Gloge (1971a) also shows that, for the lowest-order mode (i.e., the HE_{11} mode), the expression for u is

$$u(V) = \frac{(1 + \sqrt{2})V}{1 + (4 + V^4)^{0.25}} \tag{3.101}$$

By substituting Equation 3.101 into Equation 3.98, we can get an expression for $b(V)$. Figure 3.6 shows a plot of $b(V)$. From this expression, we can also obtain plots

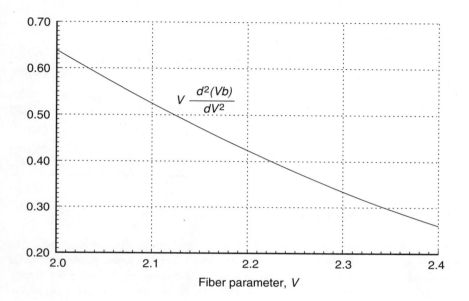

Figure 3.7 Plot of $V d^2(Vb)/dV^2$ versus V for the lowest-order fiber mode
in the region of practical single-mode fiber design.

of $d(Vb)/dV$ and $V d^2(Vb)/dV^2$ as functions of V, as also shown in the figure (see problem at end of chapter). The single-mode values of interest are from $V = 2.0$ to 2.4, as shown in Figure 3.7. The value of $V d^2(Vb)/dV^2$ decreases monotonically from 0.64 to 0.25. Hence, the waveguide dispersion is, in general, a small negative value.

Example: Calculate the waveguide dispersion in units of $ps \cdot km^{-1} \cdot nm^{-1}$ for a 9/125 single-mode fiber with $n_1 = 1.48$ and $\Delta = 0.22\%$ operating at 1300 nm.

Solution: We begin by calculating V from Equation 2.20,

$$V \approx \frac{2\pi a}{\lambda} n_1 \sqrt{2\Delta} \tag{3.102}$$

$$\approx \frac{2\pi(4.5 \times 10^{-6})}{1300 \times 10^{-9}}(1.48)\left(\sqrt{2(0.0022)}\right) \tag{3.103}$$

$$\approx 2.14 . \tag{3.104}$$

(We note that V falls within the expected range of $2.0 < V < 2.405$ for single-mode fiber.) From Figure 3.7, we find $V d^2(Vb)/dV^2 \approx 0.480$ at $V = 2.14$.

We also have $n_2 = n_1(1 - \Delta) = 1.48(1 - 0.0022) = 1.477$, so

$$\frac{\Delta \tau_{wg}}{L \, \Delta \lambda} = -\left(\frac{n_2 \Delta}{c}\right)\left(\frac{1}{\lambda}\right)\left(V \frac{d^2(Vb)}{dV^2}\right) \tag{3.105}$$

$$= -\left(\frac{(1.477)\,(0.0022)}{3 \times 10^8}\right)\left(\frac{1}{1300 \times 10^{-9}}\right)(0.48) \tag{3.106}$$

$$= -4.00 \times 10^{-6} \, s \cdot m^{-1} \cdot m^{-1} \tag{3.107}$$

$$= -4.00 \, ps \cdot km^{-1} \cdot nm^{-1}. \tag{3.108}$$

3.3.3 Total Dispersion (Single-Mode Fibers)

To minimize the total dispersion of a single-mode fiber, it is necessary to operate at a wavelength longer than 1.27 µm to allow the small positive material dispersion to cancel the small negative waveguide dispersion, causing a net dispersion of zero, as illustrated in Figure 3.8. This zero-dispersion point occurs near 1300 nm, a wavelength that, fortunately, has a fairly low attenuation (although not as low as the attenuation minimum at 1550 nm), allowing operation of high–data-rate links at this wavelength. This low dispersion at 1300 nm has provided the impetus for the development of a family of sources and receivers that operate at this wavelength.

The size of the waveguide dispersion has been found to be sensitive to the doping levels as well as the values of Δ and a. For various combinations of Δ and a, and for triangular and other doping profiles (as in Figure 2.9), it has proved possible to achieve zero dispersion at other wavelengths between 1300 nm and 1700 nm, allowing the development of fibers that combine minimum losses at 1550 nm with zero dispersion at this wavelength. This process is called *dispersion shifting* and results in so-called *dispersion-shifted fibers*.

3.3.4 Dispersion-Adjusted Single-Mode Fibers

We have found that the lowest losses occur at 1500 nm wavelength, while the lowest total dispersion occurs (in a step-index single-mode fiber) at 1300 nm wavelength. Since these wavelengths are near each other, it has become a goal to combine the two features so that optimum performance in terms of bandwidth and loss occurs at the same wavelength. Of the two parameters, dispersion is the most adjustable and is the one that is manipulated. This manipulation has led to two adjustments; one is to move the zero-dispersion wavelength (*dispersion shifting*) to

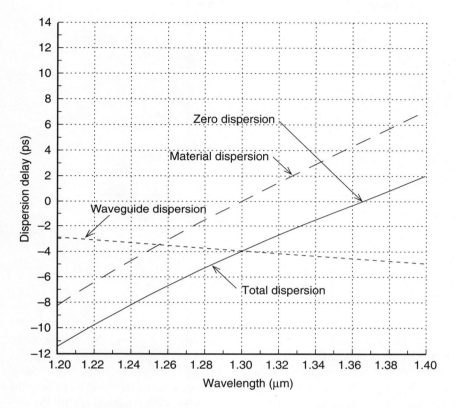

Figure 3.8 Addition of material dispersion and waveguide dispersion
to achieve zero dispersion at a value just below 1.32 μm. (This plot
is the solution of a problem at the end of this chapter.)

higher values in the vicinity of 1550 nm (as in Figure 3.9), and the other is to minimize or flatten the dispersion over a range of wavelength values (*dispersion flattening*).

Dispersion-Shifted Step-Index Fibers

The *material dispersion* of silica can be adjusted in small amounts by doping the core of the fiber with G_eO_2 or other dopants. (G_eO_2 has the greatest effect on the material dispersion.)

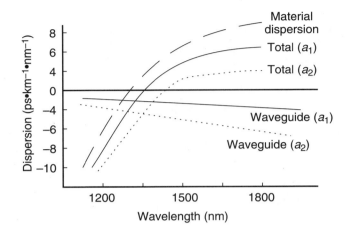

Figure 3.9 Example of a dispersion-shifted fiber. The waveguide dispersion is modified by changing *a* (while also changing Δ to keep *V* fixed) to increase the zero-dispersion wavelength.

The *waveguide dispersion* depends on the fiber-core radius, the fractional change in the fiber index of refraction Δ, and on the shape of the fiber profile *g*. A variety of fiber profile adjustments can be made to tune the zero dispersion wavelength. For the step-index profile the core radius *a* is reduced while increasing the fractional index change Δ. (We still want to keep the cutoff wavelength for the second-lowest mode, the LP_{11} mode, between 1000 nm and 1300 nm to keep the power distribution of the lowest-order mode reasonably close to the center of the fiber. This requires that the value of *V* be kept between 1.5 and 2.4.) The equation for *V* indicates that a linear decrease in *a* requires a quadratic increase in Δ in order to keep *V* constant. It has been found that the waveguide dispersion is very sensitive to changes in the fiber parameters. For example, reducing a core radius from 5.5 μm to 1.8 μm changes the zero-dispersion wavelength from 1300 nm to 1750 nm (Ainslie and Day, 1986). Experimental results quoted in Ainslie and Day (1986) indicate that dispersion-shifted step-index fibers are easily made and achieve the desired dispersion shift. Unfortunately, these fibers also exhibited an increased loss (0.30 dB/km or greater at 1550 nm) that seems to be due to the strong concentration of germania doping required in the core. While various models have been proposed (Ainslie and Day, 1986) for various single-cladding designs, none have enabled the production of dispersion-shifted step-index fibers with losses comparable to the unshifted fibers. An alternative is to consider dispersion shifting with a non–step-index profile, either a graded-index profile or a multi-layer profile.

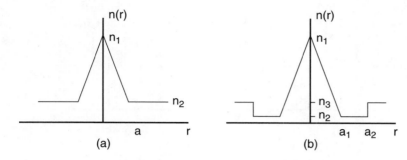

Figure 3.10 (a) Triangle profile. (b) Triangle profile with depressed cladding.

Dispersion-Shifted Graded-Index Fibers

Graded-index fibers offer more parameters for the fiber designer to manipulate (Gambling, 1979b; Paek, 1981; Bhagavatula, 1988). One additional parameter is the profile (g) of the core from the refractive index equation

$$n(r) = n_1 \sqrt{1 - 2\Delta \, (r/a)^g} \, .$$

(3.109)

Studies (Gambling et al., 1979a; White, 1982) of the propagation of modes in these structures show that the single-mode behavior exists for $V \leq 3.53$ for a parabolic profile ($g = 2$) and for $V \leq 4.38$ for a triangular profile ($g = 1$).

The triangular profile (Figure 3.10) was demonstrated (Ainslie and Day, 1986) to shift the zero-dispersion wavelength successfully without any penalty in fiber losses with a Δ of 0.115 and a core radius a of 3.2 μm. The measured losses were 0.25 dB/km at 1.56 μm. (Other experiments have achieved losses as low as 0.21 dB/km (Ainslie and Day, 1986).) The triangular profile has the advantage of the mode-field diameter staying quite small. The cutoff wavelength for the LP_{11} mode is usually fairly low (i.e., in the vicinity of 0.85 μm), a potential disadvantage in that it raises the susceptibility to microbending losses. A remedy to this low value of cutoff wavelength is to use a depressed-cladding triangle profile as in Figure 3.10(b). The extra cladding tends to guide the LP_{01} mode better. In a design cited in Ainslie and Day (1986), the ratio of a_2/a_1 was > 8.5.

Dispersion-Shifted Multi-Index Fibers

More complicated profiles give the designer ample parameters to manipulate in an attempt to optimize the fiber performance. The *double-clad fiber* (or *"W"*

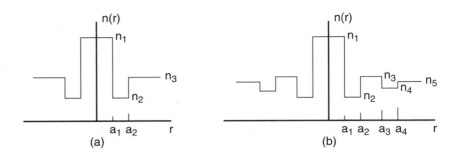

Figure 3.11 Refractive index profile of (a) a double-clad or "W" profile fiber and (b) a quadruple-clad fiber.

fiber) of Figure 3.11(a) has been used widely for dispersion-flattening, but also works for dispersion shifting. The dispersion-shifted fibers were sensitive to micro-bending (Ainslie et. al., 1986), however. Other designs incorporating a triangular center profile with a raised outer cladding have produced dispersion-shifted fibers with losses as low as 0.17 dB/km at 1550 nm (Ainslie and Day, 1986).

Dispersion-Flattened Fibers

An alternative approach to minimizing dispersion effects with less loss penalty than the step-index dispersion-shifted fibers is to attempt to reduce the dispersion to a nonzero minimum between 1300 nm and 1500 nm, as shown in Figure 3.12 for the quadruple-clad profile of Figure 3.11(b). This technique is called *dispersion flattening*. (Zero dispersion can occur toward the ends of the range in some designs.) These dispersion-flattened fibers allow the use of multiple wavelengths with reasonable loss and dispersion performance; the advantages of this are discussed in Chapter 10, on wavelength-division multiplexing. As noted above, multi-layer profiles have been used successfully to flatten the dispersion characteristics of fibers. Step-index designs can achieve dispersion flattening only with unacceptably high losses.

3.3.5 Modal Dispersion

Group delay (or *modal dispersion*) is important in multimode fibers. It is caused by the different pathlengths associated with each of the modes of a fiber.

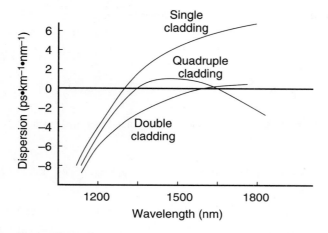

Figure 3.12 Example of a dispersion-flattened fiber. The quadruple-clad
fiber provides enough design variables to produce a flattened
small-dispersion region between 1300 nm and 1600 nm.

Modal Dispersion I: Step-Index Fiber

Assuming a multimode step-index fiber and considering each mode as a ray
of light at a slightly different angle, we can see from Figure 3.13 that each mode will
have a slightly different pathlength in traversing the fiber. For an ideal, straight step-
index fiber, the fastest mode is the one that travels straight down the fiber (i.e., at an
angle of incidence of 90 degrees to the core-cladding interface). The slowest mode is
that which is incident at the critical angle of the fiber.

The time delay between the fastest and slowest pulse is the *modal pulse
delay distortion* $\Delta\tau_{modal}$ and, for a step-index fiber, is given by Cherin (1983)

$$\Delta\tau_{modal} = \frac{L(n_1 - n_2)}{c}\left(1 - \frac{\pi}{V}\right),\tag{3.110}$$

where L is the length of the fiber and V is the parameter given by Equation 2.19.
(This expression ignores the constant delay common to all modes and represents
only the *difference* in propagation times.) We note that for most multimode fibers, V
is larger than 10, so a useful approximation is

$$\Delta\tau_{modal} \approx \frac{L\,\Delta n_1}{c}.\tag{3.111}$$

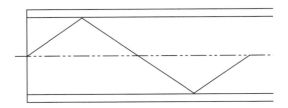

Figure 3.13 Step-index fiber showing high order mode path.

Since modal dispersion is due to the existence of various modes, it is *not* present in single-mode fibers.

Example: Consider a 50/125 step-index fiber with $n_1 = 1.47$ and $\Delta = 1.5\%$. Calculate the group delay (or modal dispersion) in units of ns $\cdot$ km^{-1} for this fiber at an operating wavelength of 850 nm.

Solution: We find the normalized dispersion as

$$\frac{\Delta\tau_{modal}}{L} \approx \frac{\Delta n_1}{c} \tag{3.112}$$

$$\approx \frac{(0.015)\,(1.47)}{3\times10^8} \tag{3.113}$$

$$\approx 7.35\times10^{-11}\ \text{s}\cdot\text{m}^{-1} \tag{3.114}$$

$$\approx 73.5\ \text{ns}\cdot\text{km}^{-1}. \tag{3.115}$$

Alternative solution: We can use the more exact formula,

$$\frac{\Delta\tau_{modal}}{L} = \frac{(n_1 - n_2)}{c}\left(1 - \frac{\pi}{V}\right). \tag{3.116}$$

Finding $(n_1 - n_2)$,

$$n_1 - n_2 \approx \Delta\, n_1 \tag{3.117}$$

$$\approx (0.015)\,(1.47) \tag{3.118}$$

$$\approx 2.21\times10^{-2}. \tag{3.119}$$

To find V, we use

$$V \approx \frac{2\pi a}{\lambda} n_1 \sqrt{2\Delta} \tag{3.120}$$

$$\approx \left(\frac{2\pi(25 \times 10^{-6})}{850 \times 10^{-9}} \right) (1.47) \left(\sqrt{2(0.015)} \right) \tag{3.121}$$

$$\approx 47.1 . \tag{3.122}$$

Hence,

$$\frac{\Delta\tau_{modal}}{L} \approx \frac{(n_1 - n_2)}{c} \left(1 - \frac{\pi}{V} \right) \tag{3.123}$$

$$\approx \frac{(2.21 \times 10^{-2})}{3 \times 10^8} \left(1 - \frac{\pi}{47.1} \right) \tag{3.124}$$

$$\approx 6.87 \times 10^{-11} \text{ s} \cdot \text{m}^{-1} \tag{3.125}$$

$$\approx 68.7 \text{ ns} \cdot \text{km}^{-1} . \tag{3.126}$$

Once we find the pulse spread $\Delta\tau$ caused by modal dispersion, we need a method of calculating the maximum data rate that the fiber can support. This maximum data rate DR_{max} can be found from

$$DR_{max} = \frac{1}{4 \, \Delta\tau} . \tag{3.127}$$

If the limiting pulse spread is caused by modal dispersion, for example, we would have

$$DR_{max} = \frac{1}{4 \, \Delta\tau_{modal}} . \tag{3.128}$$

Modal Dispersion II: Graded-Index Fiber

For the graded-index fiber, we must account for the inhomogeneous velocity of the light in the fiber as well as the sinusoidal paths as shown in Figure 2.11, Section 2.7.1. In particular, we note that, although the higher-order modes have longer pathlengths because of their further excursion from the axis, their average velocity will be higher because of the increased velocity as the ray moves away from the center axis. To a first-order approximation, the effect of the longer pathlength is canceled by the higher velocity, and modes that leave the axis at the same time will arrive at the next axis crossing at the same time. More exact analysis of the graded index fiber predicts (Keiser, 1991) that the delay time incurred by the m-th mode of the fiber will be given by

$$\tau_{\text{modal}} = \frac{LN_{g1}}{c} \times \left(1 + \frac{g - 2 - \varepsilon}{g + 2} \Delta \left(\frac{m}{N}\right)^{g/(g+2)}\right.$$

$$+ \frac{\Delta^2}{2} \frac{3g - 2 - 2\varepsilon}{g + 2} \left(\frac{m}{N}\right)^{2g/(g+2)}$$

$$\left. + \text{ other terms with } \Delta^3, \Delta^4, \text{ etc.}\right), \quad (3.129)$$

where

$$\varepsilon = -\frac{2n_1}{N_{g1}} \frac{\lambda}{\Delta} \frac{d\Delta}{d\lambda}. \quad (3.130)$$

$$N_{g1} = n_1 - \lambda\frac{dn_1}{d\lambda}, \quad \text{and} \quad (3.131)$$

$$N = a^2 \Delta k^2 n_1^2 \frac{g}{g + 2}. \quad (3.132)$$

The information on $d\Delta/d\lambda$ and $dn_1/d\lambda$ is material-dependent and wavelength-dependent and would have to be provided from a study of the materials used in fabrication of the fibers.

From Equation 3.129, we note that the term that is linear in Δ can be eliminated if the profile index is set to

$$g = g_{\text{opt}} = 2 + \varepsilon, \quad (3.133)$$

where

$$g_{\text{opt}} = 2 - \frac{2n_1}{N_{g1}} \frac{\lambda}{\Delta} \frac{d\Delta}{d\lambda}. \quad (3.134)$$

An approximation for g_{opt} that is cited frequently (Olshansky and Keck, 1976; Midwinter, 1979) is

$$g_{\text{opt}} \approx 2 - \frac{12\Delta}{5}. \quad (3.135)$$

This approximation is accurate only if the modal dispersion is considered and the material dispersion is ignored. Olshansky (Olshansky and Keck, 1976; Olshansky, 1976a) gives more involved expressions for the optimum profile (beyond the scope of this text) that include material and waveguide dispersion.

From this discussion, we see that graded-index fibers have an *optimum index profile* g_{opt} that minimizes the time delay of the modes. The net delay $\Delta\tau_{modal}$ from the lowest-order modes to the highest-order modes is given by (Gloge et al., 1979; Marcuse and Presby, 1979; and Cherin, 1983)

$$\Delta\tau_{modal} \approx \begin{cases} n_1\Delta\dfrac{(g - g_{opt})L}{(g + 2)c} & g \neq g_{opt} \\[2ex] \dfrac{n_1\Delta^2 L}{2c} & g = g_{opt} \end{cases} \tag{3.136}$$

We observe that $\Delta\tau_{modal}$ can be positive or negative depending on the size of g relative to g_{opt}. For negative $\Delta\tau_{modal}$, the interpretation is that the higher-order modes are arriving before the lower-order modes.

Example: Consider a graded-index fiber with $\Delta = 2\%$ and $g_{opt} = 2.0$. If $g = 95\%$ of g_{opt}, calculate the ratio of $\Delta\tau_{modal}$ ($g \neq g_{opt}$) to $\Delta\tau_{modal}$ ($g = g_{opt}$).

Solution: We have $g = 0.95g_{opt} = 0.95(2.0) = 1.90$, so

$$\frac{\Delta\tau_{modal}(g \neq g_{opt})}{\Delta\tau_{modal}(g = g_{opt})} = \frac{n_1\Delta\dfrac{g - g_{opt}}{(g + 2)c}L}{\dfrac{n_1\Delta^2 L}{2c}} \tag{3.137}$$

$$= \frac{2(g - g_{opt})}{\Delta(g + 2)} \tag{3.138}$$

$$= \frac{(2)(1.90 - 2)}{(0.02)(1.90 + 2)} \tag{3.139}$$

$$= -2.56 \tag{3.140}$$

$$= -256\% . \tag{3.141}$$

Hence, we see that a small difference between g_{opt} and the achieved g can cause a major change in the value of the ratio. This high sensitivity in the dispersion to values of g leads to major efforts to establish tight tolerances on the value of g. This tight tolerance is difficult to maintain, as we shall learn in the later chapter describing fiber fabrication and measurement.

We now want to compare the modal dispersion of a step-index multimode fiber with that of a graded-index fiber of the same size. A comparison of Equation 3.111 and Equation 3.136 shows a dispersion that is Δ smaller for the graded-index fiber than the equivalent step-index fiber (see the following example). This improved modal-dispersion performance is one of the primary advantages of the graded-index fiber over a multimode step-index fiber of the same diameter.

Example:

(a) Calculate the ratio of the modal delay per km in a 50/125 graded-index fiber with $n_1 = 1.46$, $\Delta = 1.5\%$, and $g = g_{opt} = 2$ to the modal delay in a step-index fiber of the same size with the same n_1 and Δ.

Solution: The time delays are given by

$$\frac{\Delta\tau_{GI}(g = g_{opt})}{L} \approx \frac{n_1\Delta^2}{2c} \tag{3.142}$$

$$\frac{\Delta\tau_{SI}(g = g_{opt})}{L} \approx \frac{n_1\Delta}{c} . \tag{3.143}$$

Taking the ratio,

$$\frac{\Delta\tau_{GI}(g = g_{opt})}{\Delta\tau_{SI}(g = g_{opt})} = \frac{\dfrac{n_1\Delta^2}{2c}}{\dfrac{n_1 \Delta}{c}} \tag{3.244}$$

$$= \frac{\Delta}{2} \tag{3.145}$$

$$= \frac{0.015}{2} \tag{3.146}$$

$$= 0.00750 . \tag{3.147}$$

Hence, we see that, when optimized, the graded-index fiber can have one to two orders of magnitude less dispersion than the equivalent step-index fiber.

(b) Consider the same question if the graded-index fiber is not optimized. Let $g = 2.1$ and $g_{opt} = 2.0$.

Solution:

$$\frac{\Delta\tau_{GI}(g \neq g_{opt})}{\Delta\tau_{SI}(g \neq g_{opt})} = \frac{\dfrac{n_1\Delta(g - g_{opt})}{(g + 2)c}}{\dfrac{n_1\Delta}{c}} \tag{3.148}$$

$$= \frac{(g - g_{opt})}{(g + 2)} \tag{3.149}$$

$$= \frac{(2.1 - 2.0)}{4.1} \tag{3.150}$$

$$= 0.0244 . \tag{3.151}$$

The dispersion of the graded-index fiber is still less than that of the step-index fiber, but the ratio has been reduced by more than one order of magnitude due to the non-optimum value of profile.

Current practical values of the total dispersion in multimode graded-index fibers are on the order of 0.2 ns $\cdot$ km^{-1} for laser sources and 1.0 ns $\cdot$ km^{-1} for LED sources. (The LED dispersion is higher because the material-dispersion effects for the wide–spectral-width LED are not negligible.) Practical data bandwidth-distance products for a laser source and a multimode graded-index fiber with an optimized profile are on the order of 0.5 to 2.5 GHz $\cdot$ km.

3.3.6 Dispersion Units

Modal dispersion is independent of the source linewidth and is usually specified for a fiber or cable by the characteristic pulse spread per kilometer length (i.e., in units of ns $\cdot$ km^{-1}). For multimode fibers dominated by modal dispersion, the specification can be alternatively given as an analog bandwidth-distance product (in units of GHz $\cdot$ km).

Both material dispersion and waveguide dispersion are dependent on the source linewidth. To remove the dependence, the specification for these quantities can also be given per nanometer of linewidth. Hence, the units are ns $\cdot$ km$^{-1} \cdot$ nm^{-1}.

3.3.7 Pulse-Spreading Approach

We have been using the time delay of waves to characterize the dispersion of a fiber. An alternative approach is to quantitatively consider the pulse spreading that is induced by the dispersion. In this approach we can imagine measuring the pulse width at the input and at the output and attributing the increase in pulse width to the dispersion.

We begin by defining the *RMS pulse width*. The *RMS pulse width* σ_s is defined as the standard deviation of the pulsewidth (or, equivalently, the square root of the variance) and is related to the first moment of the pulse M_1 and the second moment of the pulse M_2 by

$$\sigma_s^2 = M_2 - M_1^2 , \tag{3.152}$$

where

$$M_1 = \int_{-\infty}^{\infty} tp(t) \, dt \tag{3.153}$$

and

$$M_2 = \int_{-\infty}^{\infty} t^2 p(t) \, dt . \tag{3.154}$$

Many optical pulses are symmetric (or are assumed to be). The mean value M_1 of a symmetric input pulse is zero; so, for symmetric pulses, the *mean-square pulsewidth* σ_s^2 is

$$\sigma_s^2 = \int_{-\infty}^{+\infty} t^2 p(t) \, dt . \tag{3.155}$$

The pulse width at the output of the fiber is a combination of the pulse width at the source and the pulse spreading of the fiber caused by dispersion. For the mean-square pulse width, the effects are combined according to

$$\sigma_{out}^2 = \sigma_{in}^2 + \sigma_{fiber}^2 . \tag{3.156}$$

We usually determine σ_{fiber} by measuring the input and output pulses, calculating σ_{in} and σ_{out} from the measured waveforms, and computing σ_{fiber} from Equation 3.156 as

$$\sigma_{fiber} = \sqrt{\sigma_{out}^2 - \sigma_{in}^2} . \tag{3.157}$$

Similar results are also available in terms of the RMS pulse widths of the dispersion terms, where, again, the results are combined as

$$\sigma_{fiber}^2 = \sigma_{modal}^2 + \sigma_{material}^2 . \tag{3.158}$$

We now will give some expressions for the dispersion-induced pulse spreading.

Material Dispersion

The RMS pulse spread for material dispersion is given (Keiser, 1991) by

$$\sigma_{mat} = \frac{L}{c}\frac{\sigma_\lambda}{\lambda}\left(\lambda^2\frac{d^2 n_1}{d\lambda^2}\right)^2, \tag{3.159}$$

where σ_λ is the RMS spectral width of the source.

Modal Dispersion: Step-Index Fiber

For a step-index fiber, Senior (1985) gives an estimated spreading due to modal dispersion as

$$\sigma_{modal\ SI} \approx \frac{L n_1 \Delta}{2c\sqrt{3}} \tag{3.160}$$

$$\approx \frac{L\,(NA)^2}{4\sqrt{3}\,n_1 c}. \tag{3.161}$$

The maximum data rate that can be achieved in a fiber (assuming a negligible input pulsewidth) can be estimated as (Olshansky, 1979b)

$$DR_{max} \approx \frac{0.2}{\sigma_s}. \tag{3.162}$$

For the step-index fiber this gives

$$DR_{max} = \frac{0.8\sqrt{3}\,n_1 c}{L\,(NA)^3}. \tag{3.163}$$

Modal Dispersion: Graded-Index Fiber

Keiser (1991) gives a complete expression for the pulse spreading caused by modal dispersion σ_{modal} in a graded-index fiber as

$$\sigma_{\text{modal GI}} = \frac{LN_{g1}\Delta}{2c} \frac{g}{g+1} \sqrt{(g+2)/(3g+2)}$$

$$\times \sqrt{c_1^2 + \frac{4c_1c_2(g+1)\Delta}{2g+1} + \frac{16\Delta^2 c_2^2(g+1)^2}{(5g+2)(3g+2)}} \qquad (3.164)$$

where

$$N_{g1} = n_1 - \lambda\frac{dn_1}{d\lambda}, \qquad (3.165)$$

$$c_1 = \frac{g-2-\varepsilon}{g+2}, \qquad (3.166)$$

$$c_2 = \frac{3g-2-2\varepsilon}{2(g+2)}, \text{ and} \qquad (3.167)$$

$$\varepsilon = -\frac{2n_1}{N_{g1}} \frac{\lambda}{\Delta} \frac{d\Delta}{d\lambda}. \qquad (3.168)$$

While this expression is a complete solution, we would like to have a simpler estimate of the RMS pulse spread in the graded-index fiber.

To find the optimum value of g, we can use the estimate (Olshansky and Keck, 1976; Midwinter, 1979)

$$g_{\text{opt}} \approx 2 - \frac{12\Delta}{5}. \qquad (3.169)$$

Gloge et al. (1979) give the following expression for the pulse spread in a graded-index fiber.

$$\sigma_{\text{modal GI}} = \begin{cases} \dfrac{0.246LN_{g1}\Delta \,|g - g_{\text{opt}}|}{c(g+2)} & 1 > |g - g_{\text{opt}}| \gg \Delta \\ \dfrac{0.150LN_{g1}\Delta^2}{c} & g = g_{\text{opt}}. \end{cases} \qquad (3.170)$$

The dependence of the pulse spread is very sensitive to small changes in g, as seen in Figure 3.14. An alternative expression for the RMS pulse spread at $g = g_{\text{opt}}$ is (Senior, 1985)

$$\sigma_{\text{modal GI}} = \frac{n_1\Delta L}{2c\sqrt{3}} \qquad (\text{for } g = g_{\text{opt}}). \qquad (3.171)$$

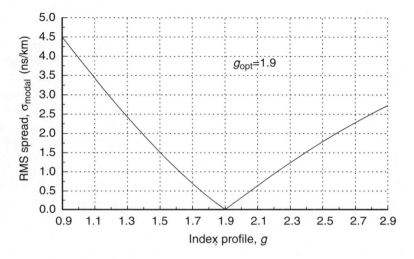

Figure 3.14 Typical modal pulse spread σ_{modal} for a graded-inde
fiber vs. the refractive index profile g.

Estimates of expected RMS pulse spread can be made by calculating the pulse spread expected when $g = g_{\text{opt}}$ and then multiplying that value by 10 to allow for expected variations.

3.3.8 Mode-Mixing Effects

In the above discussions, the delay or the pulse spread has been seen to be linearly dependent on the length of the fiber. While this is true for graded-index and single-mode fibers, another effect causes a variation from this rule for long lengths of poorly made step-index multimode fibers. Due to the long length of these fibers, the modes of a multimode fiber have ample opportunity to mix because of imperfections or geometric changes, leading to an averaging effect on the modal velocity (Keiser, 1991). The spread in modal velocities in such fibers has been found to be proportional to $\sqrt{LL_0}$, where L is the physical length of the fiber and L_0 is a characteristic *equilibrium coupling length* of the fiber. For lengths longer than L_0, it is observed that the pulse spread is proportional to $\sqrt{L}$ rather than directly proportional to the length L itself (i.e., the assumption of a linear dependence on L provides a conservative estimate of the delay) (Keiser, 1991). Typical values of L_0 are in the neighborhood of 500 to 1000 m. (This length is a function of the quality control of the fiber manufacturer and increases as fibers are made better.)

3.3.9 Data Rate–Distance Product

From the discussion of dispersion characteristics, we see that longer distances require a reduction of data rate to achieve a specified bit-error rate (BER). Since the dispersions are linearly dependent on distance, the product of the data rate and the distance is (theoretically) a constant. This tradeoff between distance and data rate is fundamental to fiber–optic-system designs. Individual fibers are specified by their data rate–distance product.

A typical single-mode fiber will have a product of several $Gb \cdot s^{-1} \cdot km$, a typical graded-index fiber product will be several 100s of $Mb \cdot s^{-1} \cdot km$, and a typical multimode step-index fiber will have a product of a few 10s of $Mb \cdot s^{-1} \cdot km$. Thus we see that, based on data rate alone, the single-mode fiber provides the best performance (especially when operated near 1300 nm), followed by the graded-index fiber, with the multimode fiber step-index fiber offering the least data handling capacity. Note, however, that other considerations (especially the cost of the fiber and system components) can dominate the fiber selection decision in any system where data rate is not the prime consideration.

Our discussion of information-handling capacity has been couched in terms of digital transmission, since that is the primary format that is used in fiber-optic systems. Fibers can also carry analog information, and the same discussion can be formulated in terms of bandwidth rather than data rate without changing the conclusions on the order of information-handling capacity among the different types of fibers.

3.4 Fiber Strength

3.4.1 Glass Fiber Breaking Strength

Glass fibers have two properties that are unusual compared to metal cables (Ritter, 1978; Kalish et al., 1979; Nagel, 1988; Kurkjian et al., 1989; Kapron, 1990; Biswas, 1991; Kapron and Yuce, 1991; Kurkjian and Innis, 1991; Roberts et al., 1991). First, the material is brittle rather than ductile (i.e., it will break rather than stretch). This "fast fracture" mode of failure leaves little margin for error in applying loads to fiber. Second, glass fiber that is under tension will eventually fail (although the time-to-failure can be very long if the tensile stress is low). This delayed time behavior leads to major efforts to reduce the stress level on the fiber within a fiber-optic cable.

The theoretical strength of glass is on the order of 1 GPa (where 1 Pa = $1 N/m^2$ and 1 GPa = 1.45×10^5 psi). Actual fiber strengths are typically two orders of magnitude smaller and are highly variable. The failure stress for optical fibers

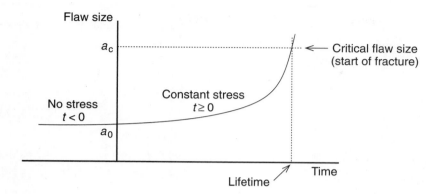

Figure 3.15 A typical time history of crack size a for a fiber under constant stress.

from the same fabrication lot can vary by more than an order of magnitude. The explanation for this variability lies in the flaws that occur on the fiber's surface. The flaws are cracks in the glass surface that are randomly distributed in size. When a fiber is placed under static stress, the flaws grow increasingly bigger (in a process called *crack propagation*) until one of the flaws exceeds a critical size and the fiber fails. Unfortunately, this crack propagation process causes delayed failure in a fiber after it has been deployed or installed. A time history of a typical crack size is shown in Figure 3.15. Here the crack size a remains constant when no stress is applied to the fiber ($t < 0$). With stress applied ($t \geq 0$), the crack size grows until it reaches the critical size, causing failure.

Three factors are required for the crack propagation to occur: an applied stress, the presence of flaws, and the presence of moisture. (Samples that have been carefully dried in an inert, moisture-free environment do not show crack growth. One of the strengthening methods under consideration is to coat the fiber with a hermetic [moisture-resistant] barrier, as discussed in Chapter 12, on fiber fabrication.) Reduction of any of the three required factors will reduce the crack propagation.

Based on models of fracture mechanics (Kalish et al., 1979; Kapron and Yuce, 1991; Kurkjian and Inniss, 1991), the predicted flaw size a_c that will cause a failure at a stress level of σ_f is

$$a_c = \left(\frac{K_{Ic}}{Y \sigma_f} \right)^2 , \tag{3.172}$$

where K_{Ic} is a mechanical strength property of the material ($K_{Ic} = 0.81$ MPa $\cdot \sqrt{\text{nm}}$ for silica glass) and Y is a parameter that depends on the shape of the flaws in the material ($Y = \sqrt{\pi}$ for a surface flaw).

Example: Calculate the critical flaw size that corresponds to the theoretical strength of glass, $\sigma_f = 14$ GPa.

Solution: The flaw size is found from

$$a_c = \left(\frac{K_{Ic}}{Y\sigma_f} \right)^2 \tag{3.173}$$

$$= \left(\frac{(0.81 \times 10^6)}{(\sqrt{\pi})\,(14 \times 10^9)} \right)^2 \tag{3.174}$$

$$= 1.065 \times 10^{-9} \tag{3.175}$$

$$= 1.065 \text{ nm} . \tag{3.176}$$

Hence, we find that a "perfect" fiber (in terms of strength) requires flaws that are less than 1 nm in size.

The *useful life* of the fiber is the time that it endures under the stress load until failure. Since the useful life of the fiber has obvious system implications, we seek some basic understanding of the processes and parameters involved. For applications requiring high mechanical stress, it should be noted that the design of the optical cable will include sufficient strength members to carry the major stresses. The fiber will be isolated from the loads in a properly designed cable and ideally will have only negligible stress applied.

In characterizing fiber strength, it is desirable to be able to predict the strength of a long piece of fiber (i.e., a kilometer or more) based on the breaking strength of short samples (i.e., a few meters). Since the distribution of flaws of certain sizes along a lengthy fiber is statistical in nature, a statistical approach must be used in this prediction. This requires that a statistically meaningful number of samples must be used to acquire sufficient confidence in the predicted strength. Usually, an empirical fit is made to apply a standard statistical curve (called the *Weibull distribution*) to the data accumulated by testing the fiber. If $F(\sigma,L)$ is the cumulative probability that a fiber of length L will fail below a stress level σ, then the surviving fibers have a probability distribution, $S = 1 - F$, that is given (Nagel, 1988) by

$$S = 1 - F \tag{3.177}$$

$$= e^{-(L/L_o)\,(\sigma/\sigma_o)^m} , \tag{3.178}$$

where L is the length of the fiber under test, σ is the breaking stress, and m is an empirical parameter fitted to the experimental data for the fiber, and L_o and σ_o are constants. (Large values of m, in the range of 25 to 35, indicate a fiber with negligible flaws.) Taking the natural logarithm of both sides of Equation 3.178 twice in succession, we can write

$$\ln \ln \left(\frac{1}{1-F}\right) = \ln L + m \ln \sigma + \text{constants} . \tag{3.179}$$

This equation represents a linear expression in terms of m. Figure 3.16 shows a typical set of data for an experimental test. Here, 50 fibers of some length (say, 20 m) have been tested under carefully controlled environmental conditions (especially the relative humidity and temperature) until they break. The breaking strengths are ordered from the lowest to the highest (i.e., from just above $4 \text{ GN} \cdot \text{m}^{-2}$ to just below $6 \text{ GN} \cdot \text{m}^{-2}$). Since there are 50 samples, it is noted that each sample represents 2% of the total sample set. The data is plotted on commercially available Weibull plotting paper, as in Figure 3.16. A straight line is fitted to the data, and the slope is computed to determine the value of m. The value of m for this sample is about 20.

The 50% failure probability is a frequently used benchmark, although it contains no information about the spread of the data. For example, 50% of our samples would have broken at a strength of approximately $5.5 \text{ GN} \cdot \text{m}^{-2}$.

To extrapolate the results to longer samples of, say, 1 km requires a scaling computation that relates the failure stress at one length, σ_{f1}, to the failure stress at the second length, σ_{f2}. From our equations we find

$$\frac{\sigma_{f1}}{\sigma_{f2}} = \left(\frac{L_2}{L_1}\right)^{1/m} . \tag{3.180}$$

In terms of a Weibull plot this computation equates with a translation of the vertical axis by an amount of $\ln(L_2/L_1)$; this axis shift often is done graphically.

Example: Consider the data sample shown in Figure 3.16. If the 50% failure strength of a set of 20 m samples is $5.2 \text{ GN} \cdot \text{m}^{-2}$ and the estimated slope of the line fitted to the data is 20, calculate the 50% failure strength of 1 km samples of the same fiber (under the same environmental conditions).

Solution: The estimated failure strength is

$$\sigma_{f1} = \sigma_{f2} \left(\frac{L_2}{L_1}\right)^{1/m} \tag{3.181}$$

$$= 5.2 \left(\frac{20}{1 \times 10^3}\right)^{1/20} \tag{3.182}$$

$$= 4.2 \text{ GN} \cdot \text{m}^{-2} . \tag{3.183}$$

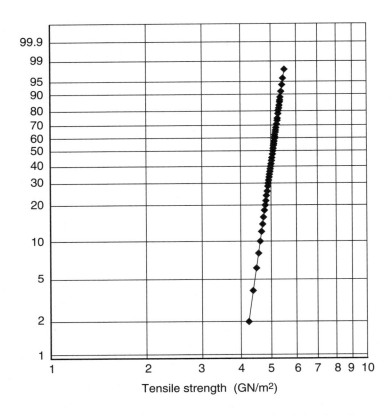

Figure 3.16 Representative experimental results of short-length
stress tests of 50 filters fitted with a Weibull distribution.

We note that longer length will cause the median fiber to break at a lower
stress level. This is because the longer fiber has a wider variety of flaws
than the shorter sample.

Fibers have actually been manufactured with flaw sizes below the critical
value for theoretical strength (Nagel, 1988; Kurkjian et al., 1989; Kurkjian and Innis,
1991), resulting in values of m in the range of 35 to 40. This low-flaw fiber is accomplished by firepolishing the glass used to make the fiber and by carefully filtering the
coating material to avoid impure particles that might induce cracks in the fiber surface. Such processing increases the cost of fibers and, even then, does not ensure that
subsequent handling of the fiber in cabling, splicing, and installation will not induce
flaw formation. (The user must be careful not to confuse data from "pristine" fibers

in the laboratory environment, just after manufacture, with performance after the fiber has been cabled and installed in the operating environment.)

The presence of water vapor at a crack location has a severe deleterious effect on the fiber strength and lifetime (Kalish et al., 1978). For example, a fiber with a static-fatigue strength of 2.5 GPa has a lifetime of several years at 2% relative humidity, but at 97% relative humidity it will last about 1 hour. The effect of the humidity is not yet quantized, but can be included empirically in the values of the parameter m by measurement under the expected service conditions. The mechanism for reducing fiber strength is thought to be a weakening of the molecular bonds of the glass at the tip of the crack because of infiltration of water vapor. (It should be noted that water vapor hastens the crack propagation process. Fiber under zero or extremely low stress will not fail just by the presence of water vapor.) One method of reducing water-vapor levels at the fiber is to hermetically seal the fiber by applying one or more metallic layers to the fiber in the drawing process before coating. These fibers are discussed in more detail in Chapter 12.

3.4.2 Time-Dependent Fiber Strength

Having seen how variable the strength of a fiber is, we now further complicate the picture by taking into account the time-dependence of the failure mechanism. Ultimately we hope to be able to predict the minimum service lifetime of a fiber when subject to certain maximum stress conditions. These lifetime predictions depend on destructive testing of samples of the fiber.

Two types of time-dependent strength testing can be done, static-fatigue testing and dynamic-fatigue testing. In both tests it is particularly important to handle the fiber carefully to avoid introduction of new flaws that can skew the results of the test.

Static-Fatigue Testing

In *static-fatigue testing* a group of fiber samples is subjected to a constant stress over a (long) period of time under carefully controlled environmental conditions. Usually, several groups of fibers are tested at differing levels of applied stress. The time-to-failure is noted for each sample as it breaks. Theory (Ritter, 1978; Kalish et al., 1979; Nagel, 1988) predicts that the time-to-failure under static testing t_{fs} is given by

$$t_{fs} = \frac{K_s}{\sigma_f^n},$$

(3.184)

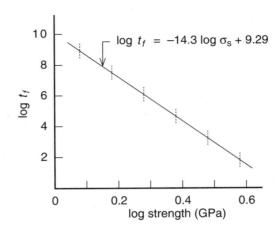

Figure 3.17 Typical data from a static fatigue test.

where K_s is a material constant, σ_f is the stress applied, and n is a fiber parameter called the *stress corrosion susceptibility parameter* (that is to be measured with either static- or dynamic-fatigue testing). Taking the logarithm of this equation we have

$$\log t_{fs} = -n \log \sigma_f + \log K_s \qquad (3.185)$$

The experiment proceeds as follows. A randomized assembly of fiber samples are tested under carefully controlled environmental conditions. Several levels of constant stress are applied to subgroups of fiber samples and the failure time of each sample is recorded. (One major disadvantage of static testing is the long time-to-failure of the strongest fibers in the sample.) The observed data is plotted as the log of failure times versus the stress level, as in Figure 3.17. The average time-to-failure of each subgroup is computed. A straight-line curve is fit to the average time-to-failure of each subgroup as a function of the log of the applied stress (as in Figure 3.17). The slope of the curve gives n. This parameter will be used to predict the useful life of the fiber.

Dynamic-Fatigue Testing

In *dynamic-fatigue testing* a linearly increasing stress (in time) $\dot{\sigma}$ is applied to the fiber samples. The time-to-failure of each sample is measured. (The primary advantage of dynamic-fatigue testing is that it is quicker to perform than static-fatigue testing.) Theory of dynamic-fatigue testing predicts a time-to-failure t_{fd} at a failure stress level σ_f (in Pa) that is given by the relation (Kalish et al., 1979; Nagel, 1988)

$$t_{fd} = \frac{K_d}{\sigma_f^n} .$$

(3.186)

The failure stress is also given by $\sigma_f = \dot{\sigma} t_{fd}$. Manipulation of these two equations gives

$$\log \sigma_f = \left(\frac{1}{n+1} \right) \log \dot{\sigma} + \left(\frac{1}{n+1} \right) \log K_d ,$$

(3.187)

where K_d is a material constant, n is the stress corrosion susceptibility parameter, and σ_f is the applied stress at failure. The material constants K_s and K_d are related by (Kalish et. al., 1979; Nagel, 1988)

$$\log K_d = \log K_s + \log (n + 1) .$$

(3.188)

The experiment proceeds as follows. The fiber samples are prepared and a stress is applied that linearly increases in time. As each fiber breaks, the time of breakage is noted (usually from video recording of the test that incorporates a clock timer in the image). The breaking stress is calculated from the measured breaking time and the known value of the linear stress rate. The log of the breaking strength of the fiber is plotted against the log of the stress rate $\dot{\sigma}$. The measured slope of the straight line fitted to the data gives the desired value of n. The experiment is repeated with different rates of linear increase (about four different rates are usually used) and subsequent values of n are found. An average value of n is found from the tests.

Lifetime Predictions

Once n is known, Equation 3.184 can be used to calculate the predicted time-to-failure t_{fs1} for a fiber when subjected to a lower stress level σ_{f1} in its application by the proportionality,

$$\frac{t_{fs1}}{t_{fs2}} = \left(\frac{\sigma_{f2}}{\sigma_{f1}} \right)^n .$$

(3.189)

3.4.3 Intercomparison of Testing Results

Values of n found from these two methods are widely scattered (Kalish et. al., 1979; Nagel, 1988; Kurkjian et. al., 1989) but usually lie between a minimum of 15 and a maximum of 45 (except for some "super-strength" fibers made with special fabrication processes). Experiments have been reported where the static-fatigue values of n are either above or below the dynamic-fatigue values. The differences are attributed to experimental setup differences and control of the environmental conditions. Within the same experimental setup, one measurement or the other is usually smaller; this smaller value is assumed as a conservative estimate of n. Efforts continue to resolve the experimental difficulties in reproducibly measuring a valid value of n. For now, however, we find that the dependence of n on experimental and environmental conditions is not well understood and that care must be taken in extrapolating results from short-length short-term fatigue data into lifetimes for long-length long-term fatigue applications.

3.4.4 Prooftesting

While Figure 3.16 represents the usual stress measurement of an optical fiber, Figure 3.18 represents the typical data after the fiber has been cabled, stored, handled, and installed. In the figure, we note that much of the fiber is high in strength, but a low-strength tail has appeared, which contains weak pieces of the fiber. Because we lack the detailed understanding to control the appearance of these weak sections of fiber, we seek a screening mechanism that can be used to remove the undesirable sections (Bogatyrjov et. al., 1991; Glaesemann and Gulati, 1991).

One technique that is used to raise the overall minimum time-to-failure is that of *prooftesting*. In prooftesting, a fiber is momentarily subjected to a stress level that exceeds the static stress that is anticipated in operation (see Figure 3.19). In the prooftest the optical fiber is taken off of its reel and passed through a friction pulley arrangement to a region of high stress (the darkened portion of the fiber in Figure 3.19), where it is subjected to a constant stress (the *proof strength*) which is several times higher than that anticipated in actual operation of the fiber. The fiber is then returned to its unloaded state through another set of friction pulleys. (The unloading of the fiber must be done rapidly to avoid any significant growth in the flaws that survived the prooftest (Bogatyrjov et al., 1991).) It should be noted that the higher the proof strength, the more capable the fiber, but at an increased cost of diminished length of the fiber. (The fiber will break more frequently at higher values of proof strength.) While this disadvantage can be partially overcome with the use of high-strength splices, the increased handling of the fiber will increase fiber costs and, perhaps, weaken the fiber after testing. Miyajima (1983) has analyzed the trade-offs

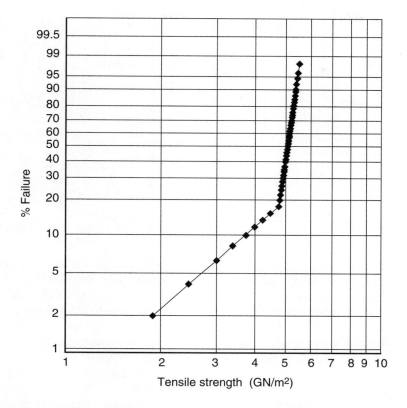

Figure 3.18 Typical data from a strength test of a fiber after normal handling.

involved in setting proof-strength levels and the costs involved in setting them too high.

All fiber flaws that exceed a certain value will break when subjected to the prooftest. Hence, prooftesting places an upper limit on the size of the flaws in the fiber (and, undesirably, ensures that small flaws will become bigger as a result of the test). By truncating the distribution of flaw sizes in this way, the surviving fibers will be stronger (on the average) than the fibers originally tested. The time-to-failure of a fiber after prooftesting can be shown to be (Bogatyrjov et al., 1991)

$$t_{fs} = \frac{B\sigma_p^{n-2}}{\sigma_s^n} \,, \tag{3.190}$$

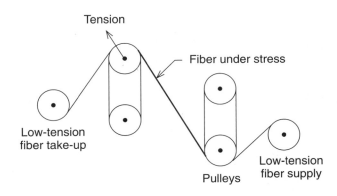

Figure 3.19 Mechanism for fiber prooftest.

where t_{fs} is the time-to-failure under static stress after prooftesting, B is a fiber constant, n is the static corrosion susceptibility factor measured from the dynamic- or static-fatigue testing, and σ_s is the static stress that is applied to the fiber in its application. The B parameter can be obtained either by destructively testing some samples of the fiber after prooftesting or from the dynamic- or static-fatigue data, along with some fiber strength information measured in an inert environment (usually gaseous nitrogen or liquid nitrogen) (Ritter, 1978; Kalish et al., 1979).

Plots of Equation 3.190 are called *lifetime prediction diagrams* (see Figure 3.20). Three curves are plotted for three different prooftest levels. The fibers with the higher prooftest strengths will have longer lifetimes at any given level of applied stress. Alternatively, the diagram can be used to find the maximum allowed stress that can be applied to a fiber that has been prooftested to a certain level to ensure a certain minimum required lifetime.

Equation 3.190 is sometimes used backwards to compute the proof stress σ_p required to test a fiber that is required to have a minimum lifetime $t_{fs\,min}$ at some specified applied stress as

$$\sigma_p = \left(\frac{\sigma_a^n\, t_{fs\,min}}{B} \right)^{1/(n-2)} . \qquad (3.191)$$

It should be noted that $t_{fs\,min}$ is a conservative estimate of the fiber lifetime. All prooftested fibers will exceed this time *(if handled properly)*. The average fiber will last considerably longer.

It should be noted that the prooftest level is frequently specified as a certain amount of *strain* of the fiber. Strain is defined as the elongation of the fiber over the

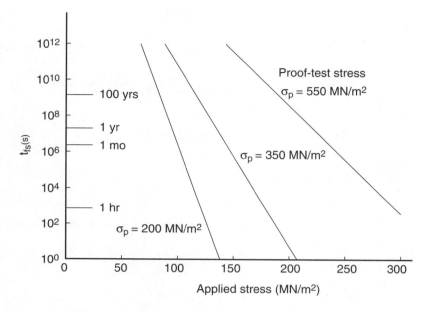

Figure 3.20 Representative optical fiber lifetime prediction diagram.

original length and is usually expressed as a percentage. While metallic conductors can suffer strains of up to 20% without breaking, allowable fiber strains are much less, typically 3% or so. Stress and strain are related by the Young's modulus of the material—a parameter associated with the elasticity of the material. For glass, the Young's modulus is approximately 68.9 GPa (10^7 psi), so a 1% strain is a stress of 689 MPa (10^5 psi). Frequently, a 1% strain is used as a standard value of prooftest.

So far, we have assumed implicitly that the fiber is under a constant stress during its application. This assumption is frequently violated by the installation and maintenance procedures. One additional question is the effect of the loading history on the life of the fiber. Consider an undersea fiber with the following stress history:

- constant low stress while on a reel,
- a different low stress on the fiber while in the hold of the cable-laying ship,
- a high stress level during cable deployment,
- a low to moderate stress level while on the ocean floor (although this can be highly variable depending on the geography of the immediate location of the cable),
- maximum stress during recovery from the ocean floor,

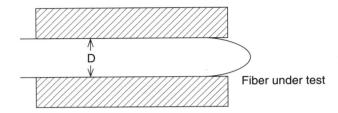

Figure 3.21 Diagram of two-point bending test.

- ■ low stress during repair, and
- ■ high stress during redeployment.

How are we to estimate the lifetime of the fiber after redeployment? The estimated lifetime is (Kalish et. al., 1978)

$$t_{min} = \frac{B\sigma_p^{n-2} - \sum_{i=1}^{m} t_i\sigma_i^n}{\sigma_a^n} \;,$$

(3.192)

where the summation term in the numerator reflects the decrease in lifetime owing to the stress-time history of the fiber. Here σ_i is the stress level and t_i is the time duration at that level of stress. We note that, while dependent on the accumulated stress history of the fiber, the order of the stress sequences is irrelevant in determining the fiber lifetime after redeployment.

3.4.5 Two-Point Bending Test

So far, we have been considering tensile strength; the fiber has been subject to an axial stress. Another scenario that stresses fibers occurs when a fiber turns around a post or is wrapped on a spool (as occurs in fiber-guided missiles). The model for the stress is beyond the scope of this text and is discussed by Roberts et al. (1991) and Mathewson et al. (1986). In testing the fiber strength for this configuration the *two-point bending test* is used. A short length of fiber is bent and placed in grooves in two plates (see Figure 3.21). One of the plates is moved steadily inward until the fiber breaks. (Usually many fibers are tested simultaneously, and the plate position at breaking is recorded for each fiber.) The breaking stress is predicted from theory as (Biswas, 1991)

$$\sigma_f = 1.1978E\left(\frac{d_f}{D-d}\right),$$

(3.193)

where σ_f is the breaking stress, E is the Young's modulus of glass, d_f is the diameter of the fiber, D is the plate separation at breaking, and d is the fiber diameter (including the coating material). Weibull plots are usually made from the sample, with a special correction to account for the "effective gauge length" (Mathewson, 1986) of the sample that is actually under stress.

3.5 Summary

In this chapter we introduced more of the parameters used to describe a fiber's performance. The losses of the fiber are due to various factors and predict a loss minimum occurring at 1550 nm for silica fibers—a wavelength that is currently being exploited by the industry. The bandwidth or data-rate limitations are because of various dispersion factors that can cause a pulse to spread. In multimode fibers, modal- and material-dispersion effects dominate. In single-mode fibers, modal dispersion is absent but material and waveguide dispersions are significant. Near 1300 nm, the combination of material dispersion and waveguide dispersion can produce zero dispersion. This region produces the minimum dispersion of the fiber.

We have seen that fiber strength is difficult to characterize because glass exhibits the peculiar characteristics of failing suddenly and always moving toward failure when under stress. The statistical nature of the surface flaws that can cause failure adds further complications to a simplified model for fiber failure. Fiber strength is statistically characterized by destructively testing short samples and extrapolating the results to longer lengths (with an appropriate skepticism about the extrapolation predictions). Lifetime predictions can be done with the aid of dynamic test results. To remove some of the variability of fiber breaking strength, prooftesting can be performed to remove the weaker pieces of fiber, at an increased cost, since the fiber will generally be available in shorter lengths after prooftesting. The user must exhibit care in handling high-strength fiber to avoid inducing larger flaws in the fiber after the prooftesting.

3.6 Problems

1. Find an expression that relates α (in units of dB/km) to α_p (in units of m^{-1}).
2. An optical fiber has a loss of 1 dB/km. Find the corresponding value of the loss coefficient α (in units of m^{-1}).
3. The optical power after propagating through a fiber that is 450 m long is reduced to 30% of its original value. Calculate the fiber loss α_p in dB/km.

4. (a) Calculate the pulse delay for a 1 km single-mode fused silica step-index fiber if $\lambda = 1.0\ \mu m$, $\Delta\lambda/\lambda = 0.12\%$, $V = 1.5$, $n_1 = 1.453$, and $n_2 = 1.450$.

   (b) Calculate the pulse delay distortion of a multimode step-index fiber that is 1 km long and has $n_1 = 1.453$ and $n_2 = 1.438$ with $V = 38$.

5. Consider a fused silica single-mode fiber that is 1 km long. Find the material dispersion component of the delay distortion if $\Delta\lambda = 3.0$ nm and the operating wavelengths are 800 nm, 900 nm, 1300 nm, and 1500 nm.

6. Calculate the material dispersion delay at $\lambda = 820$ nm, $\Delta\lambda = 3$ nm, $L = 2$ km, for a fused silica single-mode fiber.

7. Consider a 0.80 km fiber made of fused silica with a step-index. The following applies: $\lambda = 1.0\ \mu m$, $\Delta\lambda/\lambda = 0.12\%$, $V = 38$, $n_1 = 1.453$, and $n_2 = 1.438$.

   (a) Calculate the pulse spread due to group delay.

   (b) Calculate the pulse spread due to material dispersion.

8. Using a computer, plot $b(V)$, $d(Vb)/dV$, and $V\,d^2(Vb)/dV^2$ over a range of $V = 0$ to $V = 3$.

9. Consider a 9/125 single-mode fiber with $n_1 = 1.48$ and $\Delta = 0.22\%$ as described in the waveguide dispersion example calculation. Let $L = 1000$ and $\Delta\lambda = 1$ nm.

   (a) Plot $\Delta\tau_{wg}$ for wavelengths ranging from 1200 nm to 1400 nm.

   (b) Plot $\Delta\tau_{mat}$ for the same range of wavelengths.

   (c) Plot the sum of the material pulse spread and the pulse spread due to waveguide dispersion.

10. We want to compute the fraction of a tensile load that is carried by the coating of a fiber. The fraction of the total stress σ_{total} that is carried by the coating is found from

$$\frac{\sigma_{coating}}{\sigma_{total}} = \frac{E_{coating}A_{coating}}{E_{coating}A_{coating} + E_{fiber}A_{fiber}}, \qquad (3.194)$$

where $E_{coating}$ and E_{fiber} are the Young's modulus of the coating and fiber material, respectively, and $A_{coating}$ and A_{fiber} are the cross-section areas of the coating and fiber respectively.

Consider a 62.5/125 fiber that has a polymer coating that is 0.05 mm thick surrounding it. The Young's modulus of the coating polymer is 350 MPa and of glass is 71.9 GPa.

   (a) Calculate the fraction of the total applied stress that is carried by the coating.

(b) Calculate the fraction of the total applied stress that is carried by the fiber.

(c) Suppose that the fiber were double-coated with two different concentric coatings. What do you think would be the expression for the fraction of the total stress carried by the interior coating?

References

Agrawal, G. P., *Nonlinear Fiber Optics*. New York: Academic Press, 1989.

Ainslie, B. J. and C. R. Day, "A review of single-mode fibers with modified dispersion characteristics," *J. Lightwave Technology*, vol. LT-4, no. 8, pp. 967–979, 1986.

Artiglia, M., G. Coppa, P. Di Vita, H. Kalinowski, and M. Potenza, "Simple and accurate microbending loss evaluation in generic single-mode fibers," *Proc. 12th ECOC (Barcelona, Spain)*, vol. 1, pp. 341–344, 1986.

Artiglia, M., G. Coppa, P. Di Vita, H. Kalinowski, and M. Potenza, "Bending loss characterization in single-mode fibers," *CSELT Technical Report*, vol. XV, no. 6, pp. 411–415, 1987.

Artiglia, M., G. Coppa, P. DiVita, M. Potenza, and A. Sharma, "Mode field diameter measurements in single-mode optical fibers," *J. Lightwave Technology*, vol. 7, no. 8, pp. 1139–1152, 1989.

Bendow, B., H. Rast, and O. H. El-Bayoumi, "Infrared fibers: an overview of prospective materials, fabrication methods, and applications," *Optical Engineering*, vol. 24, no. 6, pp. 1072–1080, 1985.

Bhagavatula, V., J. Lapp, A. Morrow, and J. Ritter, "Segmented-core fiber for long-haul and local-area-network applications," *J. Lightwave Technology*, vol. 6, no. 10, pp. 1466–1469, 1988.

Biswas, D. R., "Characterization of polyimide-coated fibers," *Optical Engineering*, vol. 30, no. 5, pp. 772–775, 1991.

Bogatyrjov, V., M. Bubnov, E. Dianov, S. Rumyantzev, and S. Semjonov, "Mechanical reliability of polymer-clad and hermetically coated fibers based on proof testing," *Optical Engineering*, vol. 30, no. 6, pp. 690–699, 1991.

Cherin, A., *Introduction to Optical Fibers*. New York: McGraw-Hill, 1983.

100

Chraplyvy, A. R., "Limitations on lightwave communications imposed by optical-fiber nonlinearities," *J. Lightwave Technology*, vol. 8, no. 10, pp. 1548–1557, 1990.

Gambling, W., H. Matsumura, and C. Ragsdale, "Zero total dispersion in graded-index single-mode fibers," *Electronics Letters*, vol. 15, pp. 474–476, 1979a.

Gambling, W., H. Matsumura, and C. Ragsdale, "Mode dispersion and profile dispersion in graded-index single-mode fibres," *Microwaves, Optics, and Acoustics*, vol. 3, pp. 239–246, 1979b.

Gartside, C. H. III, P. D. Patel, and M. R. Santana, "Optical fiber cables," in *Optical Fiber Telecommunications II*, (S. E. Miller and I. P. Kaminow, eds.), pp. 217–261, New York: Academic Press, 1988.

Glaesemann, G. and S. Gulati, "Design methodology for the mechanical reliability of optical fiber," *Optical Engineering*, vol. 30, no. 6, pp. 709–714, 1991.

Gloge, D., "Dispersion in weakly guiding fibers," *Applied Optics*, vol. 10, no. 9, pp. 2442–2445, 1971a.

Gloge, D., "Weakly guiding fibers," *Applied Optics*, vol. 10, no. 10, pp. 2252–2258, 1971b.

Gloge, D., E. A. Marcatili, D. Marcuse, and S. D. Personick, "Dispersion properties of fibers," in *Optical Fiber Telecommunications*, (S. E. Miller and A. G. Chynoweth, eds.), pp. 101–124, New York: Academic Press, 1979.

Kaiser, P. and D. B. Keck, "Fiber types and their status," in *Optical Fiber Telecommunications II*, (S. E. Miller and I. P. Kaminow, eds.), pp. 29–54, New York: Academic Press, 1988.

Kalish, D., B. Tariyal, and H. Chandan, "Effect of moisture on the strength of optical fibers," in *Proc. of the 27th International Wire and Cable Symposium*, pp. 331–341, 1978.

Kalish, D., P. L. Key, C. R. Kurkjian, B. K. Tariyal, and T. T. Wang, "Fiber characterization—mechanical," in *Optical Fiber Telecommunications*, (S. E. Miller and A. G. Chynoweth, eds.), pp. 401–433, New York: Academic Press, 1979.

Kapron, F. P., "Fiber-optic test methods," in *Fiber Optics Handbook for Engineers and Scientists*, (F. C. Allard, ed.), pp. 4.1–4.54, New York: McGraw-Hill, 1990.

Kapron, F. P. and H. H. Yuce, "Theory and measurement for predicting stressed fiber lifetime," *Optical Engineering*, vol. 30, no. 6, pp. 700–708, 1991.

Keiser, G., *Optical Fiber Communications,* 2nd Edition. New York: McGraw-Hill, 1991.

Kurkjian, C. B., J. T. Krause, and M. J. Matthewson, "Strength and fatigue of silica optical fibers," *J. Lightwave Technology*, vol. 7, no. 9, pp. 1360–1370, 1989.

Kurkjian, C. and D. Inniss, "Understanding mechanical properties of light-guides: a commentary," *Optical Engineering*, vol. 30, no. 6, pp. 681–689, 1991.

Lemaire, P. J., "Reliability of optical fibers exposed to hydrogen: prediction of long-term loss increase," *Optical Engineering*, vol. 30, no. 6, pp. 780–789, 1991.

Marcuse, D. and H. Presby, "Effects of profile deformations on fiber bandwidth," *Applied Optics*, vol. 18, pp. 3758–3763, 1979.

Mathewson, M. J., C. R. Kurkjian, and S. T. Gulati, "Strength measurement of optical fibers by bending," *J. American Ceramics Soc.*, vol. 69, no. 11, pp. 815–821, 1986.

Midwinter, J., *Optical Fibers for Transmission*. New York: Wiley, 1979.

Miyajima, Y., "Studies on high-tensile proof tests of optical fibers," *J. Lightwave Technology*, vol. LT-1, no. 2, pp. 340–346, 1983.

Miyamoto, M., M. Sakai, R. Yamauchi, and K. Inada, "Bending loss evaluation of single-mode fibers with arbitrary core index profile by far-field pattern," *J. Lightwave Technology*, vol. 8, no. 5, pp. 673–677, 1990.

Nagel, S.R., "Fiber material and fabrication methods," in *Optical Fiber Telecommunications II*, (S. E. Miller and I. P. Kaminow, eds.), pp. 121–215, New York: Academic Press, 1988.

Olshansky, R. and D. Keck, "Pulse broadening in graded-index optical fibers," *Applied Optics*, vol. 15, no. 12, pp. 483–491, 1976.

Olshansky, R., "Multiple-α index profiles," *Applied Optics*, vol. 18, pp. 683–689, 1979a.

Olshansky, R., "Propagation in glass optical waveguides," *Review of Modern Physics*, vol. 51, pp. 341–367, 1979b.

102

Paek, U., G. Peterson, and A. Carnevale, "Dispersionless single-mode light-guides with α index profiles," *Bell Sys. Technical J.*, vol. 60, pp. 583–598, 1981.

Petermann, K. and R. Kuhne, "Upper and lower limits for the microbending losses in arbitrary single-mode fibers," *J. Lightwave Technology*, vol. LT-4, no. 1, pp. 3–7, 1986.

Ritter, J. E., Jr., "Probability of fatigue failure in glass fibers," *Fiber and Integrated Optics*, vol. 1, no. 4, pp. 387–389, 1978.

Roberts, D. R., E. Cuellar, M. T. Kennedy, and John E. Ritter, Jr., "Calculation of the static fatigue lifetime of an optical fiber," *Optical Engineering*, vol. 30, no. 5, pp. 716–727, 1991.

Sakaguchi, S. and S. Takahashi, "Low-loss fluoride optical fibers for mid-infrared optical communication," *J. Lightwave Technology*, vol. LT-5, no. 9, pp. 1219–1228, 1987.

Schick, G., K. A. Tellefsen, A. J. Johnson, and C. J. Wieczorek, "Hydrogen sources for signal attenuation in optical fibers," *Optical Engineering*, vol. 30, no. 6, pp. 790–801, 1991.

Senior, J., *Optical Fiber Communications: Principles and Practice*. Englewood Cliffs, NJ: Prentice Hall, 1985.

Stolen, R., "Nonlinear properties of optical fibers," in *Optical Fiber Telecommunications*, (Miller, S.E. and Chynoweth, A.G., eds.), pp. 125–150, New York: Academic Press, 1979.

Stolen, R., "Nonlinearity in fiber transmission," *Proc. IEEE*, pp. 1232–1236, 1980.

Tran, D. C., George H. Sigel, Jr., and B. Bendow, "Heavy metal fluoride glasses and fibers: a review," *J. Lightwave Technology*, vol. LT-2, no. 5, pp. 566–586, 1984.

White, K., "Design parameters for dispersion-shifted triangular profile single-mode fibers," *Electronics Letters*, vol. 18, pp. 725–727, 1982.

Part Two

Fiber-Link Components

Chapter 4

Optical Sources

4.1 Introduction

In this chapter we will consider the optical sources currently used in fiber-optic systems, describe their operational parameters, and consider coupling between the source and the fiber.

The sources currently used with fiber optics are semiconductor light sources (Kressel and Butler, 1977; Burrus et al., 1979; Bergh and Copeland, 1980; Selway et al., 1980; Kressel, 1981; Bowers and Pollock, 1988; Lee et al., 1988; Shumate, 1988; Yu et al., 1990; Lee, 1991), *light emitting diodes* (LEDs) or *semiconductor lasers*. These sources enjoy a combination of usable properties in size, wavelength availability, power, linearity, simplicity of modulation, low cost, and reliability that make them suitable for this application. Sources are currently classified according to wavelength. *Short-wavelength sources* produce light from 500 to 1000 nm and are typically a ternary blend of semiconductors such as GaAlAs. *Long-wavelength sources* operate in the region of 1200 to 1600 nm in an effort to minimize fiber losses and dispersion effects and, for reasons dealing with the physical properties of semiconductor materials, are made up of quaternary semiconductor blends such as InGaAsP.

The devices are forward-biased semiconductor *pn* junctions. The materials for making the diodes are selected from semiconductors allowing direct transitions and are typically doped much more heavily than an electronic diode. The mechanism for producing light requires current densities of relatively high value compared to most other electronic devices. Charges in the current flow tend to spread themselves widely apart as they cross the junction, lowering the current density. To overcome this spreading effect, current flow is confined to a small area. To help confine the light to a low-loss portion of the device, other doped layers are added on either side of the *pn* junction. The resulting junctions, called *homojunctions* or *heterojunctions* depending on their material composition, have served to increase the overall operating efficiency of the devices. Because temperature effects are important in semiconductors, this overall increase in the efficiency of the sources has led to increased stability and reliability as well.

4.2 Light Generation by Semiconductors

Light can be generated by a radiative recombination of an electron and a hole within the semiconductor. Electrons and holes can also recombine nonradiatively, producing heat within the semiconductor but no light. The fraction of the total recombinations that occur radiatively is expressed by the *internal quantum efficiency* η_i of the device. An efficient semiconductor light generator (i.e., one with a high internal quantum efficiency) will have many more radiative recombinations than nonradiative recombinations. This is accomplished by choosing the semiconductor material properly and by flooding the light-generating region with charge carriers. This flooding with charge carriers is done by using energy barriers to confine the carriers in the vicinity of a forward-biased junction.

When a *pn* junction is forward biased, holes will be injected from the *p* material into the *n* material and electrons will be injected from the *n* material into the *p* material. After crossing the *pn* junction, the injected *minority carriers* will find themselves in the presence of majority carriers of the opposite charge and will recombine. The energy possessed by the charge-carrier pair before the recombination process can be converted into either electromagnetic radiation (a *radiative transition*) or into heat (a *nonradiative transition*). The energy E produced by the recombination is approximately equal to the band-gap energy E_g of the material. If the energy E of a radiative transition corresponds to an optical frequency ($E = h\nu$), then light is generated. (Some materials produce rf and microwave radiation.) For an optical source, we want the radiative transitions to totally dominate the nonradiative emissions.

In general, there are two types of radiative recombinations. In a *spontaneous emission*, the light produced is incoherent, randomly polarized, and randomly directed. These recombinations are used in light-emitting diodes. The second radiative recombination requires the presence of a stimulating light wave to trigger the recombination. The light produced by this *stimulated emission* is in phase with the stimulating light (i.e., coherent light is produced), of the same polarization, and in the same direction. Lasers make use of these stimulated emissions to produce their light.

Recombination of a charge-carrier pair requires that the vector momentum of the carriers be conserved. (The momentum of a particle is mv; the momentum of a photon is hk, where $|k|$ is $2\pi/\lambda$ and the direction of k is the direction of emission.) In silicon and germanium, it is difficult to meet the momentum conservation requirement without a third momentum vector from an intermediary phonon (i.e, some of the heat that is within the semiconductor). This is a so-called *indirect transition* and is so inefficient that we do not make optical sources from either silicon or germanium.

Some semiconductors that are made from alloys of elements in columns III and V of the periodic chart, and columns II and VI, do not require this third momentum term and produce a *direct transition*. The direct transition in these materials is much more probable than the indirect transition in silicon and germanium. Some examples of the alloys used to make sources for fiber-optic communications are mixtures of gallium (Ga), aluminum (Al), and arsenic (As), and mixtures of indium (In), gallium (Ga), arsenic (As), and phosphorous (P). Other materials can be used to make sources for non-fiber applications.

The diode devices are made by growing layers of semiconductor material on top of one another, using the process of liquid-phase epitaxy. In this process, a substrate crystal is exposed to a liquid solution containing the proper mixture of components. Under precise temperature control, the liquid precipitates on the substrate crystal. If done properly, the crystalline structure of the layer mimics that of the substrate or layer below it. This requires that the spacing of the atoms of the grown layer (the so-called *lattice spacing*) be close to the lattice spacing of the lower layer. (Too much difference in the lattice spacing leads to mechanical strain between the layers and a subsequent large increase in the nonradiative transitions.) Layers are grown on top of each other, with the thickness of each layer being precisely controlled by the temperature of the media and the time duration of the growth process. Other techniques, including vapor-phase epitaxy, molecular-beam epitaxy, and metal-organic chemical vapor beam deposition are also being studied and used to construct optical-diode sources (Long et al., 1988; Yu et al., 1990).

4.2.1 Wavelength and Material Composition

The wavelength of the emitted light is determined by the *bandgap energy* E_g of the material. (The bandgap energy is the energy required to create a hole-electron pair in the material. This energy is recovered when the hole and the electron recombine.) The relationship between bandgap energy and nominal wavelength is

$$\lambda = \frac{hc}{E_g}, \tag{4.1}$$

where λ is the nominal emitted light wavelength, h is Planck's constant (6.63×10^{-34} joules $\cdot$ sec), and E_g is the bandgap energy (usually specified in eV, where 1 eV $= 1.6 \times 10^{-19}$ joules). If the wavelength is specified in units of μm and the bandgap energy is in units of eV, the relationship reduces to

$$\lambda(\text{in } \mu\text{m}) = \frac{1.240}{E_g(\text{in eV})}. \tag{4.2}$$

Since the bandgap energy depends on the material composition, the wavelength of the emitter is determined by the material composition of the emitter. (The wavelength is also a function of operating temperature of the device. The wavelength is shifted toward a longer wavelength at a rate of 0.6 nm/C by temperature increases (Yu et al., 1990).)

Some short-distance, inexpensive systems use LEDs emitting in the visible region of the spectrum, typically at the red color of 665 nm. These sources are made of GaP. Other sources work in the infrared region between 800 and 930 nm. These sources are made out of alloy semiconductors with the composition $Ga_{1-x}Al_xAs$ where the subscript indicates a variable concentration in the alloy. Long-wavelength sources operating at the minimum-dispersion wavelength of 1300 nm and at the minimum-loss wavelength (for silica fibers) of 1550 nm are made of InGaAsP.

Long-Wavelength Sources

An alloy of $In_xGA_{1-x}As_yP_{1-y}$ is used for operation between 1000 and 1700 nm. The designer must decide the values of x and y. The choice is constrained by two factors, the desired wavelength and the spacing of atoms (the *lattice spacing*) in each of the materials. This latter constraint is important as semiconductor fabrication demands that the lattice spacing be equal (within a tolerance of 0.1% (Lee, 1991)) on either side of a semiconductor junction. This is so that junctions can be grown that will have the necessary mechanical and thermal properties to allow crystal growth without introducing numerous defect sites at the interface. Mismatched lattice spacings will usually cause unacceptable strains and dislocations across the junction, resulting in low optical generation efficiency or low lifetime. (Recently, a new family of lasers, the *strained quantum well laser* (Ohtoshi and Chinone, 1989; Koren et al., 1990; Tanbun-Ek et al., 1990), has been developed that purposely slightly mismatches the atomic lattice spacing across a junction to achieve an operating advantage in output power, operating efficiency, or spectral purity.)

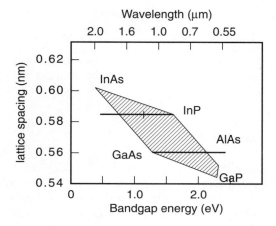

Figure 4.1 Composition relations for InGaAsP and GaAlAs emitters.

The goal of the designer in choosing the device's semiconductor alloys is to achieve operation at the desired wavelength while matching the lattice spacings across the junctions. Figure 4.1 shows the regime of possible alloys of gallium (Ga), aluminum (Al), arsenic (As), and phosphorous (P) that can be used in fabricating some optical sources. The vertical axis is the lattice spacing of the alloy molecules; the horizontal axis is the bandgap energy of the alloy or, equivalently, the emitted wavelength. The boundaries of the shaded area comprise ternary semiconductor alloys (i.e., three-element alloys). For example, the line between the corners labeled InAs and GaAs is the alloy $In_xGa_{1-x}As$ for values of x ranging from 0 to 1. The interior region shows the quaternary alloys of $In_xGa_{1-x}As_yP_{1-y}$.

Most long-wavelength quaternary devices are built on a substrate of InP. We find the InP corner of the shaded figure on the right side and observe that the lattice spacing is slightly larger than 0.58 nm. The lattice spacing constraint means that we are required to build our junctions on a line of constant lattice spacing (i.e., a horizontal line through the InP point as drawn in the figure). Material science studies (Yu et al., 1990) show that the bandgap energy in an $In_xGa_{1-x}As_yP_{1-y}$ alloy along this line is determined by the alloy fraction y as

$$E_g(\text{in eV}) = 1.34 - 0.72y + 0.12y^2 . \tag{4.3}$$

and that, for this relationship, the alloy fraction x is related to y by

$$x = \frac{0.4526}{1 - 0.031y} \, . \tag{4.4}$$

For an operating wavelength of 1300 nm, we find (see the problems at the end of this chapter) that $y = 0.589$ and $x = 0.461$ and that the emitting alloy should be constructed as $In_{0.461}Ga_{0.539}As_{0.589}P_{0.411}$.

Short-Wavelength Sources

Also shown on Figure 4.1 is the line for GaAlAs, extending from the GaAs corner of the shaded area to the AlAs point noted in the figure. Fortuitously, this line is almost horizontal, indicating little change in the lattice parameter. For $Ga_xAl_{1-x}As$, the bandgap energy [E_g (in eV)] depends on the alloy fraction x as (Keiser, 1991)

$$E_g(\text{in eV}) = 1.424 + 1.266x + 0.266x^2 \tag{4.5}$$

The value of x is constrained to the region $0 < x < 0.37$ because the transition becomes indirect for higher values of x.

4.2.2 Typical Device Structure

The typical light emitting semiconductor is fabricated with four primary layers, grown on top of a substrate. (Other layers may be added for other purposes.) As shown in Figure 4.2 for a long-wavelength source, the substrate is usually InP and the layers consist of

- an n-type InP buffer layer,
- a thin active region of p-type InGaAsP,
- a p-type cladding layer of InP, and
- a heavily-doped p^+-type layer of InGaAs.

(Typical dimensions of these layers are given later, in discussions of particular sources.) The layers are added to help confine the current carriers and the generated light in the vertical direction. (Techniques to confine the current carriers and light in the horizontal direction are discussed later.)

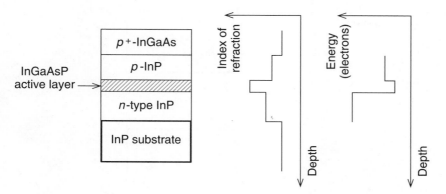

Figure 4.2 Typical diode structure consisting of four-layers on a substrate, index of refraction variation through the layers, and energy well for electrons.

Current-Carrier Vertical Confinement

Current-carrier confinement is desirable because the radiative recombination efficiency increases with current density. To consider the confinement of the carriers, we want to focus our attention on the active layer and its adjacent layers. As seen in the right side of Figure 4.2, electrons will be injected from the n-type InP material into the active region (seen as a drop in energy on the diagram). The electrons face an energy barrier at the interface between the active region and the p-type material below it. Few electrons will have the energy to climb this barrier and, so, the electrons will be confined to the active region. A similar energy barrier between the active layer and the n-type material exists for the holes that are injected from the p-type material into the active layer; the holes are again trapped in the active region. This boundary structure on either side of the active layer is called a *double heterostructure*.

Light-Wave Vertical Confinement

Light confinement in the vertical direction is desired because the efficiency of the stimulated emission process depends on the strength of the optical field. Without any confinement mechanism, the optical field will quickly diverge out into the inactive material and lose its power to the absorption in the material. The light wave can be confined by shaping the indices of refraction of the layers on either side of the

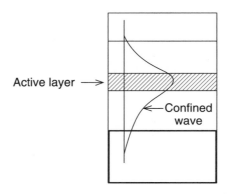

Figure 4.3 Side view of an optical field confined in the vertical direction.
The portion of the field lying outside of the active region is less than it
would be in the absence of any confinement.

active region. As seen in the center of Figure 4.2, the index of refraction of the active region is greater than the index of the two adjacent layers. The two outer layers (the p^+ layer and the InP substrate) have lower indices of refraction than their adjacent layers. This step-wise variation in the refractive indices forms a light guide that serves to confine the light field to the active region and its adjacent layers. Figure 4.3 shows a side view of the fundamental spatial mode in the presence of the confining structure. While an appreciable portion of the field extends beyond the active region, the field would be considerably wider in the absence of any confinement. The fraction of the field that is within the active region is a function of the thickness of the active layer d and the height of the index mismatches at each interface (Kressel and Butler, 1977).

4.3 Light Emitting Diodes

4.3.1 LED Configurations

Two configurations (Marcuse, 1977; Botez and Ettenberg, 1979; Lee et al., 1988; Yu et al., 1990) of light emitting diodes have become popular, surface emitters and edge emitters. The surface emitters are widely used in multimode fiber systems since the wide-angle beams that they produce are more efficiently coupled into multimode fibers; the edge emitters are used in both single-mode and multimode systems, since they provide a tighter emission pattern that is more efficiently coupled into fibers with low numerical apertures.

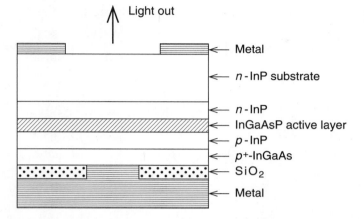

Figure 4.4 Representative InGaAsP long-wavelength surface emitter.

4.3.2 Surface-Emitting LEDs

The *surface-emitting LED* (also known as a *SLED* or *Burrus emitter*) is illustrated in Figure 4.4 and Figure 4.5.

In the InGaAsP device, the four layers grown on the InP substrate consist of (Yu et al., 1990)

1. an *n*-type InP buffer layer that is 2 to 5 μm thick,

2. a *p*-type InGaAsP active layer that is 0.4 to 1.5 μm thick,

3. a *p*-type InP layer that is 1 to 2 μm thick, and

4. a p^+-type InGaAs "cap" layer that is about 0.2 μm thick. (This latter layer is to help reduce the metal-to-semiconductor contact resistance.)

The device consists of the double heterojunction structure around the active region. The light is emitted from a circular planar region of the active layer, usually 20 to 50 μm in diameter (Yu et al., 1990). As mentioned earlier, it is desirable to keep the current density as high as possible in the active region. This is done vertically through the use of heterojunctions, as explained earlier. Confinement of the current carriers in the horizontal dimension can be done by a variety of techniques, which include:

- including a layer of dielectric insulation, such as SiO_2, (Figure 4.4) with a hole etched through to allow current flow in a limited area,

- using proton bombardment to create a high-resistivity region outside of the boundaries of the active region to minimize current flow through this region (Figure 4.5),

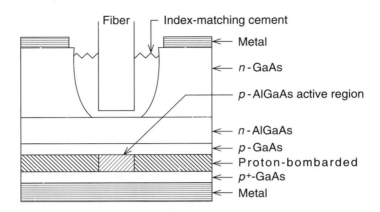

Figure 4.5 Representative GaAlAs short-wavelength surface emitter.

- etching away the surrounding material to form a *mesa structure* (Figure 4.6) that isolates the active region, or
- diffusing some zinc into the material to form a low-resistivity region that provides a channel for the current flow (Figure 4.7).

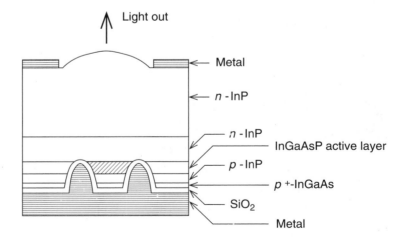

Figure 4.6 Representative GaAlAs short-wavelength surface emitter
with a mesa current-confinement structure and a dome at the
output face to reduce light loss due to total internal reflection.

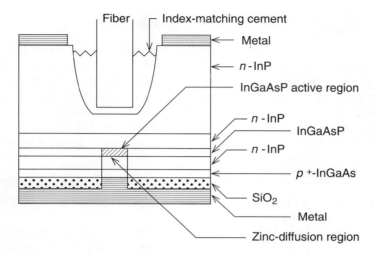

Fiber   — Index-matching cement
← Metal
← n-InP
— InGaAsP active region
— n-InP
— InGaAsP
— n-InP
← p^+-InGaAs
— SiO$_2$
— Metal
— Zinc-diffusion region

Figure 4.7 Representative InGaAsP long-wavelength surface emitter incorporating current confinement using a low-resistivity region produced by diffusion of zinc into the structure.

Light from the active light-emitting region can be collected from either side of the device (i.e., through the substrate or through the other side). In a GaAlAs device, the GaAs substrate will absorb appreciable light, so a well is etched out of the substrate to allow the fiber to approach the active region more closely (Figure 4.5). In InGaAsP devices, the InP substrate does not appreciably absorb the light, so the well can be omitted.

Because of the high refractive index of the semiconductor materials ($n = 3.4$ for InP, $n = 3.6$ for GaAs), there is a high reflection loss if the light is coupled into air. Additionally, the critical angle for a semiconductor-air interface is only 15 degrees, thereby causing significant power to be reflected back into the semiconductor. One potential solution to both problems is the use of an index-matching epoxy (as in Figure 4.5) to join a fiber pigtail to the source. A solution to the refraction problem in surface-emitting devices is the formation of a domed output surface on the device, which has less refractive losses due to the geometry of the interface (Figure 4.6). For a semiconductor-air interface, the coupling efficiency is approximately $1/[n(n + 1)^2]$; while for an LED with a hemispherical output dome, the coupling efficiency is approximately $[2n/(n + 1)^2]$—an improvement of $2n^2$. For a fiber, however, this improvement can be transitory, since the effective area of the source is also magnified by n^2. Since any increase in effective emitter area beyond the core area of the fiber is ineffective, improvement in the output light level occurs only when the original source area is smaller than the fiber core. (The improved efficiency is real, however, for LEDs used in non-fiber applications.)

Output Beam Pattern: Surface-Emitting LEDs

Since the emitting region of the surface emitting LED is circularly symmetric, the emitted *beam pattern* will also be symmetric with a 60-degree (typical) half-cone angle beam divergence. Here, *beam divergence* is defined as the angular spread of the emitted beam as measured in the far-field of the beam. The angular spread is measured at the points where the power is decreased to one-half the maximum on-axis power (i.e., the −3 dB optical power point). Both the full-angle and the half-angle can be used to describe the beam divergence; the user must take note of the specified test conditions.

Example: An optical beam is found to be −10 dB from its on-axis value of power at a measured angle of 75 degrees.

(a) If the angular power dependence is $P(\theta) = P_0 \cos^n \theta$ where θ is the angle measured from the on-axis position and P_0 is the on-axis power, find the value of n.

Solution:

$$\frac{P(\theta)}{P_0} = \cos^n \theta \tag{4.6}$$

$$10^{-\frac{10}{10}} = \cos^n 75° \tag{4.7}$$

$$0.1 = (\cos 75°)^n \tag{4.8}$$

$$\log(0.1) = n \log(\cos 75°) \tag{4.9}$$

$$n = \frac{-1}{-0.587} \tag{4.10}$$

$$n = 1.703 . \tag{4.11}$$

(b) Calculate the full-angle beam divergence of this source.

Solution: We need the angle where P/P_0 is reduced to a value of 1/2.

$$\frac{P}{P_0} = \cos^{1.703} \theta \tag{4.12}$$

$$\frac{1}{2} = (\cos \theta)^{1.703} \tag{4.13}$$

$$\cos \theta = (0.5)^{\frac{1}{1.703}} \tag{4.14}$$

$$= 0.665 \tag{4.15}$$

$$\theta = 48.3° \text{ (half–angle divergence)} \qquad (4.16)$$
$$\theta = 96.6° \text{ (full–angle divergence)} . \qquad (4.17)$$

4.3.3 Edge-Emitting LEDs

Edge-emitting LEDs remove the light along an axis transverse to the current flow, as shown in Figure 4.8.

Representative edge-emitting LED structures are shown in Figure 4.9. The four layers grown on the substrate are similar to the surface-emitting devices except that the active layer is much thinner, being 0.05 to 0.25 μm thick (compared to 0.4 to 1.5 μm for the surface-emitting device).

The insulating SiO_2 layer has a stripe hole in it that guides the current down into the active region, laterally confining the current. The typical width of the active

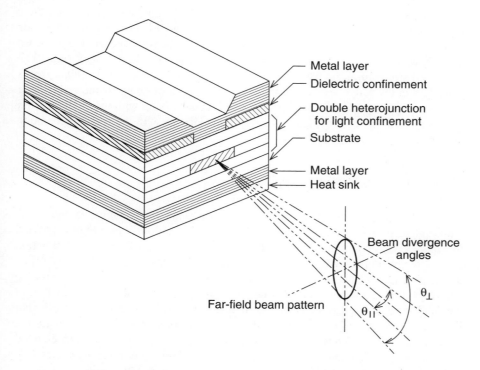

Figure 4.8 Representative edge emitter and far-field beam pattern.

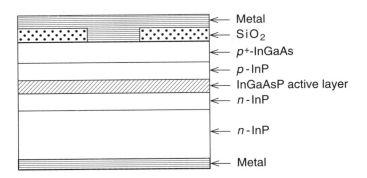

Figure 4.9 Representative InGaAsP edge-emitting LED structures.

region of an edge emitter (chosen to match the core size of typical fibers) is 50 to 70 μm (Keiser, 1991). The length of the region is typically 100 to 150 μm.

The heterojunctions on either side of the active region play an additional role in the edge emitter; they act as waveguides to help confine the light. The active region is designed to have a high index of refraction with the materials on either side of the active layer having a lower index of refraction. The substrate and the p^+-doped layer, lying further away from the active layer, have an even lower index value. The combined structure of the five layers makes an optical waveguide. The light that is generated in the active region is confined to the region made up by the active layer and its two adjacent layers. This optical confinement, along with the charge carrier confinement of the heterojunction, increases the efficiency of the optical generation process.

Beam Pattern: Edge-Emitting LEDs

Because of the asymmetry of the rectangular-shaped active region at the emitting end of the device, the far-field pattern will be elliptical (see Figure 4.8) and two angles will be required to describe the beam divergence. Because of diffraction effects, the beam perpendicular to the junction will have the larger divergence (i.e., the larger divergence is aligned along the shortest dimension). A typical value for the half-angle in this direction is 60 degrees. The angle parallel to the junction will have the smaller value, typically a 30-degree half-angle. Besides their asymmetric output beam, edge emitters will typically produce less power (approximately 1/2 to 1/6 the power) than surface emitters. We will find, however, that the amount of light coupled into a fiber is comparable for both devices, due to offsetting differences in coupling efficiency.

4.3.4 LED Output Power Characteristics

Typically, LEDs used in fiber communications produce power levels of several microwatts or tens of microwatts in the fiber. For low drive currents, the output power is a linear function of current, as shown in Figure 4.10. As the drive current becomes larger, eventually the output power saturates (i.e, levels off). The *linearity* of the diode refers to the linearity of the curve that relates the output power to the drive current. Linearity is important to allow faithful analog modulation of the output power by direct current modulation.

The amount of power that can be produced for any given drive current is a function of the *operating efficiency* of the LED. The operating efficiency of a diode is the measure of its ability to convert electrical input power into optical output power. This efficiency is determined by three factors.

- The first factor is the ratio of radiative transitions to nonradiative transitions that occur in the active layer of the material. Nonradiative transitions rob the optical generation process of carriers. Nonradiative recombination tends to occur at crystal defects, especially at the boundaries between the active layer and its two surrounding layers. The ratio of radiative recombination to nonradiative recombination is maximized by having a high current density in the active region, as was previously discussed, and by careful layer growth to remove boundary defects.

- The second factor is the amount of light absorption that occurs in the active region of the emitter. This is *self-absorption*. It can be important when the nonradiative transitions described in the previous factor are at minimum

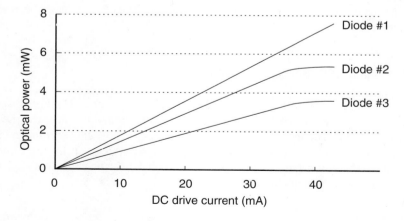

Figure 4.10 Output power versus drive current for a typical LED source.

and when the doping levels of the active region are high (Keiser, 1991). This absorption is minimized by keeping the active layer of the device thin, as is done in most modern double-heterojunction devices. Studies (Keiser, 1991) indicate that an optimum thickness can be calculated for both surface emitters and edge emitters as a function of doping levels and nonradiative recombination rates.

■ The third factor is the reflection losses (i.e., Fresnel losses) and refraction losses that occur at the semiconductor output face, as described previously.

4.3.5 LED Spectral Width

The *spectral width* of the source is important because it determines the contribution to material dispersion, as seen in Chapter 3. Lower spectral width will allow increased data rate if material dispersion effects are the limiting factor. The spectral width $\Delta\lambda$ of LEDs can be approximated by (Saleh and Teich, 1991)

$$\Delta\lambda(\text{in }\mu\text{m}) \approx 1.45\lambda^2(\text{in }\mu\text{m})\, kT\,(\text{in eV}) , \tag{4.18}$$

where λ is the nominal LED operating wavelength *in units of* μm and kT is the product of Boltzmann's constant and the junction temperature *in units of eV*. We note that the linewidth generally increases as the square of the wavelength; long-wavelength LEDs will have appreciably more linewidth than short-wavelength LEDs. Typical LED spectral widths for an GaAlAs source are on the order of a few to several 10s of nm. The spectral width of a InGaAsP surface-emitting LED can be about 100 nm, while that of an equivalent edge emitter is typically 60 to 80 nm (Yu et al., 1990). Heating the device or operating at increased drive currents can increase the spectral width of a source.

4.3.6 LED Modulation Bandwidth

The speed of response of an LED (i.e., how fast an LED can respond to changes in its driving current) depends on the time duration of the charge carriers in the active region. The length of time that the minority charge carriers remain in the active region is determined by the *lifetime* of the carriers τ_{lifetime}. If the drive current is modulated by a sinusoidal signal (added to a DC bias), the power output P_{out} of the laser follows a modulation frequency response of

$$P_{\text{out}} = \frac{P_0}{1 + \omega^2\tau_{\text{lifetime}}^2} , \tag{4.19}$$

where P_0 is the unmodulated power and ω is the modulation frequency.

We now want to measure the modulation response of the LED. An optical detector converts the incident optical power into a current that is proportional to the optical power. The *electrical* power P_{elec} from the detector is proportional to the square of the electrical current out of the detector and, so, it is proportional to the square of the optical power (i.e., $P_{elec} \propto i^2 \propto P_{optical}^2$). So, to measure the modulation response of the LED, we measure the frequency response of the electrical power out of the detector (assuming that the detector bandwidth exceeds the LED bandwidth). If we record the 3-dB frequency $\omega_{3-dB\ electrical}$ of the detected electrical power, we will be measuring the frequency where $P_{out}^2/P_0^2 = 1/2$. Hence, we find

$$f_{(3-dB\ electrical)} = \frac{1}{2\pi\tau}.$$ (4.20)

(Another definition of modulation bandwidth has been taken at the frequency when $P(\omega)/P_0 = 1/2$; the user must determine which definition has been used to describe the bandwidth.)

The lifetime $\tau_{lifetime}$ depends on several parameters. For a lightly doped active layer (doping levels on the order of 2×10^{17} atoms $\cdot$ cm^{-2}) and a thin active-region thickness d, the nonradiative recombination processes are dominated by the nonradiative recombinations that occur at the layer interfaces. The radiative recombination component of the lifetime depends on the carrier density in the active region.

- When the drive current is low, the radiative lifetime is independent of the drive current level and linearly dependent on the doping level of the active region.

- When the drive current is high, the radiative carrier lifetime is proportional to $\sqrt{d}$ and inversely proportional to $\sqrt{J}$, where d is the active layer thickness and J is the current density in the active region (Keiser, 1991).

Increasing the modulation bandwidth requires reducing the carrier lifetime. For low–drive-current LEDs the modulation bandwidth can be improved by increasing the doping level. At high current levels the bandwidth can be improved, assuming that nonradiative recombinations are not dominating the process, by reducing the value of d (i.e., decreasing the thickness of the active region), increasing the current density in the device (i.e., increasing the drive current, decreasing the active area of the device, or both), or all of these.

LED Rise-Time

An alternative to the frequency-domain approach to device response is to work in the time domain. The *rise-time* of a device can determine the maximum modulation rate of the source. If the fall-time of the LED is negligible compared to the rise-time, then the maximum modulation rate is given approximately by the inverse of the rise-time. Several factors combine to determine the rise-time capability of an LED.

The rise-time can be limited by capacitance C_s associated with the active region. Since charge is present in this region, it represents a capacitive charge-storage medium. (We will neglect other parasitic capacitances that might be present.) The capacitance associated with this charge storage in an LED is in the range of 350 to 1,000 pF (Keiser, 1991). Keiser (1991) gives an expression for the rise-time t_r (defined as the time required for the device to transition from 10% of its final light-output value to 90% of its final value) as

$$t_r = 2.20 \left(\frac{2kTC_s}{qI_p} + \tau_{\text{lifetime}} \right), \tag{4.21}$$

where k is Boltzmann's constant, T is the device temperature, C_s is the junction space-charge capacitance, and I_p is the size of the current step function used to drive the device. This relationship illustrates that the rise-time can be reduced by reducing the capacitance and by maximizing the drive current. (The high-current limit is seen to be $2.20\tau_{\text{lifetime}}$.) In fact, the rise-time of the device can be minimized by over-driving the diode with a current waveform that momentarily exceeds its pulsed value. The fall-time can also be minimized by providing a momentary negative bias to the device when it is first turned off. (The negative bias clears the space charge from the active region.)

LED Power-Bandwidth Tradeoff

Keiser (1991) also shows that the power-bandwidth product for an LED can be expressed as

$$P\,\Delta f = \frac{hc}{2\pi q\lambda} \frac{1}{\tau_{\text{lifetime}}} J, \tag{4.22}$$

where τ_{lifetime} is the radiative lifetime of the minority carriers in the active layer and J is the current density. At a given drive current, therefore, the user must trade-off speed and power (i.e., a fast LED is a low-power LED).

4.3.7 LED Summary

Due to the combination of their operating properties (especially their wide beam divergence, relatively low power coupled into a fiber and relatively large spectral width), LEDs are typically suitable for use in systems using multimode fibers requiring less than 50 Mb $\cdot$ s^{-1} of information rate.

4.4 Laser-Diode Sources

Laser-diode sources produce more power than an LED, have a narrower spectrum, and can couple more power into a fiber (Panish, 1976; Casey and Panish, 1978; Selway et al., 1980; Thompson, 1980; Bowers and Pollock, 1988; Dutta and Zipfel, 1988; Lee et al., 1988; Shumate, 1988; Cartledge, 1990; Oshiba and Tamura, 1990; Yu et al., 1990; Lee, 1991). The structure of a laser diode is much like that of an edge-emitting LED. (While surface-emitting lasers are under investigation, the majority of present diode lasers are edge-emitters.) The principal difference between an edge-emitting LED and the edge-emitting laser is that, in the laser, the active region is thinner vertically and narrower horizontally. In addition, multilayer reflectors are added to the ends of the structure to provide optical feedback. (This feedback raises the optical-field strength to ensure that the stimulated emissions dominate the spontaneous emissions in the laser. The mirror structure also serves to reduce the beam divergence of the emitted pattern and to narrow the spectrum of the light output.) Double heterojunctions are used to confine both the charge carriers and the optical fields in the vertical direction. Additional structures are incorporated to confine the current and the light laterally.

4.4.1 Gain-Guided Lasers

Several techniques are incorporated into the structure of the devices to confine horizontally the electrical carriers to a narrow region. Figure 4.11 shows a typical *stripe-geometry* laser diode. Here a layer of SiO$_2$ has been fabricated and a narrow stripe has been etched through the SiO$_2$ layer, followed by the deposition of the metal. The stripe can range from 5 to 20 μm in width and from 150 to 300 μm in length. The emitting region in the active layer is formed with a width that is slightly larger than the stripe (allowing for current-spreading mechanisms). In the emitting region under the stripe, the index of refraction is slightly higher than in the laterally adjacent regions, because of the presence of the current carriers in that region. This slight rise in the refractive index forms a lateral waveguiding structure. (The vertical waveguide structure is formed by the heterojunction materials.) In this way, the gen-

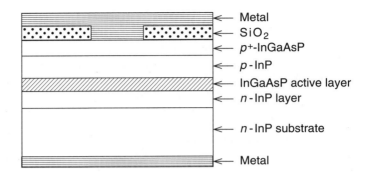

Figure 4.11 Representative stripe-geometry laser.

erated light is guided both vertically and laterally and is more confined as it propagates down the length of the diode. The confinement minimizes absorption in the non-active region of the active layer. The mechanism of providing lateral waveguiding with the change in the refractive index caused by the current carriers is called *gain-guiding*.

Other techniques can be used to provide a narrow current stripe. Figure 4.12 shows a device that uses proton bombardment to create a high-resistivity region outside of the desired current-flow region. (The protons break bonds in the semiconductor material and raise its resistivity. The lateral extent and depth of the bombardment are externally controllable.) Figure 4.13 shows a stripe laser that uses a "V"-shaped groove that has been etched in the laser material to control the current flow. (The thinner cap layer on top of the active region has lower resistance and, so, the current flow is most intense at the bottom of the groove.)

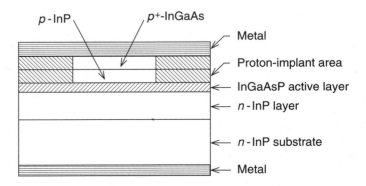

Figure 4.12 Stripe laser that uses proton bombardment to form the stripe region.

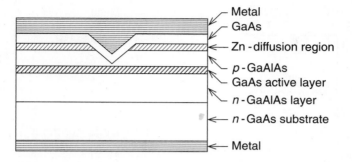

Figure 4.13 Stripe laser using V-groove technique to form the stripe region.

4.4.2 Index-Guided Lasers

An alternative laser-diode structure incorporates a deliberate change in material in the lateral direction across the active layer to form a waveguide structure. The change in the index of refraction of the materials forms a lateral waveguide that confines the light to a narrower region. Several structures have been devised to perform this function:

- Figure 4.14 shows a *buried heterostructure laser* where the *n*-type InGaAsP active emitting region is surrounded to the left and right by *p*-type InP. The change in material is accompanied by a step-change in the refractive index providing the lateral waveguide. The vertical waveguide structure is again done by the heterojunctions.

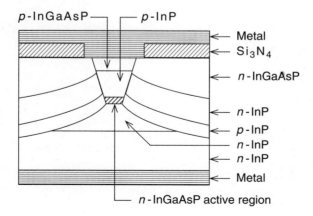

Figure 4.14 Index guiding of light in a buried-heterostructure laser.

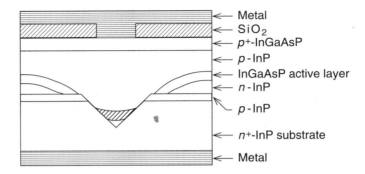

Figure 4.15 Index-guided laser with a buried-channel substrate.

- Figure 4.15 shows a *buried-channel substrate laser*. Here the active region has "slumped" into an etched groove, isolating the light-emitting portion of the active layer.

- Figure 4.16 illustrates a *double-channel planar buried-heterostructure laser* (Yu et al., 1990), where the light-emitting region of the active layer has a channel on either side of it that isolates it and provides the light-guiding structure.

All of these lasers have a step change in the refractive index. It is also possible to guide light with a more gradual, tapered change in the index of refraction. (These structures lose part of their carrier confinement, however.) Figure 4.17 illustrates an *inverted-rib laser*. In this laser the first layer boundary below the active layer has a channel etched into it. Because of the narrow layer-spacing, the effects of this change in geometry are coupled into the active region as a gradual change in the

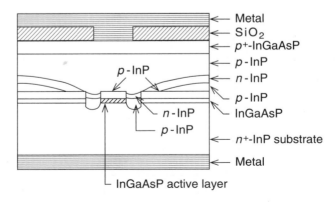

Figure 4.16 Index-guided double-channel planar buried-heterostructure laser.

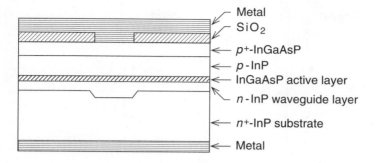

Figure 4.17 Index-guided inverted-rib laser.

index (Yu et al., 1990). This index change provides a lateral-light guiding structure. Figure 4.18 shows a *ridge-guide laser* (Yu et al., 1990). In this geometry the shaped region is above the active region and, again, induces a waveguiding change in the index of refraction, due to the coupling of the ridge effects into the active region.

Generally, index-guided lasers are superior to gain-guided lasers. They typically have a lower threshold current (i.e., the current drive where lasing begins), have better mode stability under pulsed operation, and have a narrower frequency spectrum than the gain-guided lasers.

4.4.3 Beam Patterns

The typical diode laser emitting region at the output face is 250 to 500 μm long by 5 to 15 μm across by 0.1 to 0.2 μm high (Keiser, 1991). These dimensions give the beam an asymmetric far-field pattern with a perpendicular beam divergence

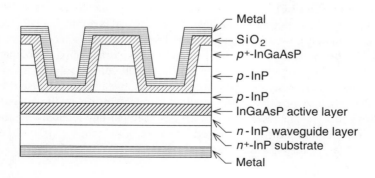

Figure 4.18 Ridge-guide laser.

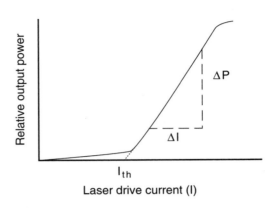

Figure 4.19 Output characteristics of a diode laser.

of 30 degrees to 50 degrees (measured perpendicular to width of the emitting region) and a parallel beam divergence of 5 degrees to 10 degrees. This latter value is about one-fifth the comparable value for an edge-emitting LED and implies that the laser-beam pattern is more directional than the LED. This directionality is beneficial in trying to couple the light into optical fibers.

4.4.4 Laser Power Characteristics

The drive mechanism that operates the laser is the current through the forward-biased device. A typical plot of the output power versus the driving current amplitude is shown in Figure 4.19. Observe that there are two linear regions of operation. When the drive current is below the threshold current, the diode is not lasing. It is operating as an LED, emitting a small amount of incoherent light. When the drive current is above threshold, the device is lasing and is producing coherent light. Increases in output power in this region are linearly related to increases in driving current. The *threshold current* is obtained by linearly extrapolating the output power curve down to the zero level of power. The threshold-current value is a key diode-laser parameter. Obviously, it is desirable to have as low a value of threshold current as possible, because the drive power required to reach threshold will only reduce the overall efficiency of the laser. Much effort has been put into reducing the threshold current. Typical values for index-guided lasers are in the range of 10 to 30 mA; for gain-guided lasers, they are in the range of 60 to 150 mA (Yu et al., 1990).

Conversion Efficiency

Two types of conversion efficiency can be defined for the diode laser or LED. The first is the *overall electrical conversion efficiency* (or *efficiency*), given by

$$\eta = \frac{P_{\text{out optical}}}{P_{\text{in electrical}}} \qquad (4.23)$$

$$= \frac{P_{\text{out}}}{V_f I_f}, \qquad (4.24)$$

where η is the efficiency, P_{out} is the optical output power of the source, and P_{in} is the electrical input power. Since the input power is the product of the forward voltage across the device V_f and the forward current through the device I_f, Equation 4.24 follows.

Example: A diode laser produces 1 mW of light when driven at a forward current of 100 mA. The nominal forward voltage across the diode is 2 volts. Calculate the overall electrical conversion efficiency.

Solution:

$$\eta = \frac{P_{\text{out}}}{V_f I_f} \qquad (4.25)$$

$$= \frac{1 \times 10^{-3}}{(2.0)\,(100 \times 10^{-3})} \qquad (4.26)$$

$$= 5 \times 10^{-3} \qquad (4.27)$$

$$= 0.5\% . \qquad (4.28)$$

The second conversion efficiency is the *incremental efficiency* η_i, given by

$$\eta_i = \frac{dP}{dI}, \qquad (4.29)$$

where dP/dI is the slope of the output characteristics measured *above threshold*. The slope of the output characteristics is a measure of the efficiency of converting charge carriers into photons of light. A related efficiency is called the *external quantum efficiency* η_{ext}, given by

$$\eta_{\text{ext}} = \frac{q}{E_g} \frac{dP}{dI} , \qquad\qquad (4.30)$$

where q is the electron charge (1.6×10^{-1} coulombs) and E_g is the bandgap energy of the material (determined by the alloy composition).

Linearity

Another property of the output-power curve that affects the performance of an optical device in the operation of an analog communication system is the *linearity*. Nonlinearities associated with saturation of the output power at high drive currents or changes in the output-power level associated with jumps in lasing wavelength cause harmonic distortion and intermodulation of multiple signals carried on the link. Generally speaking, LEDs are quite linear in operation while laser diodes can suffer some linearity deficiencies in comparison. If the nonlinearities are intolerable in the desired application, several compensation techniques have been devised to ameliorate their effect. These techniques are based on feedback and other circuit procedures that sense the nonlinearity and attempt to cancel it.

4.4.5 Laser Resonator

The *optical resonator* is required to allow the light to make the equivalent of several passes through the low-gain active region. In a semiconductor laser, this optical resonator is a Fabry-Perot resonator, made up of parallel front and rear surfaces of the diode as shown in Figure 4.20. The parallel alignment is easily achieved by cleaving the crystal along its crystalline planes. In GaAs, the power reflectivity at the crystal interface is 32% ($n = 3.63$); in InGaAsP it is 33% ($n = 3.71$). This reflectivity is normally high enough to eliminate the need for mirrors; however, the back surface of the laser is frequently coated with a multilayer 100%-reflecting surface, resulting in emission from the front surface only. For higher power lasers the front surface is also coated with a partially transmitting coating to protect the front surface of the laser from forming defects due to ambient moisture.

4.4.6 Laser Modes

Just like the fiber, the introduction of the reflecting surface in the optical resonator introduces *electromagnetic modes* within the resonator. The modes are geometrical descriptions of the waves that fit the boundary conditions imposed by

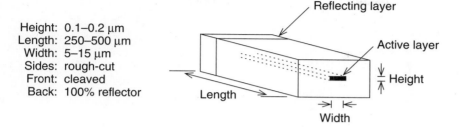

Height: 0.1–0.2 μm
Length: 250–500 μm
Width: 5–15 μm
Sides: rough-cut
Front: cleaved
Back: 100% reflector

Figure 4.20 Physical structure of a laser diode.

the resonator boundaries. Each mode, in general, oscillates at a different frequency, as (simplistically) shown in Figure 4.21. Each frequency is determined by the spacing ofthe resonator mirrors (the *longitudinal modes*) and the width and height of the waveguiding region of the diode (the *lateral modes*). The spacing of the longitudinal mode frequencies Δv is determined by the separation of the mirrors and is given by

$$\Delta v = \frac{c}{2L},$$
(4.31)

where c is the velocity of light, and L is the separation of the mirrors. The spacing of the frequencies of the transverse modes is a more difficult expression and will not be displayed here.

It is noted that, as the diode current is increased, the central mode grows faster than the side modes, thereby increasing the so-called *side mode suppression ratio (SMSR)* of the diode laser. This results in a spectral-power output that is more single-frequency at high drive currents. This effect is illustrated in Figure 4.22.

There are several disadvantages to multi-mode lasers.

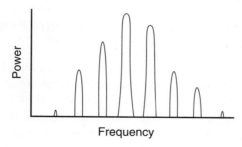

Figure 4.21 Spectral output of a multimode diode laser.

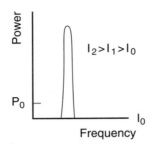

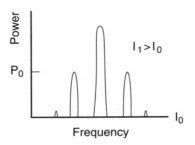

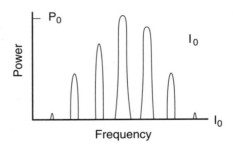

Figure 4.22 Reduction of the number of modes as drive current is increased.

- The linewidth of the source is broader than the linewidth of any single line. This has adverse implications for the material and waveguide dispersion of a fiber link with a subsequent reduction in the bandwidth.

- A multimode diode laser will randomly jump from mode to mode, causing the power output to vary with time. This leads to a noise, called *mode partition noise* in the output. (This source noise is described in more detail later in this chapter).

■ Coherent detection techniques, as described in Chapter 9, have inherent advantages. They require a single-frequency source, rather than a multi-mode laser.

Single-Mode Diode Lasers

The first step in reducing the number of modes is to reduce the number of transverse modes to one in both the vertical and horizontal directions. This is a *single-mode laser*. The higher-order modes can be eliminated by making the difference between the propagation losses of the lowest-order mode and the higher-order modes as large as possible. In this way, preference is given to the lowest-order mode in establishing oscillation.

■ Higher-order transverse modes in the vertical dimension are eliminated by reducing the height of the active light-emitting region and the waveguiding layer structure.

■ Similarly, to reduce the number of lateral modes, the width of the current stripe and/or the waveguiding horizontal region can be minimized. (For this reason, index-guided lasers, with their narrowly defined guiding region, tend to have fewer modes than gain-guide lasers, with their less-constrained guiding region.) A single transverse mode will be supported when the width of the emitting region is smaller than about 2 μm (Lee, 1991).

Another technique to eliminate the higher-order modes in the lateral direction is to incorporate regions with high optical loss into the edges of the emitting region. Since the higher-order modes extend more into these lossy regions, they will have a higher loss and will be unable to sustain oscillation.

The number of longitudinal modes can be reduced by shortening the resonator length. This has the adverse effect of decreasing the laser power, since less gain in incurred in the shorter distance. Other techniques have become more useful in reducing the number of longitudinal modes.

4.4.7 Single-Frequency Diode Lasers

Once a laser has been made to operate in a single mode, it desirable to reduce the number of longitudinal modes. Such a laser is a *single-frequency* laser and would have a spectrum as shown in Figure 4.23.

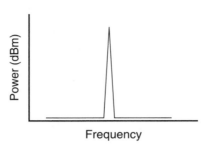

Figure 4.23 Single-frequency operation of a laser.

Once a laser is made to operate at a single frequency under steady-state conditions, it becomes important to maintain that state under pulse conditions. The charge carriers in the current pulse and the photons in each of the modes are coupled together. Under a current pulse, the light in the modes typically undergoes oscillatory behavior (called *relaxation oscillations*) causing the frequency content of the light to vary in time. In addition, the index of refraction of the lasing medium is a function of the gain of the medium. The time-varying gain leads to a time-varying index of refraction which, in turn, leads to a change of the center operating frequency of the laser. For a step-increase in current, the frequency is observed to shift in a *frequency chirp*. Techniques to reduce the observed chirp include biasing the laser in its "OFF" state just above threshold, reducing the active region volume, and running the laser continuously and using an external modulator (usually an electro-optic modulator) to modulate the laser beam after it has left the laser.

Once achieved, single-frequency laser operation can be easily upset by light reflected from any interface in the system back into the laser resonator. While it is possible to eliminate reflections by carefully matching indices of refraction at all interfaces or using polarization-rotating devices as reflection isolators, both techniques require user awareness of the reflection problem and careful design to remove the effect.

Techniques have been developed to produce single-frequency lasers with superior mode characteristics, including the distributed-feedback laser and the distributed Bragg reflector laser. Other techniques that use composite cavities (i.e., combinations of more than one resonator) are discussed in Chapter 9.

Distributed-Feedback Laser

A diode laser does not need reflecting surfaces at the ends of the medium in order to lase. Instead of concentrating the reflectivity at the ends of the laser, it can

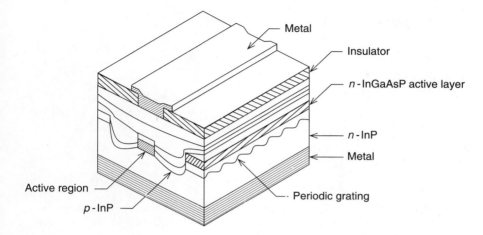

Metal

Insulator

n-InGaAsP active layer

n-InP

Metal

Active region

p-InP

Periodic grating

Figure 4.24 Physical structure of a distributed feedback (DFB) laser.
(Note: for simplicity, not all layers are labeled.)

be continuously distributed throughout the lasing medium, resulting in a *distributed feedback (DFB) laser* (Lee, 1991; Yariv, 1991). Such a laser, shown in Figure 4.24, uses a corrugated surface as the interface between two layers in the heterostructure. The periodicity of the structure Λ determines the wavelength of maximum interaction λ_B by

$$\lambda_B = \frac{2n\Lambda}{k}, \tag{4.32}$$

where n is the index of the mode in the laser and k is an integer that indicates the grating order. (Usually $k = 1$, but sometimes $k = 2$ is used.) Although physically removed from the active region, the periodic grating affects the propagation properties of the active region and results in an output at (Lee, 1991)

$$\lambda = \lambda_B \pm \frac{\lambda^2_B}{2nL_g} (m + 1), \tag{4.33}$$

where m is an integer and L_g is the grating length. Usually the $m = 0$ modes are the ones to oscillate. While theory predicts that two modes will oscillate (i.e., one for each value of the sign), random imperfections in the cleaving at the end facets of the device introduce asymmetric phase differences and result in only one mode oscillating.

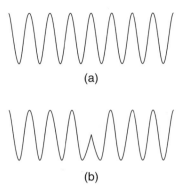

(a)

(b)

Figure 4.25 Two versions of grating structures for distributed feedback lasers.
(a) Periodic grating structure. (b) Grating structure with a $\pi/2$ phase shift in its center.

　　　　The stability of the single-frequency DFB laser under pulsed operation can be improved by modifying the grating. Figure 4.25 shows a periodic grating structure and a grating structure that has a phase shift of $\pi/2$ in the middle of it. (This latter grating design is called a *λ/4-shifted grating*.) It has been shown (Lee, 1991) that this is the optimum phase shift to improve the performance of the laser. These lasers will operate at a single frequency corresponding to the wavelength, λ_B. Phase-shifted grating lasers have superior frequency stability and a narrower linewidth than non-shifted DFB lasers.

　　　　DFB lasers exhibit good single-frequency operation and have little sensitivity to drive-current pulses and temperature changes.

Distributed Bragg Reflector Laser

　　　　The distributed Bragg reflector (DBR) laser also uses a corrugated interface to provide reflection at a wavelength λ_B that is determined by the grating period Λ as

$$\lambda_B = \frac{2n_e\Lambda}{l} , \qquad (4.34)$$

where n_e is the mode's propagation constant and l is an integer that indicates the order of the grating (usually $l = 1$). The corrugated sections providing the reflections are located outside of the current-pumped active region. Figure 4.26 compares the basic structures of the DFB laser (where the corrugation region overlaps the current-pumped region) and the DBR laser. The DBR laser will oscillate at the single frequency corresponding to λ_B. These lasers typically have a higher threshold current than the DFB lasers and are more susceptible to temperature variations (Yu et al., 1990). They are also more susceptible to frequency chirp when pulsed.

Quantum Well Lasers

Improvements in fabrication technology now allow the thickness in the active layer of the diode structure to be as small as 5 to 10 nm with smooth, defect-free interfaces. In such thin layers the electronic behavior of the electron charge carriers can no longer be modeled by their behavior in bulk material, but their quantum mechanical states must be used. The behavior of these electrons gives rise to a family of devices known as *quantum well lasers* (Lee, 1991; Yariv, 1991). Early devel-

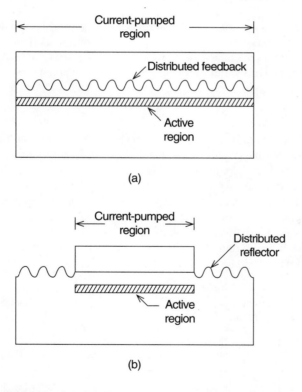

Figure 4.26 Side views of slices through two lasers. (a) A distributed feedback (DFB) laser. (b) A distributed Bragg reflector (DBR) laser. (For simplicity, the layer structures are not shown.)

opmental models of these lasers have shown nice combinations of desirable properties, including low threshold current, higher output power (and gain), narrow linewidth, frequency stability under pulsed operation, and low noise. Quantum well lasers have been demonstrated in GaAlAs for some time (Lee, 1991); work on InGaAsP devices is proceeding. Usually the long-wavelength devices are *multiple quantum well (MQW) lasers* where thin layers of active GaInAs are alternated with thin barrier layers of GaInAsP (forming the walls of the energy "well").

Quantum well structures that are built with a purposeful mismatch of lattice constant (a "strained" lattice) offer the promise of improved performance over unstrained devices (Lee, 1991). These *strained layer multiple quantum well lasers* show promising results in the laboratory and continue to be explored.

4.5 Laser Temperature Dependence

One major problem with laser diodes is the temperature dependence of the threshold current. Long-wavelength sources made of InGaASP exhibit a greater temperature sensitivity than the short-wavelength sources made of GaAlAs. The temperature dependence of the threshold current I_{th} can be expressed as

$$I_{th} = I_0 e^{T/T_0},$$ (4.35)

where I_0 is a constant (established at a reference temperature) and T_0 is an empirical constant fit to the measured data. Typical values of T_0 are found in Table 4.1.

Laser diodes are usually elaborately heat-sinked to provide a constant operating temperature, or they incorporate a thermoelectric cooler to remove heat from the diode. The cooler capacity needs to be large enough to maintain a constant operating temperature while the diode is operating.

As an alternative, laser diode drive circuits can include temperature compensation circuitry to minimize the effects of temperature changes. A photodetector samples a small portion of the output light (either at the back facet of the laser or by

Table 4.1: Representative range of values of empirical parameter T_0 for temperature dependence of threshold current.

Material	Wavelength	T_0
InGaAsP	1.3 µm	60–70K
InGaAsP	1.5 µm	50–70K
GaAlAs	0.85 µm	110–140K

splitting off a small fraction at the output) and provides a voltage proportional to the laser power. Through a feedback mechanism, the drive current can be adjusted to maintain a constant output power level. (The same circuit provides compensation for power changes due to aging of the laser diode.) Further details on temperature-compensating circuitry are contained in the discussion on electronic drive circuits, later in this chapter.

4.6 Source Reliability

Reliability of source devices plays an important role in determining overall system reliability (Ettenberg and Kressel, 1980; Hirao et al., 1986; Goodwin et al., 1988; Nakajima et al., 1987; Dutta and Zipfel, 1988; Fukuda, 1988). This is particularly true since early diode lasers had low lifetimes of a few hundreds of hours.

The failure rate for any device is modeled by the *bathtub curve* (Figure 4.27), so named because its shape resembles a transverse profile of a bathtub. After a short period of operation, certain devices will fail (called "infant failures") due to fabrication problems and other factors. Those that survive will typically be long-lived, with few failures. At the end of the device lifetime, the failure rate will increase due to accumulated effects. The problems for the manufacturer and user are to minimize the infant failures (or at least be certain that they are screened out and not used), to maximize the useful lifetime of the device under the required operating conditions, and to allow for a graceful degradation in performance at the end of useful life. The optical sources are typically characterized by their useful predicted lifetime.

First we will discuss laser reliability and then LED reliability.

4.6.1 Laser Reliability

A typical diode laser will have a power output that decreases exponentially in time, following a relation of the form

$$P_{out}(t) = P_i \, e^{-t/\tau_m} , \qquad (4.36)$$

where $P_{out}(t)$ is the time-dependent output power, P_i is the initial power of the laser, and τ_m is the exponential lifetime. This lifetime can be found by measuring the average time that it takes the power of a sample batch of lasers to reach a predetermined fraction of its initial value. Knowledge of τ_m allows us to extrapolate the power behavior to any amount of time.

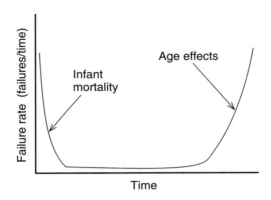

Figure 4.27 "Bathtub curve" for device failure rates.

Example: It is predicted that a certain diode laser will have its power decrease to 90% of its initial value in 3 years. How many years will be required for the power to decrease to 10% of its initial power?

Solution: We begin by finding τ_m from the initial information,

$$\frac{P_{out}}{P_i} = e^{-t/\tau_m} \tag{4.37}$$

$$0.90 = e^{-3/\tau_m} \tag{4.38}$$

$$-\frac{3}{\tau_m} = \ln(0.90) \tag{4.39}$$

$$\tau_m = -\frac{3}{\ln(0.90)} \tag{4.40}$$

$$= 28.5 \text{ years} . \tag{4.41}$$

To find the time required to reduce the power to 10% of the original power, we use

$$\frac{P_{out}}{P_i} = 0.10 = e^{-t/28.5} \tag{4.42}$$

$$\ln(0.10) = -\frac{t}{28.5} \tag{4.43}$$

$$t = -28.5 \ln(0.10) \tag{4.44}$$

$$t = 65.6 \text{ years} . \tag{4.45}$$

Example: Suppose that we want the laser in the previous example to reach 10% of its original power at 20 years of life. What fraction of the output power should be present at the end of 3 years to ensure this performance?

Solution: We need to find the required τ_m first.

$$\frac{P_{out}(20 \text{ years})}{P_i} = 0.10 = e^{-20/\tau_m} \qquad (4.46)$$

$$-\frac{20}{\tau_m} = \ln(0.10) \qquad (4.47)$$

$$\tau_m = -\frac{20}{\ln(0.10)} \qquad (4.48)$$

$$\tau_m = 8.69 \text{ years} . \qquad (4.49)$$

Finding the power after 3 years,

$$\frac{P_{out}(3 \text{ years})}{P_i} = e^{-3/8.69} \qquad (4.50)$$

$$= 0.708 \qquad (4.51)$$

$$= 70.8\% . \qquad (4.52)$$

Reliability in diode lasers is an issue because of the harsh electrical and optical environment encountered inside the laser's active region. A typical diode laser is subject to severe operating conditions that include current densities of 2000 to 5000 A/cm^2 within the active region and optical power densities of 10^5 to 10^6 W/cm^2. For the purpose of testing lasers it has been observed that continuous (CW) operation is the most strenuous environment and is used for most testing. The definition of when a laser has reached the end of a useful life is open to some interpretation. One method of determining this is to hold the driving current fixed and to define the end of life as that time when the output power falls below a usable amount (e.g., 1.25 mW). Another method monitors the output power and, as the power starts to fall, adjusts the drive current to a (larger) value required to maintain the original power level. For this mode of operation, the end of life occurs when the drive current reaches the rated maximum of the device. Other definitions are also used, making it difficult to compare test data. Sometimes device performance is measured below threshold to characterize the laser. This is particularly true when tests are run at high temperatures, where the amount of current required to achieve lasing is extreme enough to introduce new causes of degradation.

Laser Degradation Mechanisms

Three sources of degradation have been identified in diode lasers (Ettenberg and Kressel, 1980; Hirao et al., 1986; Goodwin et al., 1988; Nakajima et al., 1987; Dutta and Zipfel, 1988; Fukuda, 1988):

- *Facet damage*—This is physical damage to the reflecting surfaces of the laser due to operating at high optical power densities. Improved cleaving techniques and the addition of protective passivating layers on the laser facets have removed this problem.

- *Ohmic contact degradation*—All semiconductor devices operated at high current densities and elevated temperatures exhibit deterioration in the ohmic contacts at the metal-semiconductor interface. Improved solders and heat-sinking have remedied this problem.

- *Internal damage formation*—The least understood and hardest to control degradation mechanism is that due to the formation of internal lattice defects in the semiconductor crystal. These nonradiating defects initially form along the lines of crystal dislocations and have been named *dark lines* due to their appearance under a microscope. These defects were found in GaAlAs short-wavelength lasers, where they led to severe lifetime problems in early devices, resulting in a major effort to improve the fabrication of these lasers. Improved fabrication techniques have reduced the effect of these defects, and any devices with these defects are usually found by preliminary testing of the lasers. The sources that have dislocation defects are identified by running a *burn-in test* of approximately 100 hours for each device. Sources suffering from degraded performance in this test are either sufficiently defective to justify rejection or will typically be susceptible to degradation by the propagation of dark lines in the future and are also rejected. It should be noted that InGaAsP long-wavelength devices do not show these defects and have superior reliability performance over short-wavelength lasers.

4.6.2 Laser Testing

The testing of lasers (Ettenberg and Kressel, 1980; Hirao et al., 1986; Goodwin et al., 1988; Nakajima et al., 1987; Dutta and Zipfel, 1988; Fukuda, 1988) is not a standardized process. Manufacturers have different testing conditions, different sampling techniques, and different definitions of the end of useful life. Results are quoted in statistical form, which can also be misleading, or, worse yet, by anecdotes about particular lasers under test, as if such lasers were the norm rather than statistical oddities on the outer fringe of the distribution. As the room-temperature lifetime of laser sources becomes longer, techniques are required to accelerate the degradation processes. Two possibilities occur, to increase the drive current level or to increase the operating temperature.

High-Current Testing

Increasing the operating current level has been observed to decrease the operating life τ by the proportionality

$$\tau \propto J^{-n}, \tag{4.53}$$

where J is the current density and n is an empirically fit parameter with values typically ranging from 1.5 to 2.0. This mechanism accelerates the facet degradation more than high-temperature testing and so it is not used very often.

Example: Consider a laser with a predicted lifetime of 20 years at an operating current of 100 mA.

(a) Assuming that $n = 1.75$, what would be its lifetime if the current were doubled?

Solution:

$$\tau \propto J^{-n} \tag{4.54}$$

$$\frac{\tau_1}{\tau_2} = \frac{J_1^{-n}}{J_2^{-n}} \tag{4.55}$$

$$= \left(\frac{J_1}{J_2}\right)^{-n} \tag{4.56}$$

$$= \left(\frac{I_1}{I_2}\right)^{-n} \tag{4.57}$$

$$= \left(\frac{100}{200}\right)^{-1.75} \tag{4.58}$$

$$= 3.36 . \tag{4.59}$$

Hence,

$$\tau_2 = \frac{\tau_1}{3.36} \tag{4.61}$$

$$= \frac{20}{3.36} \tag{4.62}$$

$$= 5.95 \text{ years} . \tag{4.63}$$

(b) If the current was halved?

Solution:

$$\frac{\tau_1}{\tau_2} = \left(\frac{I_1}{I_2}\right)^{-n} \tag{4.64}$$

$$= \left(\frac{100}{50}\right)^{-1.75} \tag{4.65}$$

$$= 0.297 \tag{4.66}$$

$$\tau_2 = \frac{\tau_1}{0.297} \tag{4.67}$$

$$= \frac{20}{0.297} \tag{4.68}$$

$$= 67.3 \text{ years} . \tag{4.69}$$

Note that the nonlinear behavior of this dependence favors operation of lasers requiring long lifetimes at the lowest current level consistent with required link performance. Modest reductions in drive current can produce large increases in lifetime.

High-Temperature Testing

Just as the lifetime is sensitive to changes in drive current, it is also sensitive to changes in temperature. High temperatures can cut the lifetime significantly. The general form of the operating life dependence on temperature is

$$\tau \propto e^{E/kT} , \tag{4.70}$$

where k is Boltzmann's constant, T is the operating temperature, and E is an empirically determined *activation energy* parameter.

Increased temperature operation is currently the preferred method of acceleration rather than increased current operation. Since high-current operation is not desirable, testing is done at moderate lasing current levels or, in some cases, at current levels below threshold with intermittent testing at lasing current drives.

Example: Consider a laser diode with a predicted lifetime of 10 years at room temperature (300K).

(a) Calculate the predicted lifetime if the operating temperature is reduced by 10 degrees C.

Solution:

$$\frac{\tau_1}{\tau_2} = \frac{e^{E/KT_1}}{e^{E/KT_2}} \tag{4.71}$$

$$= e^{\frac{E}{K}\left(\frac{1}{T_1} - \frac{1}{T_2}\right)}. \tag{4.72}$$

Lacking any other information, we will use $E = 0.7\ eV = 0.7(1.6 \times 10^{-19}) = 1.12 \times 10^{-19}$ joules.

$$\frac{\tau_1}{\tau_2} = \exp\left[\frac{1.12 \times 10^{-19}}{1.38 \times 10^{-23}}\left(+\frac{1}{300} - \frac{1}{290}\right)\right] \tag{4.73}$$

$$= \exp\left[\left(\frac{1.12 \times 10^{-19}}{1.38 \times 10^{-23}}\right)\left(-1.149 \times 10^{-4}\right)\right] \tag{4.74}$$

$$= 0.393 . \tag{4.75}$$

$$\tau_2 = \frac{\tau_1}{0.393} \tag{4.76}$$

$$= \frac{10}{0.393} \tag{4.77}$$

$$= 25.4 \text{ years} . \tag{4.78}$$

(b) Calculate the expected lifetime if the temperature is *raised* by 10 degrees C.

Solution:

$$\frac{\tau_1}{\tau_2} = \exp\left[\frac{1.12 \times 10^{-19}}{1.38 \times 10^{-23}}\left(+\frac{1}{300} - \frac{1}{310}\right)\right] \tag{4.79}$$

$$= 2.39 \; . \tag{4.80}$$

$$\tau_2 = \frac{\tau_1}{2.39} \tag{4.81}$$

$$= \frac{10}{2.39} \tag{4.82}$$

$$= 4.18 \text{ years} \; . \tag{4.83}$$

Figure 4.28 shows some typical results of laser lifetime tests. The power output of the test lasers is plotted as a function of time, showing a history of the power variation of the laser. Typical median lifetimes for AlGaAs short-wavelength lasers under continuous wave (i.e., DC operation) are about 10^5 hrs; InGaAsP laser have much longer median lifetimes (Yu et al., 1990).

4.6.3 LED Reliability

Failure mechanisms in LEDs are similar to those in lasers (except that there is no facet damage in LEDs), but, because of more benign current density levels and optical power levels, degradation of LED output power is much more gradual. However, because LEDs produce less power than lasers, there is a tendency to run them at high drive-current levels. The same processing and passivating improvements that apply to lasers have also brought about improvements in LED performance. Burn-in testing can be used to remove the devices that exhibit infant failure. Generally, LED lifetime due to normal aging degradation is not an issue in system design. (There can

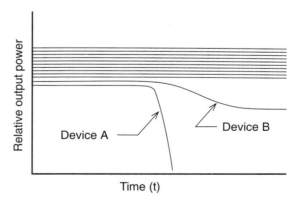

Figure 4.28 Typical laser diode lifetime test data. (The curves have been displaced for clarity.) Device A has failed; device B has lost significant output power but has not failed yet.

still be, however, freak failures of atypical devices or failure due to user errors.) Typical predicted lifetimes of LEDs, based on high-temperature testing, are in the range of 10^5 to 10^8 hours, depending on their design and fabrication properties.

4.6.4 Laser Modulation Response

Laser diodes can be intensity modulated (IM) by modulating the drive current of the device; the light can be frequency modulated (FM) by pulsing the drive current, and can be phase modulated (PM) with the use of an external modulator. Since frequency modulation and phase modulation often require coherent detection techniques, discussion of these latter two modulation formats is deferred to Chapter 9. We will consider only intensity modulation here.

We have seen from the output power characteristics of the laser that, once past threshold, the output power is linear with further increases in current. To pulse modulate the laser intensity, we pulse the drive current from the threshold level (or just above that value) to some larger value; the output power of the laser jumps from a small value to a larger value. For continuous bipolar signals, we need to establish a bias point partway up the power curve and allow the drive current to deviate around that bias value.

4.6.5 Modulation Bandwidth

The speed of response of the source plays a role in determining the maximum data rate that can be transferred if the fiber does not present a limit.

Laser Intensity Modulation Bandwidth

One of the important characteristics of the source for high-speed data links is the maximum modulation rate of the source. This maximum modulation rate is characterized by the modulation frequency response (i.e., a sine-wave modulating signal frequency is increased until the source response decreases). This upper limit of the frequency response is determined by the RC time constant of the source (where R is the source and circuit resistance and C is determined by the parasitics of the device) and by the interaction of the light and charge carriers in the active region of the device. Figure 4.29 shows a representative response of an intensity-modulated diode laser. This response is characterized by a low-frequency flat portion and a high-frequency resonant peak. The exact frequency location and magnitude of this peak is dependent on the operating point and device parameters of the diode (Tucker, 1985; Yariv, 1991). The frequency width is described by the 3-dB frequency (i.e., the frequency where the response is 1/2 of the low-frequency value). While it is possible

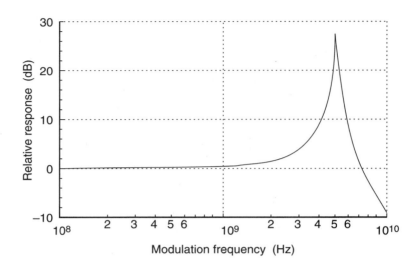

Figure 4.29 Representative frequency response of an intensity-modulated diode laser.

to use compensation techniques to utilize the full frequency response of the device, including the resonant peak, most applications confine themselves to the flat portion of the response to the left of the peak. Current wide-bandwidth lasers have a modulation bandwidth of more than 10 GHz.

    Another effect is observed when diode lasers are driven with a pulse of current. Due to the influence of the charge carriers on the index of refraction of the active region, the optical length of the resonator changes with current density. This change in optical width appears as a linear shift in operating frequency of the source (Tucker, 1985; Yariv, 1991). This *frequency chirp* can have the detrimental effect of increasing the linewidth of the source, an effect that is discussed further in Chapter 9. On the other hand, the same frequency shift can be used as a method of frequency modulating the laser. As discussed earlier, distributed feedback (DFB) lasers and distributed Bragg reflector (DBR) lasers are more resistant to this chirp effect.

4.7 Source Noise

4.7.1 LED Source Noise

    The LED is considered to be essentially a noise-free source, i.e., the source does not contaminate the signal with additional noise.

4.7.2 Laser Source Noise

The laser has three potential noise sources.

1. The first noise is that of *spontaneous emission* added to the coherent light of the laser output. Such a noise is called *relative intensity noise*, commonly abbreviated as RIN. For a relatively powerful laser, this small amount of spontaneous light is overpowered by the large amount of coherent light and the noise is negligible. For weaker sources, such as can occur in portions of an analog signal transmission, the noise can be appreciable and can affect the analog link operation. For coherent communication links where the information is encoded in the frequency or phase of the optical signal, this noise can limit the link performance. The noise is discussed further in Chapter 9.

2. The second noise is *partition noise*. A pulsed multimode laser (or even some supposedly single-frequency lasers operated at high current levels) will operate at several frequencies simultaneously, as seen in Figure 4.21. The power is "partitioned" among the various longitudinal spectral modes. While the *total* power can be kept constant, the power in each mode is not a steady function of time; the power distribution among modes changes with time. Each time the distribution changes, the power output undergoes fluctuation leading to a noise term on the nominally stable output. One technique for removing this noise is to use a true single-frequency laser, such as the distributed feedback (DFB) laser.

3. The third noise is associated with a multimode source when combined with a multimode fiber in a system that modifies the fiber-mode power distribution. This can occur at splices and connectors. When viewed at the end of a fiber, a coherent source will form an interference pattern consisting of the constructive and destructive interference of the modes in the fiber. For an ideal fiber, with all modes excited by the source, there is no fluctuation in power as time progresses. One of the potential bad effects of splices or connectors (or an imperfect fiber) in a fiber link, however, is to redistribute the power in the higher modes as light passes through the splice or connector. Usually some of the higher modes are removed by passage through the connector or imperfection. The multimode source changes its spectral distribution of power as a function of time. The laser, in turn, excites the various fiber modes with more or less power, depending on the instantaneous power distribution. Since the power in the higher modes is removed by the splice or connector, the total source power minus that removed is time-

varying. This time-varying amplitude represents noise. This noise (actually due to the combined fiber, splice or connector, and source) is attributed to the source and is called *modal noise*. Since the noise mechanism requires coherent light to form the interference pattern, one solution is to use a high power source of low coherence (or *short coherence length*, in the jargon). The superradiant LED is such a source.

4.8 Electronic Driving Circuits

As noted in the specifications, optical diode sources are operated at 1–2 volts of forward bias with variable current drive, depending on the device and the output power required. Since light is only unipolar, bipolar electrical signals must be DC-shifted to a unipolar representation before driving the optical source. The DC level can be electronically removed after detection. Both analog and digital modulation can be accomplished.

In analog modulation the DC bias point is placed on the linear portion of the output transfer characteristics, as shown in Figure 4.30. The time-varying signal is then applied about the bias point. Obviously, the amplitude of the signal must be kept small enough to avoid the nonlinear portion of the transfer characteristics to avoid distorting the signal.

For digital modulation, the device is biased at a low value of optical output (usually, but not always, at zero power) for a logical **0** and pulsed to higher output power for a logical **1**. The ratio of the lower power to the higher power is the *extinction ratio* of the modulation scheme. Nonzero extinction ratios incorporate a link performance penalty, usually in the maximum data rate of the link.

4.8.1 LED Driving Circuits

For low data rates and compatibility with digital circuits, LED sources and lasers with low threshold currents can frequently be driven directly from the output of a digital gate. The TTL family, with its ability to source and sink current, is particularly appropriate for direct drive. Other logic families and certain applications requiring larger drive currents use driver circuits designed to provide appreciable drive current. Figure 4.31 illustrates one of these driving circuits.

Another frequently used technique is to place the source in the collector or emitter arm of a bipolar transistor circuit to increase the drive current, as shown in Figure 4.32. Current-limiting resistors of the proper magnitude keep the current

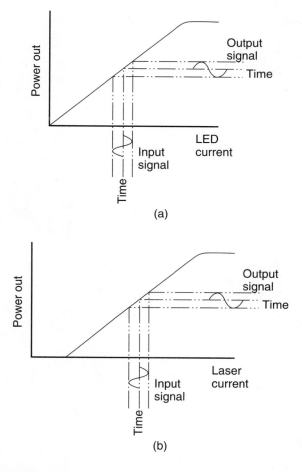

Figure 4.30 DC bias with AC modulation for (a) an LED and (b) a laser.

below the maximum current rating of the diode. Switching speeds are limited by the switching speed of the transistor. The same technique is also useful in logic gates with an open-collector output stage allowing the user to tailor the design of the diode driving circuit.

For digital modulation, the device is operated at zero volts for a logical **0** and at some appropriate forward bias for a logical **1**. As described earlier in the section on LED rise time, a momentary reverse bias and a momentary overdrive can speed up the device performance.

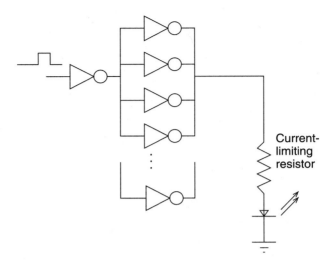

Figure 4.31 LED and laser drive using logic circuits.

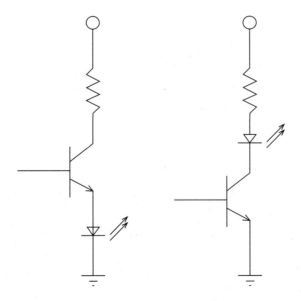

Figure 4.32 Light source drive using bipolar transistors.

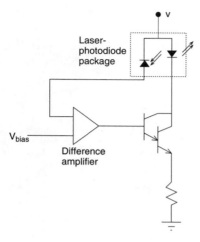

Figure 4.33 Optical feedback circuit for bias point establishment and stability.

4.8.2 Laser Driving Circuits

The primary problem in analog operation of laser diodes is maintaining the stability of the operating point in the face of temperature changes, both ambient changes and localized device heating due to its operation. This stabilization can be done by sampling the output power of the laser from a photodiode mounted in the same package as the laser and using a feedback circuit to stabilize the operating point (see Figure 4.33). In this circuit, the scaled replica of the photodiode output is applied to the input of a difference amplifier to be compared with a reference voltage that determines the desired bias point. The output of the amplifier controls the DC laser current supplied by the Darlington configuration of the two transistors. Any change, in temperature or aging effects, that decreases or increases the output power is sensed by the photodiode and difference amplifier. The DC drive current is adjusted to oppose the change, and the desired operating point is maintained. Part of the design process is to have the control circuit ignore the fast signal variation due to the AC signal and react only to the slow variation in the average power due to aging or temperature effects.

A modified circuit to allow analog modulation of the source is shown in Figure 4.34. Here a current summation is done by the difference amplifier made up of the matched transistors (facing each other in the figure) to allow the addition of an

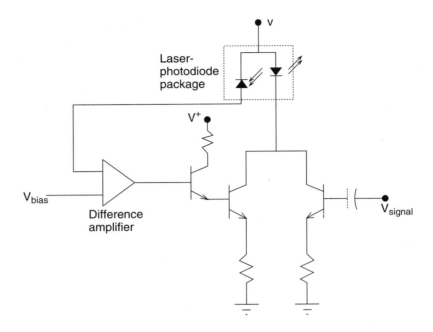

Figure 4.34 Circuit combining DC bias stability with AC modulation.

AC modulation signal on top of the bias current. Modulation of up to 50 MHz is possible with this circuit.

Open-loop stabilization of the laser source is also possible and has proved to be a popular technique. Figure 4.35 depicts a circuit that stabilizes the source's operating heat sink temperature with a *thermoelectric cooler* that has sufficient capacity to overcome the heating effects of the diode. A thermistor sensor built into the thermoelectric-cooler module monitors the heat sink temperature and controls the cooler accordingly to maintain the desired operating temperature.

Some high-threshold diode lasers require appreciably more drive current than can be provided by a transistor-amplified drive current. These lasers are usually pulsed to avoid dissipating a relatively large amount of continuous power. A block diagram of a pulsed diode-laser driving circuit is shown in Figure 4.36. The capacitor is charged to a fairly high voltage (typically 100 to 1000 volts) through the charging circuit with the diode portion of the circuit removed by an open-circuit thyristor. The thyristor is closed by a small-voltage trigger pulse generated by the signal. (Some waveshaping might be required.) The charging circuit is usually removed from the circuit by the application of the trigger pulse to an isolation circuit (not shown). When the thyristor fires, the capacitor discharges through the low-resistance path that includes the diode laser. The risetime of the pulse is determined by the RC time constant of the discharge path. The resistance includes the anode-cathode resistance of the SCR (typically 1 to 10 ohms), the forward resistance of the laser diode (obtained from the forward I-V characteristics of the diode), and any parasitic resistance. The inductance of the discharge path also lengthens the discharge pulse and causes ringing and undershoot. A short discharge path will have a typical inductance of 20 nanohenries and affects the circuit performance. A 2 Ω resistor in series with the discharge path, although slightly increasing the RC time constant, has the more beneficial effect of suppressing the ringing and undershoot in the circuit. The voltage developed across this resistor can also be used to monitor the driving-current pulse. The clamping diode shown in the figure also helps to suppress the undershoot and ringing of the circuit. The peak current of the pulse depends on the capacitor value, the voltage of that capacitor at firing (usually determined by the charging circuit), and the pulse width, since the conservation of charge says that the charge placed on the capacitor during the charging interval is removed during the discharge. The pulse-repetition rate is determined by the ability of the charging circuit to recharge the capacitor. To minimize this time, the RC time constant of the charging circuit should be as low as possible.

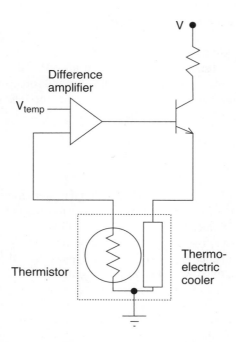

Figure 4.35 Temperature stabilization by operation of a thermoelectric cooler.

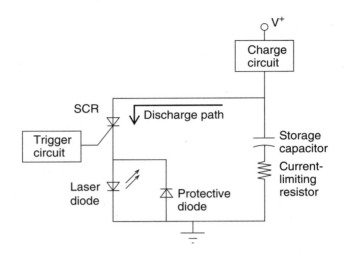

Figure 4.36 Block diagram of laser pulser circuit.

4.9 Source-to-Fiber Coupling

The source-to-fiber coupling problem (Kawano et al., 1986; Keiser, 1991) is important because substantial power is lost at this interface. We wish to find the dependence of the coupling efficiency η on the fiber and source parameters that have been discussed. We define the *coupling efficiency* as the fraction of source power that is coupled into the fiber. Mathematically, this is

$$\eta = \frac{P_f}{P_s},$$

(4.84)

where P_f is the power in the fiber and P_s is the total power emitted by the source.

It should be noted that most sources used in fiber-optic applications come with a short length of fiber-optic cable attached. (This length of cable is called a *pigtail*.) The manufacturer specifies the optical power from the source *in the pigtail* and has optimized the coupling into the fiber. Typical optical powers coupled into the pigtail range from microwatts to a few milliwatts.

The spatial emission pattern from the source plays a key role in determining the value of efficiency. Generally, this pattern is two-dimensional and asymmetric (as we have seen) for the edge-emitting LED and laser diode. The surface-emitting LED is symmetric in output.

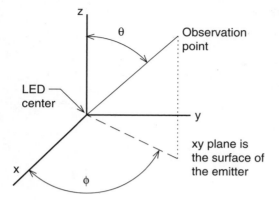

Figure 4.37 Emitter coordinate system.

4.9.1 Coupling Model

A hypothetical emitter with a well behaved symmetric emission pattern is the *Lambertian source*. Such a device has a spherical emission pattern that is symmetric and is given mathematically by

$$B(\theta,\varphi) = B_0 \cos\theta, \tag{4.85}$$

where θ and φ are the polar angles as shown in Figure 4.37, $B(\theta,\varphi)$ is the radiance of the source, and B_0 is the radiance measured along the normal to the emitting surface. The *radiance* is a measure of optical-power emission into a solid angle oriented around the axis defined by θ and φ from an infinitesimal area on the surface of the emitter. Its units are $W \cdot sr^{-1} \cdot m^{-2}$ where a steradian (abbreviated sr) is a measure of solid angle. As seen in Equation 4.85, the measured radiance from a Lambertian source is a function of the angle from the source. For a more directional source, we frequently use a modified Lambertian approximation to the emission pattern measured. A dependence of

$$B = B_0(\cos\theta)^m \tag{4.86}$$

is fitted to the data where m is a value higher than 1. Figure 4.38 illustrates such a distribution.

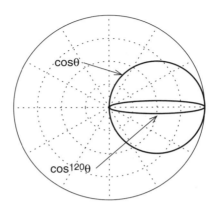

Figure 4.38 Radiance patterns for a Lambertian ($\cos \theta$) and a directive emitter $\left((\cos \theta)^{120}\right)$.

Example: A diode laser has a measured full-angle beam divergence of 30 degrees at the −3 dB power points. Calculate the value n that will allow a $(\cos \theta)^n$ spatial power distribution.

Solution: The −3 dB power points are the half-power points that occur at ± 15 degrees (since the *full*-angle beam divergence is specified). We want to fit

$$\frac{P}{P_0} = (\cos \theta)^n \tag{4.87}$$

$$0.5 = (\cos 15°)^n \tag{4.88}$$

$$\log 0.5 = n \log (\cos 15°) \tag{4.89}$$

$$n = \frac{\log (0.5)}{\log (\cos 15°)} \tag{4.90}$$

$$n = 19.9 \ . \tag{4.91}$$

An asymmetric emission distribution can also be modeled as Lambertian sources or sources with a higher dependence. Such a distribution is given by

$$B(\theta,\varphi) = \frac{1}{\left(\dfrac{\sin^2\varphi}{B_0 \cos^T\theta} + \dfrac{\cos^2\varphi}{B_0 \cos^L\theta} \right)} \ . \tag{4.92}$$

where T and L are the exponents that best fit the transverse and lateral emission patterns. Typically, for edge emitters, $T = 1$ (i.e., Lambertian), and L is a higher value. For laser diodes, L can take on values of 100 or larger because of the increased directionality of the pattern. Although many practical sources have such asymmetric emission patterns, we usually estimate the losses by using formulas based on symmetric emission assumptions. Such assumptions simplify the evaluation of integrals and provide conservative results, but a more accurate solution can be obtained with the more complicated models.

In coupling the light from the source to the fiber, we can have a fiber butted up close to optical source. (This technique is called *butt-coupling*.) The source emits with a specific angular radiation pattern; only some of the rays of light from the source will be accepted by the fiber—those within the *fiber acceptance cone* with a half-angle given by θ_{max}. (Here θ_{max} is found as $\sin^{-1}NA$, as we have seen earlier.) The power coupled into the fiber is found by the evaluation of the integral

$$P_f = \int_0^{r_u} \int_0^{2\pi} \left(\int_0^{2\pi} \int_0^{\theta_{max}} B(\theta,\varphi) \sin\theta \, d\theta \, d\varphi \right) d\varphi_s \, r \, dr . \tag{4.93}$$

Here the double integral within the parentheses is the amount of light accepted by the fiber from an infinitesimal area located at position r, φ_s. The outer integral sums up the power contributions from all of the infinitesimal emitting areas on the source. For a source smaller than the fiber core, the upper limit on the first integral would be $r_u = r_s$, the radius of the source. If the source happens to be larger than the fiber core, then the upper limit of the integral is $r_u = a$, the fiber core radius. For a step-index fiber, the value of θ_{max} is a constant. For a graded-index fiber, the value of θ_{max} will vary with the radial position r of the source, complicating the evaluation of the integral. The most general case is for a graded-index fiber with a variable exponent g of the radial dependence of index of refraction variation.

Lambertian Source

Using our assumption of having a Lambertian emitter, the radiance is $B(\theta,\varphi) = B_0 \cos\theta$ (one of the problems at the end of the chapter considers the case of a $\cos^n\theta$ source), and the integral becomes

$$P_f = \pi B_0 \int_0^{r_u} \int_0^{2\pi} \sin^2\theta_{max} \, d\varphi_s \, r \, dr . \tag{4.94}$$

We recognize $\sin^2\theta_{max} = (NA)^2$, giving

$$P_f = \pi B_0 \int_0^{r_u} \int_0^{2\pi} (NA)^2 \, d\varphi_s \, r \, dr \, . \tag{4.95}$$

Step-Index Fiber Coupling

For a step-index fiber the NA is independent of position r, so it comes outside the integrals and we evaluate the double integral as

$$P_f = \pi^2 B_0 (NA)^2 r_s^2 \quad (r_s < a) \, . \tag{4.96}$$

The power from the source is

$$P_s = \pi^2 B_0 r_s^2 \, , \tag{4.97}$$

giving the coupling efficiency

$$\eta = (NA)^2 \quad (r_s < a, \text{ Step index}) \, . \tag{4.98}$$

Revaluation of Equation 4.95 for the case of $r_s > a$ gives

$$P_f = \pi^2 B_0 (NA)^2 a^2 \quad (r_s > a) \, , \tag{4.99}$$

and, since the source power is still the same, the coupling efficiency for this case will be

$$\eta = (NA)^2 \left(\frac{a}{r_s}\right)^2 \quad (r_s > a, \text{ Step index}) \, . \tag{4.100}$$

Graded-Index Fiber Coupling

For the graded-index fiber, the NA is a function of position r and is given by

$$NA(r) = NA(0)\sqrt{1 - (r/a)^g} \, . \tag{4.101}$$

The integral for power coupled into the fiber (Equation 4.95) becomes

$$P_f = \pi B_0 \int_0^{r_u} \int_0^{2\pi} \left(NA(0)\right)^2 \left(\sqrt{1-(r/a)^g}\right)^2 d\varphi_s \, r \, dr \, . \qquad (4.102)$$

Evaluating this integral (for $r_s < a$),

$$P_f = \underbrace{\pi^2 B_0 r_s^2 \left(NA(0)\right)^2}_{P_s} \left(1 - \frac{2}{g+2}\left(\frac{r_s}{a}\right)^g\right) \quad (r_s < a) \, . \qquad (4.103)$$

The resulting coupling efficiency is

$$\eta = \left(NA(0)\right)^2 \left(1 - \frac{2}{g+2}\left(\frac{r_s}{a}\right)^g\right) \quad (r_s < a, \text{ Graded index}) \, . \qquad (4.104)$$

For the case of $r_s > a$, the power into the fiber is

$$P_f = \frac{\pi^2 B_0 \left(NA(0)\right)^2 a^2 g}{g+2} \quad (r_s > a) \, , \qquad (4.105)$$

and the coupling efficiency for this case becomes

$$\eta = \left(NA(0)\right)^2 \left(\frac{a}{r_s}\right)^2 \left(\frac{g}{g+2}\right) \quad (r_s > a, \text{ Graded index}) \, . \qquad (4.106)$$

Equations 4.98 and 4.100 and Equations 4.104 and 4.106 are the coupling-efficiency equations applicable to the step-index fiber and the graded-index fiber, respectively, and are summarized in Table 4.2.

Table 4.2: Summary of coupling efficiencies (Lambertian emitter).

	$r_s \leq a$	$r_s > a$
Step-index	NA^2	$NA^2\left(\frac{a}{r_s}\right)^2$
Graded-index	$NA^2\left[1 - \left(\frac{2}{g+2}\right)\left(\frac{r_s}{a}\right)^g\right]$	$NA^2\left(\frac{a}{r_s}\right)^2\left(\frac{g}{g+2}\right)$

Example: Consider a circular LED source with a 62.5 μm diameter. Calculate the coupling efficiencies

(a) ... into 50/125, 62.5/125, and 100/140 step-index fibers with NA = 0.20.

Solution: For a 50/125 SI fiber, $r_s > a$, so

$$\eta = NA^2 \left(\frac{a}{r_s}\right)^2 \tag{4.107}$$

$$= (0.2)^2 \left(\frac{25}{31.25}\right)^2 \tag{4.108}$$

$$= 0.0256 \tag{4.109}$$

$$= 2.56\% \tag{4.110}$$

$$= -15.92 \text{ dB} . \tag{4.111}$$

For a 62.5/125 fiber, $r_s = a$, so

$$\eta = (NA)^2 \tag{4.112}$$

$$= (0.2)^2 \tag{4.113}$$

$$= 4\% \tag{4.114}$$

$$= -13.9 \text{ dB} . \tag{4.115}$$

For a 100/140 fiber, $r_s < a$, so

$$\eta = (NA)^2 \tag{4.116}$$

$$= (0.2)^2 \tag{4.117}$$

$$= 4\% \tag{4.118}$$

$$= -13.9 \text{ dB} . \tag{4.119}$$

(b) ... into three graded-index fibers of the same dimensions with NA(0) = 0.2 and $g = 1.8$.

Solution: For a 50/125 GI fiber, $r_s > a$, so

$$\eta = NA^2(0) \left(\frac{a}{r_s}\right)^2 \left(\frac{g}{g+2}\right) \tag{4.120}$$

$$= (0.2)^2 \left(\frac{25}{31.25}\right)^2 \left(\frac{1.8}{1.8+2}\right) \tag{4.121}$$

$$= 0.01213 \tag{4.122}$$
$$= 1.213\% \tag{4.123}$$
$$= -19.16 \text{ dB} . \tag{4.124}$$

For a 62.5/125 fiber, $r_s = a$, so

$$\eta = NA^2(0) \left(1 - \frac{2}{g+2} \right) \tag{4.125}$$

$$= (0.2)^2 \left(1 - \frac{2.0}{1.8+2} \right) \tag{4.126}$$

$$= 0.01895 \tag{4.127}$$
$$= 1.895\% \tag{4.128}$$
$$= -17.22 \text{ dB} . \tag{4.129}$$

For a 100/140 fiber, $r_s < a$, so

$$\eta = (NA)^2(0) \left[1 - \left(\frac{2}{g+2} \right) \left(\frac{r_s}{a} \right)^2 \right] \tag{4.130}$$

$$= (0.2)^2 \left[1 - \left(\frac{2}{1.8+2} \right) \left(\frac{31.25}{50} \right)^2 \right] \tag{4.131}$$

$$= 0.0318 \tag{4.132}$$
$$= 3.18\% \tag{4.133}$$
$$= -14.98 \text{ dB} . \tag{4.134}$$

To compare equivalent step-index and graded-index fibers, we can use the approximation for the NA or NA(0),

$$NA \approx n_1 \sqrt{2\Delta}. \tag{4.135}$$

For a step-index fiber $(g = \infty)$ *and* $r_s = a$, we have

$$\eta(SI) = 2n_1^2\Delta. \tag{4.136}$$

For a parabolic-index fiber $(g = 2)$ *and* $r_s = a$, the efficiency is

$$\eta(GI) = n_1^2\Delta. \tag{4.137}$$

This leads to the conclusion that, all else being equal, a step-index fiber is twice as efficient in coupling light as a graded-index fiber (an improvement of 3 dB). This improved coupling is one of the advantages of a step-index fiber over a graded-index fiber.

Non-Lambertian Source

For a *non-Lambertian source* that is still symmetric and approximated by $(\cos \theta)^m$, the coupling efficiency for a *step-index fiber* with $r_s < a$ turns out to be

$$\eta_{\text{non-Lambertian}} = \frac{m+1}{2} NA^2 , \qquad (4.138)$$

showing $(m + 1)/2$ greater efficiency due to the shaping of the source. This accounts for the laser diode's superiority in delivering optical power into a fiber. Note, however, that, because of the small value of Δ or NA, all fibers typically will have significant losses at the source-fiber interface.

4.9.2 Reflection Effects

The coupling efficiencies calculated above assume a perfect match of refractive index at the core–light source interface. The lack of such a match leads to additional losses due to the *Fresnel reflection losses* at the interface. Assuming perpendicular incidence, the power transmittance T at the interface between the medium and the core of a step-index fiber is

$$T = 1 - \left(\frac{n_1 - n}{n_1 + n} \right)^2 , \qquad (4.139)$$

where n_1 is the index of refraction of the core and n is index of refraction of medium outside of the fiber core. Frequently a drop of index matching liquid or gel is placed at this interface to minimize these losses. Another technique to reduce the reflection losses is to cut the fiber ends at a non-perpendicular angle to avoid retroreflections.

4.9.3 Lens-Coupled Fiber

As indicated in the coupling efficiency results for the butt-coupled fiber, mismatching of the source area and the area of the fiber core wastes power. The

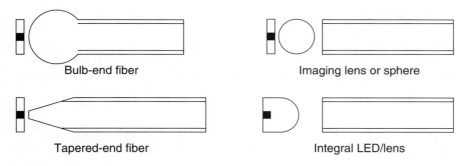

Figure 4.39 Source-fiber coupling using lenses.

obvious solution is to physically match the size of the source and the size of the fiber. If the size of the source is *smaller* than the size of the fiber (a case that typically exists with a laser), a lens can be used to optically match the apparent size while increasing the directivity of the source by the magnification factor. (Unfortunately, no remedy exists if the size of the source exceeds the size of the fiber.) Since the source and fiber sizes are so small and one does not want to have to use a lens much bigger than this size, there arises a need for small *microlenses* to perform this coupling task. Several lens geometries have been attempted in the past; some are illustrated in Figure 4.39. The goal of each of these schemes is to magnify the effective emitter area to match the area of the fiber core. The "bulb-end" fiber is easily made by heating the end of the fiber and letting it cool. The integral LED/lens incorporates a lens into the structure of the LED source at the time of fabrication.

The incorporation of the microlens has the potential disadvantages of adding fabrication steps and requiring fine resolution alignment to implement. If properly aligned, the improvement in coupling efficiency is M, the magnification of the source where M_{max} is d_{fiber}/d_{source}. (Improper alignment adds small, but tolerable, decreases in the coupling efficiency.)

Example: Consider an LED source with a 50 μm diameter operating with a 100/140 SI fiber with NA = 0.15. Calculate the coupling efficiency both with and without a coupling lens assuming a Lambertian beam pattern. Calculate the optimum magnification of the lens.

Solution: Without a lens, $r_s < a$, so

$$\eta = NA^2 \qquad (4.140)$$
$$= (0.15)^2 \qquad (4.141)$$
$$= 2.25\% \qquad (4.142)$$
$$= -16.48 \text{ dB} . \qquad (4.143)$$

With a lens, $r_s = a$, and

$$\eta = M(NA)^2 ,$$ (4.144)

$$M_{max} = M_{opt} = \frac{d_f}{d_s}$$ (4.145)

$$= \frac{100}{50}$$ (4.146)

$$= 2 ,$$ (4.147)

$$\eta = 2(0.15)^2$$ (4.148)

$$= 4\%$$ (4.149)

$$= -13.47 \text{ dB} .$$ (4.150)

Hence we find a 3 dB improvement in coupling that corresponds to the magnification of 2.

4.9.4 LED Coupling

The coupling efficiency of an LED depends on whether it is a surface-emitting LED or an edge-emitting LED.

The surface-emitting LED produces the closest approximation to our ideal Lambertian emitter. Such an LED matches the theory quite well when coupled into a large-core large-NA fiber. The match is worse when the fiber has a small core and a small NA. The coupling for such fibers can be greatly improved by reducing the size of the source area during manufacture (which, in turn, requires improved heat sinking for a given drive current) and using a lens to couple the light. Increases in coupling efficiency of 3 to 5 times have been achieved with bulb-end fibers and from 18 to 20 times with tapered-end fibers (Yu et al., 1990).

The edge-emitting LED is more difficult to model due to its asymmetric emission pattern. The narrower emission pattern (in one dimension) allows greater coupling efficiency for butt-coupled fibers. Optical techniques using cylindrical lenses to improve the coupling on each axis have been attempted (Yu et al., 1990), but the improvement in coupling has been marginal in light of the more complicated lens arrangement. Bulb-end fibers and tapered fibers can improve the coupling efficiency with these edge-emitting LEDs.

4.9.5 Laser Coupling

Laser sources, like edge-emitting diodes, have asymmetric beam patterns (although the patterns are much narrower than the LED patterns). An additional difficulty is incurred when trying to couple the light into a small-core single-mode fiber. (Core diameters can range from 5 to 9 μm.) The small size of the core increases the sensitivity of the coupling efficiency to misalignment. Lens elements are frequently used to match the small active source area to the size of the fiber core. Both microlenses and *graded-index lenses* (also called "GRIN lenses") have been used (Yu et al., 1990), as well as cylindrical lenses. While the use of lenses helps to match the source and fiber core areas, it also has the effect of increasing the sensitivity of the coupling coefficient to lateral misalignment errors. In addition, the presence of reflections from the lens elements and the fiber back into the laser can upset the frequency stability of single-frequency single-mode fiber.

4.10 Summary

We have found that the laser holds significant operational advantages over the LED except for cost and temperature sensitivity. The spectral width of the laser source is appreciably lower for the laser diode; hence, this source is more suitable for use with a single-mode fiber where dispersion is to be minimized. The optical power delivered into a pigtail is higher for the laser, due to the higher coupling efficiency enjoyed by the laser because of its narrower emission pattern. The laser also enjoys an order-of-magnitude more speed, thereby increasing the potential data rate of an optical link. The cost ratio is about a factor of two in favor of the LED source, due to the increased fabrication quality requirements needed to produce today's low-threshold, high-reliability devices. In the short-wavelength region, then, LEDs offer substantial cost advantages for links operating at modest data rates (<50 Mb $\cdot$ s^{-1}) using multimode fibers (either step-index or graded-index). At higher data rates or with single-mode fibers, laser diodes are the source of choice.

In the long-wavelength region of the spectrum, InGaAsP sources are available as both LEDs and lasers. Since most sources at the long wavelengths are to operate with single-mode fibers, early work concentrated on laser diode emitters. Long-wavelength LEDs are now commercially available and are finding applications for short-distance moderate–data-rate links, such as the proposed Fiber Distributed Data Interface (FDDI) (described in Chapter 8). While InGaAsP sources have shown slower degradation than GaAlAs devices, their temperature sensitivity is greater. Since the spectral width of a source increases as the square of the central wavelength, the spectral width of these long-wavelength sources is about 2.5 times that of an 850 nm source. Single-frequency operation is important to minimize dispersion. Such techniques as distributed feedback and multiple cavity combinations have been investigated to produce single-frequency operation. *Mode hopping,* or the jumping of the emitted light from one mode to another, is also an undesirable effect, caused by multimode operation and reflections of the light from interfaces in front of the source. Single-mode operation by control of the width of the emitter has been effective in reducing multimode effects. (Another technique to reduce these effects is to reduce the coherence of the source, using a low-coherence wide-frequency source called a *superluminescent LED*.) Polarization rotating and blocking devices called *optical isolators* are frequently used to avoid the reflection effects.

The current challenges in laser sources for fiber-optic applications are in achieving narrow spectral width, stabilizing the operating wavelength against the frequency chirp when the laser is pulsed, lowering the threshold current to values of a few to several milliamperes, and rapidly tuning the wavelength.

4.11 Problems

1. Find . . .

   (a) . . . the minimum wavelength of a GaAlAs source.

   (b) . . . the maximum wavelength of a GaAlAs source.

2. Using the alloy fraction formulas (Equations 4.3 and 4.4), find the material composition for . . .

   (a) . . . a 1.3 µm source.

   (b) . . . a 1.55 µm source.

3. Consider two $Ga_{1-x} Al_x As$ laser sources with $x = 0.02$ and $x = 0.09$ respectively. Find the bandgap energy and peak wavelength for these devices.

4. We want to estimate the transmission at air-material interfaces for high refractive-index materials like InP and GaAs. For a plane wave at perpendicular incidence, the transmissivity of a planar interface between a material and air ($n = 1$) is

$$T = 1 - \left(\frac{n-1}{n+1}\right)^2 , \tag{5.151}$$

where n is the index of refraction of the material. Find T in percent and in dB for . . .

   (a) . . . InP ($n = 3.4$).

   (b) . . . GaAs ($n = 3.6$).

5. Show that the LED bandwidth is given by Equation 4.20.

6. An optical source is selected from group of devices specified as requiring a mean time of 5×10^4 hours for the output power to degrade by -3 dB. If the device emits 5 mW at room temperature at the start of a test, what will be its emission power after (a) 1 month, (b) 1 year, (c) 5 years, and (d) 10 years?

7. A group of laser devices have operating lifetimes of 3.5×10^4 hours at 60 C and 6700 hours at 90 C. Find the activation energy for these devices and calculate the expected lifetime at 20 C.

8. A laser diode has a lateral beam divergence of 30 degrees (full angle) and a perpendicular (to the emitting junction) beam divergence of 60 degrees. What are the values of L and T associated with this beam pattern?

9. An LED has a circular emitting region with a radius of 20 µm. The pattern is assumed Lambertian with an axial radiance of $80 \text{ W} \cdot \text{cm}^{-2} \cdot \text{sr}^{-1}$ at a 100 mA drive current.

   (a) Calculate the coupling efficiency and the optical power coupled into a step-index fiber with a 140 µm core diameter and NA = 0.20.

   (b) Repeat the calculation for a 50 µm diameter graded-index fiber with a parabolic-index profile, a center index of 1.48, and $\Delta = 1\%$.

10. On the same graph plot the coupling efficiency as a function of r_s (from 0 to 50 µm) for the following step-index fibers:

    (a) ... core diameter of 50 µm and NA = 0.15

    (b) ... core diameter of 100 µm and NA = 0.20.

    In what regions could a lens improve the coupling efficiency?

11. Using a computer, plot the emission patterns (on a polar plot) of a Lambertian emitter and a laser emitter with $m = 10$.

12. (a) Show that the power coupled into a step-index optical fiber from a source (with $r_s < a$) with a radiance of $B(\theta) = B_0(\cos \theta)^m$ is given by

$$P_f = \frac{2\pi A_s B_0 \left[1 - (\cos \theta_{max})^{m+1}\right]}{m+1}. \tag{4.152}$$

   (b) Show that the total power from the source is given by

$$P_s = \frac{2\pi A_s B_0}{m+1}. \tag{4.153}$$

   (c) Calculate the coupling efficiency and show that for small NA the answer becomes

$$\eta = \frac{(m+1)\,\mathrm{NA}^2}{2}. \tag{4.153}$$

Note: For a graded-index fiber the coupling efficiency is

$$\eta = \left(\frac{(m+1)\,\mathrm{NA}^2}{2}\right)\!\left(\frac{g}{g+2}\right). \tag{4.154}$$

13. A step-index fiber is excited by a laser with an emission pattern approximated by $(\cos\theta)^7$. Calculate the improvement factor relative to a Lambertian source.

14. Consider a fiber ($a = 30\ \mu\text{m}$, $\Delta = 1.5\%$, $g = 1.95$, and $n_1 = 1.45$) excited by a surface-emitting LED ($B_0 = 2.0 \times 10^2\ \text{W} \cdot \text{cm}^{-2} \cdot \text{sr}^{-1}$ and radius of 50 μm). Calculate. . .

    (a) . . . the power emitted by the source,

    (b) . . . the power coupled into the fiber, and

    (c) . . . the coupling losses (in dB).

15. Consider an LED with a radius of 20 μm and a circularly symmetric Lambertian emission pattern with an on-axis radiance of $100\ \text{W} \cdot \text{cm}^{-2} \cdot \text{sr}^{-1}$ at a drive current of 100 mA.

    (a) Calculate the power coupled into a step-index fiber with a core diameter of 100 μm and a numerical aperture of 0.22.

    (b) Calculate the change in coupling efficiency (*in dB*) if the source size is tripled, the drive current is reduced to 50 mA, and the NA of the fiber is reduced to 0.11 while keeping the fiber diameter constant.

16. (a) Calculate the loss due to reflections in an interface that has the light go from GaAs ($n = 3.5$), through air ($n = 1.0$), and into a fiber ($n = 1.5$).

    (b) Calculate the transmission if the air is replaced with a medium (such as epoxy) that matches the index of the glass.

References

Bergh, A. and J. Copeland, "Optical sources for fiber transmission systems," *Proc. IEEE*, vol. 68, no. 10, pp. 1240–1246, 1980.

Botez, D. and M. Ettenberg, "Comparison of surface and edge emitting LEDs for use in fiber optic communications," *IEEE Trans. Electron Devices*, vol. ED-26, pp. 1230–1238, 1979.

Bowers, J. and M. Pollock, "Semiconductor lasers for telecommunications," in *Optical Fiber Telecommunications II*, (S. E. Miller and I. P. Kaminow, eds.), pp. 509–568, New York: Academic Press, 1988.

Burrus, C. A.; C. Casey, Jr.; and T. Li, "Optical sources," in *Optical Fiber Telecommunications*, (S. E. Miller and A. G. Chynoweth, eds.), pp. 499–556, New York: Academic Press, 1979.

Cartledge, J. C., "Theoretical performance of multigigabit-per-second lightwave systems using injection-locked semiconductor lasers," *J. Lightwave Technology*, vol. 8, no. 7, pp. 1017–1022, 1990.

Casey, H. and M. Panish, "Heterostructure lasers: Part A-Fundamental principles and Part B-Materials and operating characteristics," in *Heterostructure Lasers*, New York: Academic Press, 1978.

Dutta, N. and C. Zipfel, "Reliability of lasers and LEDs," in *Optical Fiber Telecommunications II*, (S. E. Miller and I. P. Kaminow, eds.), pp. 671–687, New York: Academic Press, 1988.

Ettenberg, H. and H. Kressel, "The reliability of (AlGa)As cw laser diodes," *IEEE J. Quantum Electronics*, vol. QE-16, no. 2, pp. 186–196, 1980.

Fukuda, M., "Laser and LED reliability update," *J. Lightwave Technology*, vol. 6, no. 10, pp. 1488–1495, 1988.

Goodwin, A., I. Davies, R. Gibb, and R. Murphy, "The design and realization of a high reliability semiconductor laser for single-mode fiber-optical communications links," *J. Lightwave Technology*, vol. 6, no. 9, pp. 1424–1434, 1988.

Hirao, M., K. Mizuishi, and M. Nakamura, "High-reliability semiconductor lasers for optical communications," *IEEE J. on Selected Areas in Communications*, vol. SAC-4, no. 9, pp. 1494–1514, 1986.

Kawano, K., H. Miyazawa, and O. Mitomi, "New calculations for coupling laser diode to multimode fiber," *J. Lightwave Technology*, vol. LT-4, no. 3, pp. 368–374, 1986.

Keiser, G., *Optical Fiber Communications, 2nd Edition*. New York: McGraw-Hill, 1991.

Koren, U., M. Oron, M. Young, B. Miller, J. de Miguel, G. Raybon, and M. Chien, "Low threshold highly efficient strained quantum well lasers at 1.5 μ wavelength," *Electronics Letters*, vol. 26, p. 465, 1990.

172

Kressel, H. and J. Butler, *Semiconductor Lasers and Hetrojunction LEDs.* New York: Academic Press, 1977.

Kressel, H., "Electroluminescent sources for fiber systems," in *Fundamentals of Optical Fiber Communications*, (M. F. Barnoski, ed.), pp. 187–255, New York: Academic Press, 1981.

Lee, T. P., C. Burrus, Jr., and R. Saul, "Light-emitting diodes for telecommunications," in *Optical Fiber Telecommunications II*, (S.E. Miller and I.P. Kaminow, eds.), pp. 467–507, New York: Academic Press, 1988.

Lee, T., "Recent advances in long-wavelength semiconductor lasers for optical communication," *Proc. IEEE*, vol. 79, no. 3, pp. 253–276, 1991.

Long, J., R. Logan, and R. Kerlicek, Jr., "Epitaxial growth methods for lightwave devices," in *Optical Fiber Telecommunications II*, (S.E. Miller and I.P. Kaminow, eds.), pp. 631–670, New York: Academic Press, 1988.

Marcuse, D., "LED fundamentals: Comparison of front- and edge-emitting diodes," *IEEE J. Quantum Electronics*, vol. QE-13, no. 10, pp. 819–827, 1977.

Nakajima, Y., H. Higuchi, Y. Sakakibara, S. Kakimoto, and H. Namizaki, "High-power high-reliability operation of 1.3- μm p-substrate buried crescent laser diodes," *J. Lightwave Technology*, vol. LT-5, no. 9, pp. 1263–1268, 1987.

Oshiba, S. and Y. Tamura, "Recent progress in high-power GaInAsP lasers," *J. Lightwave Technology*, vol. 8, p. 1350, 1990.

Panish, M., "Heterojunction injection lasers," *Proc. IEEE*, vol. 64, no. 10, pp. 1512–1540, 1976.

Saleh, B. E. and M. C. Teich, *Fundamentals of Photonics.* New York: John Wiley and Sons, 1991.

Selway, P., A. R. Goodwin, and P. A. Kirby, "Semiconductor laser light sources for optical communications," in *Optical Fibre Communications Systems*, (C. Sandbank, ed.), pp. 156–183, New York: Wiley, 1980.

Shumate, P., "Lightwave transmitters," in *Optical Fiber Telecommunications II*, (S. E. Miller and I. P. Kaminow, eds.), pp. 723–757, New York: Academic Press, 1988.

Tanbun-Ek, T., R. Logan, N. Olsson, H. Temkin, A. Sargent, and K. Wecht, "High output power 1.48–1.51 μm continuously graded index separate confinement strained quantum well lasers," *Applied Physics Letters*, vol. 57, p. 224, 1990.

Thompson, G., *Physics of Semiconductor Laser*. New York: Wiley, 1980.

Tucker, R. S., "High-speed modulation of semiconductor lasers," *J. Lightwave Technology*, vol. LT-3, no. 6, pp. 1180–1192, 1985.

Yariv, A., *Optical Electronics, 4th Edition*. New York: Holt Rinehart and Winston, 1991.

Yu, P. K., L. Yu, and K. Li, "Optical sources for fibers," in *Fiber Optics Handbook for Engineers and Scientists*, (F. C. Allard, ed.), pp. 5.1–5.61, New York: McGraw-Hill, 1990.

Chapter 5

Optical Receivers

5.1 Introduction

The *optical receiver* is a combination of the optical detector, electronic pre-amplifier, and electronic processing elements that recover information sent on the optical signal. The design and implementation of the receiver portion of the system is most difficult, because the receiver could be working with the weakest optical signal and we do not want to contaminate the signal with noise.

Figure 5.1 shows a block diagram of a representative optical digital data receiver. The *optical detector* converts the modulated optical input into an electronic signal for further processing. Because the optical signal is typically weak, the next step is to amplify the signal with the *preamp*. It is crucial to minimize the noise added by this amplifier. We will find that the lowest-noise preamplifiers lack the bandwidth to handle the high–data-rate signals used in fiber communications; there-fore, the *equalizer* works in combination with the preamp to restore the required bandwidth. The equalizer can also be used to help alleviate the problems caused by data spilling out into adjacent bit periods because of pulse spreading. Following the equalizer, the signal is boosted further with the *postamplifier*, frequently with some

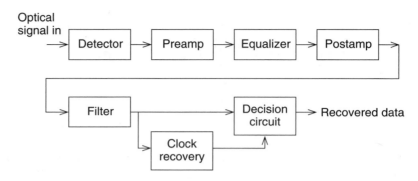

Figure 5.1 Block diagram of a representative optical receiver for a digital data link.

sort of automatic gain control that adjusts the gain subject to the strength of the signal. The *filter* following the postamp removes unwanted frequency components that might have been generated by the signal processing to this point. In some low–data-rate optical receivers, the detection is done asynchronously, in which a comparator is used to decide whether a pulse is present or not. This type of data recovery assumes that the pulses have sharp rise times and fall times. For optimum performance in high–data-rate links, the data clock is encoded into the transmitted signal and is recovered at the receiver by the *clock recovery* circuit. The recovered clock is fed into the data *decision circuit,* where the decision is made whether the voltage represents a logical **1** or a logical **0** (at the optimum sampling time provided by the recovered clock signal). Based on the results of the decision, the output is the recovered data stream, perhaps containing some errors.

In signal recovery, the issue of noise is of paramount importance, since the presence of noise leads to errors in the recovered data. In communications systems the noise can be introduced by the transmitter (as already discussed in the previous chapter), the channel, the detector, and the electronic signal processing elements. In fiber transmission, the channel noise is assumed to be zero due to the fiber's imperviousness to electromagnetic interference. The noise from the optical detector is different from that of radio and other electronic detectors in that it is signal-dependent. The noise of the electronic signal processing elements is the same as other electronic applications, and the results of prior studies have been adapted to the fiber-receiver application.

In this chapter we will first consider the optical detectors used in fiber-optic links, the pin photodiode and the avalanche photodiode. For each detector, we discuss the physical operating principles of the devices, define the parameters used to characterize them, explore their noise performance, and consider limitations on operating speed.

We then consider digital receiver design in terms of the amplifier noise, optimization of the output signal-to-noise ratio, and the requirements for an equalization amplifier.

The sensitivity of an optical receiver is a function of the noise properties of the detector *and* the preamplifier. We will then express the receiver performance in terms of the noise performance of the detector and the noise figure of the preamplifier.

5.2 Optical Detectors

A photodetector converts the optical input power into a current output. The ideal photodetector would be highly efficient, add no noise to the signal, respond uniformly to all wavelengths, would not limit the signal speed, and would be perfectly linear. Additionally, it would be small, electronically compatible with integrated circuits, reliable, and inexpensive. From the range of available photodetectors including photoconductors, phototransistors, vacuum photoemissive devices, and pyroelectric devices, only the semiconductor photodiode (Lee and Li, 1979; Plumb, 1980; Smith, 1980; Brain, 1985; Bowers, 1988; Forrest, 1988; Kasper, 1988) meets the latter set of properties sufficiently well to be considered for use in fiber-optic links.

5.2.1 Photodiodes

Physical Principles

The photodiode is a *pn* junction operated in a reverse-biased circuit, as shown in Figure 5.2. The incident light must penetrate into the *depletion region* of

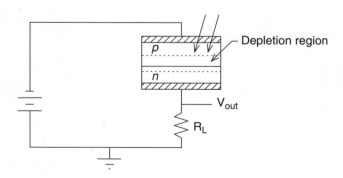

Figure 5.2 Schematic of *pn* photodiode.

Table 5.1: Bandgap energies (in eV) at 300K for representative photodiode materials.

Material	Bandgap Energy (eV) at 300K
GaAs	1.43
GaSb	0.73
GaAs$_{0.88}$Sb$_{0.12}$	1.15
Ge	0.67
InAs	0.35
InP	1.35
In$_{0.53}$Ga$_{0.47}$As	0.75
In$_{0.14}$Ga$_{0.86}$As	1.15
Si	1.14

the p and n material where the free carriers have been removed. The light is absorbed in this depletion region and delivers its energy to the material. If the absorbed energy is sufficient, a hole-electron pair can be created. This pair will be separated by the electric field existing across the depletion region, and swept to opposite sides of the depletion region. The motion of this pair of charge carriers is sensed by the outside circuitry, and the net effect is the motion of one charge quantum (1.6×10^{-19} coulombs) through the external load. The number of hole-electron pairs per second so freed is dependent linearly on the power of the optical field and, hence, the electric current is proportional to the optical power.

Spectral Response

The amount of energy required to free a hole-electron pair is the *bandgap energy* of the material. Table 5.1 gives the bandgap energy for a variety of detector materials.

Since the energy of a *photon* of light is $h\nu$ (where h is Planck's constant $[6.63 \times 10^{-34}$ joule $\cdot$ sec$]$ and ν is the optical frequency), the energy condition requires

$$h\nu \geq E_g \qquad (5.1)$$

$$\frac{hc}{\lambda} \geq E_g . \qquad (5.2)$$

From this condition we observe that any given material will exhibit a *long-wavelength cutoff*, since the photons will not have sufficient energy to free a hole-electron pair whenever the wavelength exceeds λ_{max}, where

$$\lambda_{max} = \frac{hc}{E_g} . \qquad (5.3)$$

From this relation we expect all materials to show an abrupt cutoff in performance for wavelengths exceeding λ_{max}. Since silicon has λ_{max} of 1.13 micrometers ($E_g = 1.1$ eV), it is a suitable detector material for the short-wavelength sources previously discussed, but not for the long-wavelength sources. Although other materials are also suitable (e.g., Ge, InGaAs, GaSb, GaAlSb, and others), silicon is used primarily in the short-wavelength region because of its superior properties. (Detectors made of GaAs are also used in this region since the higher carrier velocity of this material can produce higher speed devices and the detector can be integrated with high-speed GaAs electronic devices to make an integrated electronic receiver.) In the long-wavelength portion of the spectrum, silicon is ineligible, so detectors are primarily made of InGaAs.

The penetration of the light into the depletion region of the device is governed by the absorption properties of the material. The *absorption coefficient, $\alpha(\lambda)$*, for a few materials is shown in Figure 5.3. Note the effect of the cutoff wavelength, causing a negligible absorption coefficient at long wavelengths. At short wavelengths, the value of the absorption coefficient increases dramatically. The amount of incident power that is absorbed in the depletion region of depth w, assuming that the layer begins at a depth d below the device surface, is given by

$$P(w) = P_i e^{-\alpha d} \left(1 - e^{-\alpha w}\right) \left(1 - R_f\right) , \qquad (5.4)$$

where $P(w)$ is the power absorbed in the depletion region, P_i is the incident power, α is the absorption coefficient at the wavelength of interest, w is the depletion layer depth, and R_f is the power reflectivity of the detector's surface. The high value of α at short wavelengths causes little power to penetrate into the depletion region. At short wavelengths, therefore, we find a diminished response from the diode, because of this increased surface absorption.

Example: Consider a layer 1 μm thick located 3 μm below a planar air-silicon surface. Calculate the fraction of the incident power absorbed in the layer . . .

(a) . . . if $n_{silicon} = 3.5$ and $\alpha = 10^3$ cm^{-1} ($= 10^5$ m^{-1}) at the wavelength of interest.

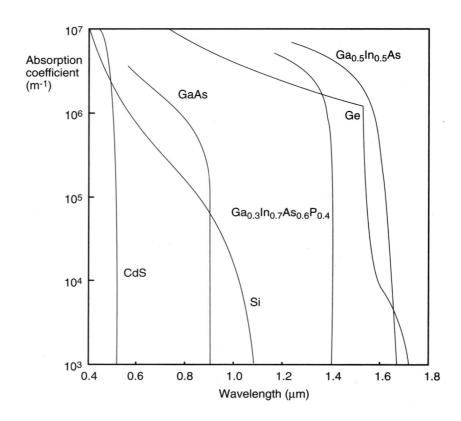

Figure 5.3 Absorption coefficient vs. wavelength for representative materials. (After Lee, T.P. and Li, T., "Photodetectors," in *Optical Fiber Telecommunications*, [S.E. Miller and A.G. Chynoweth, eds.], pp. 593–626, New York: Academic Press, 1979.)

Solution: We begin by finding the reflection coefficient

$$R_f = \left(\frac{n-1}{n+1}\right)^2 \qquad (5.5)$$

$$= \left(\frac{3.5-1}{3.5+1}\right)^2 \qquad (5.6)$$

$$= 0.309 . \qquad (5.7)$$

We then find the fraction of the incident power that is absorbed as

$$\frac{P(w)}{P_i} = e^{-\alpha d}\left(1 - e^{-\alpha w}\right)\left(1 - R_f\right) \tag{5.8}$$

$$= \left(e^{-10^5(3 \times 10^{-6})}\right)\left(1 - e^{-10^5(10^{-6})}\right)(1 - 0.309) \tag{5.9}$$

$$= 4.87\% \, . \tag{5.10}$$

(b) ... if $\alpha = 10^4 \, \text{cm}^{-1} \, (10^6 \, \text{m}^{-1})$?

Solution: For the second case we have

$$\frac{P(w)}{P_i} = e^{-10^6 \, (3 \times 10^{-6})}\left(1 - e^{-10^6(10^{-6})}\right)(1 - 0.309) \tag{5.11}$$

$$= 2.18\% \, . \tag{5.12}$$

In an effort to increase the effective depth w of the depletion region, a layer of lightly doped material (i.e., *intrinsic material*) can be formed between the p and n layers of the diode. This layer adds its depth to the depletion layer to increase the *active region* of the diode considerably. This increased depth also has the secondary benefit of reducing the capacitance of the diode. With the addition of this layer, the diode becomes a *pin diode*, as shown in Figure 5.4. The increased efficiency of these devices has made them more popular as optical receivers than are *pn* photodiodes.

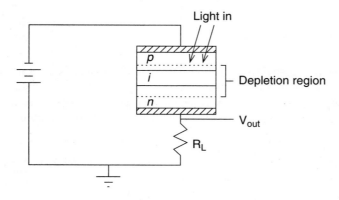

Figure 5.4 Structure of a pin photodiode.

Sensitivity

Optical detectors convert the input optical power into an output current from the detector. The sensitivity of a diode detector is measured by the *responsivity* $\mathcal{R}$ of the device, given by

$$\mathcal{R} = \frac{I_{\text{out}}}{P_{\text{in}}} , \qquad (5.13)$$

where I_{out} is the current out of the detector and P_{in} is the incident optical power. Another parameter used to describe the detector's sensitivity is the *quantum efficiency* η which is the number of hole-electron pairs generated per photon. A value less than 1 indicates that not every photon is generating carriers. Since the number of incident photons per second is $P_i\lambda/hc$ and the number of carriers flowing per second is given by I/q, we have

$$\eta = \frac{\text{carrier pairs} \cdot \text{s}^{-1}}{\text{photons} \cdot \text{s}^{-1}} \qquad (5.14)$$

$$= \frac{\dfrac{I}{q}}{\dfrac{P}{h\nu}} \qquad (5.15)$$

$$= \frac{Ihc}{qP_i\lambda} . \qquad (5.16)$$

The responsivity is then

$$\mathcal{R} = \frac{\eta q\lambda}{hc} . \qquad (5.17)$$

Figure 5.5 shows the responsivity of three typical devices plotted against wavelength. Equation 5.16 is shown as dotted lines for constant quantum efficiency. The long-wavelength falloff in the responsivity is due to the energy deficiency of the photons; the short-wavelength falloff is due to the increased absorption effects.

Example: A detector operating at 850 nm produces 80 μA of output current for a 500 μW input light beam.

(a) Calculate the responsivity of the detector.

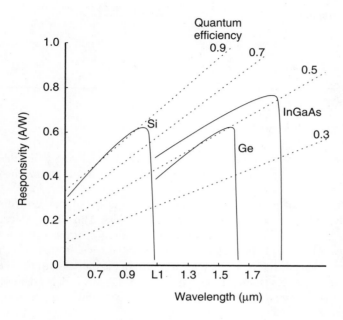

Figure 5.5 Responsivity of typical devices.

Solution:

$$\mathcal{R} = \frac{I}{P_i} \qquad (5.18)$$

$$= \frac{80 \times 10^{-6}}{500 \times 10^{-6}} \qquad (5.19)$$

$$= 0.16 \text{ A/W} \qquad (5.20)$$

$$= 160 \text{ mA/W} . \qquad (5.21)$$

(b) Calculate the quantum efficiency.

Solution:

$$\eta = \frac{Ihc}{qP_i\lambda} \qquad (5.22)$$

$$= \frac{hc\mathcal{R}}{q\lambda} \qquad (5.23)$$

$$= \frac{(6.63 \times 10^{-34})\,(3.0 \times 10^{8})\,(0.16)}{(1.6 \times 10^{-19})\,(850 \times 10^{-9})} \qquad (5.24)$$

$$= 0.234 \qquad (5.25)$$

$$= 23.4\% \; . \qquad (5.26)$$

Sensitivity—An Alternative Approach

Since the noise of an optical device plays a major role in the design of an optical receiver, it is instructive to understand in more detail the mechanisms and description of the generation of this noise.

One type of possible noise is due to the quantized nature of the light or, equivalently, the discrete generation of charge carriers. This discrete generation of carriers means that, with a weak incident signal, the output current will not be an exact replica of the ideally predicted current, but will deviate from it by a (hopefully) small amount. This deviation is one type of noise in the output current. The interesting feature of this noise (found in optical devices) is that it is dependent on both the signal level (i.e., *signal dependent*) and the environment (e.g., temperature and device dependent).

The random generation of the charge-carrier pairs by the photons has been modeled (Personick, 1981a) by the classical *Poisson random process* with a time-varying rate. In this model, the probability $\rho(t)\,dt$ that a charge-carrier pair is generated in a time interval dt, between t and $t + dt$, is given by

$$\rho(t)\,dt = \frac{\mathcal{R}p(t)}{q}\,dt \qquad (5.27)$$

$$= \frac{\eta\lambda}{hc}p(t)\,dt \;, \qquad (5.28)$$

where η is the quantum efficiency. (Implicit in the use of this model are the assumptions that the generation of each pair of carriers is independent of all other carrier generations and that only one pair of carriers is generated in the interval dt.)

From these Poisson process assumptions, some conclusions follow. The total number of carriers generated in a finite interval of time from t to $t + T$ is a random variable. The average number of carriers $\overline{N}$ generated in this time interval is given by

$$\overline{N} = \frac{\eta\lambda}{hc}\int_{t}^{t+T} p(t)\,dt \qquad (5.29)$$

$$= \frac{\eta\lambda}{hc} E, \tag{5.30}$$

where E is the total energy of the pulse in the interval between t and $t + T$ (given by the integral in Equation 5.29). The probability that the number of charges created N will equal a specific number n in this interval is predicted by (Saleh and Teich, 1991)

$$P(N = n) = \frac{\overline{N}^n e^{-\overline{N}}}{n!} \tag{5.31}$$

$$= \frac{\overline{N}^n e^{-\frac{\eta\lambda E}{hc}}}{n!}, \tag{5.32}$$

where $P(N = n)$ is the probability that a total of n charge-carrier pairs will be generated in an interval T seconds long.

Example: To illustrate the application of this probability, suppose that we wish to know the amount of energy required in the pulse to have a probability of 10^{-9} or smaller that a logical **0** will be detected when we have transmitted a logical **1**.

Solution: Assuming that a logical **0** is the creation of zero charge carrier pairs, we want to calculate the energy required to make $P(n = 0) < 10^{-9}$. From Equation 5.32, we want to solve the inequality,

$$P(n = 0) = \exp\left(-\frac{\eta\lambda E}{hc} \right) < 10^{-9}. \tag{5.33}$$

Solving for E gives

$$E > 21 \frac{hc}{\eta\lambda}. \tag{5.34}$$

From this calculation we see that the required pulse must have $21/\eta$ photons to avoid being mistaken for a logical **0** (with a probability of 10^{-9}).

Example: Calculate the number of photons N required for a detector with a quantum efficiency of 50% to have an error probability of 10^{-12}.

Solution:

$$P(n = 0) = \exp\left(-\frac{\eta\lambda E}{hc}\right) < 10^{-12} \qquad (5.35)$$

$$-\frac{\eta\lambda E}{hc} = -27.6 \qquad (5.36)$$

$$E \geq \frac{hc}{\eta\lambda}(27.6) \qquad (5.37)$$

$$\geq \left(\frac{27.6}{0.5}\right)\left(\frac{hc}{\lambda}\right) \qquad (5.38)$$

$$\geq 55.3\frac{hc}{\lambda}. \qquad (5.39)$$

We recognize that hc/λ is the energy of a photon and that dividing E by the energy per photon will give the number of photons. Also we know that the number of photons must be an integer, so

$$N = \frac{E}{\frac{hc}{\lambda}} \qquad (5.40)$$

$$\geq 56 \text{ photons}. \qquad (5.41)$$

Consider our case of the previous example, where $21/\eta$ photons are required for a BER of 10^{-9}. If the number of logical **1**s and **0**s in the data stream is equal, in an average sense, and each bit period is T_B seconds long, then the average power P_{av} in the detected light must be

$$P_{av} = \frac{21(hc/\eta\lambda)}{2T_B} \qquad (5.42)$$

to achieve the probability of error of 10^{-9}. Since this power level was derived assuming only shot noise, which is due to the quantum nature of light or charge, the resulting calculation is the *quantum limit* of the detection process (for the particular value of error probability under consideration).

To generalize our results to any desired value of probability, we will represent the probability as the *bit-error rate* or *BER*. We can then rewrite Equation 5.30 to be

$$\exp\left(-\frac{\eta\lambda E}{hc}\right) \le \text{BER} . \tag{5.43}$$

Solving for the energy E gives

$$E \ge \frac{hc}{\eta\lambda} \ln\left(\frac{1}{\text{BER}}\right). \tag{5.44}$$

The minimum average power will be

$$P_{av} = \frac{E_{min}}{2T_B} . \tag{5.45}$$

Example: Calculate the minimum power required to achieve a BER of 10^{-12} for the detector of the previous example with a 50% quantum efficiency at a data rate of 10 Mb · s^{-1}. The wavelength is 1300 nm.

Solution: The bit period T_B is given by

$$T_B = \frac{1}{\text{DR}} \tag{5.46}$$

$$= \frac{1}{10 \times 10^6} \tag{5.47}$$

$$= 10^{-7} \text{ s} . \tag{5.48}$$

The minimum power P_{min} is

$$P_{min} = \frac{E_{min}}{2T_B} \tag{5.49}$$

$$= \frac{1}{2T_B} \frac{hc}{\eta\lambda} \ln\left(\frac{1}{\text{BER}}\right) \tag{5.50}$$

$$= \frac{(6.63 \times 10^{-34})\,(3.0 \times 10^8)}{2(10^{-7})\,(0.5)(1300 \times 10^{-9})} \ln(10^{12}) \tag{5.51}$$

$$= 4.22 \times 10^{-11} \text{ W} \tag{5.52}$$

$$= 42.2 \text{ pW} . \tag{5.53}$$

5.2.2 The Avalanche Photodiode

Physical Principles

The *avalanche photodiode* (Kasper and Campbell, 1987) dopes the p and n material higher (denoted by the p^+ and the n^+ in Figure 5.6) and incorporates a narrow region of p-type material between the intrinsic region and the n^+ region. As shown in the figure, the electric field in this region is larger than in the remainder of the depletion region because most of the applied reverse bias is dropped across this region. (This particular structure is called a *reach-through avalanche photodiode* [RAPD].) In this high-field region, the field accelerates the charge carriers to velocities high enough to create more hole-electron pairs through collisions (*impact ionization*). (Field strengths in excess of 3×10^5 V/cm are required.) In the structure shown, the light enters through the p^+ region and is (ideally) absorbed in the i region. The generated carriers separate and drift across the i region. When the electrons enter the p region, they are accelerated and impact other atoms, creating more carriers. These carriers are then accelerated and, in turn, create more carriers, leading to the *avalanche effect*.

The efficiency of a hole or electron creating a new hole-electron pair is expressed by the *hole ionization rate* or the *electron ionization rate*. To minimize noise generation, it is desirable to have as large a difference as possible in these ionization rates (i.e., to have only one type of carrier responsible for most of the avalanche process). Silicon exhibits this property with an electron ionization rate that is one hundred times larger than the hole ionization rate. Other materials such as Ge,

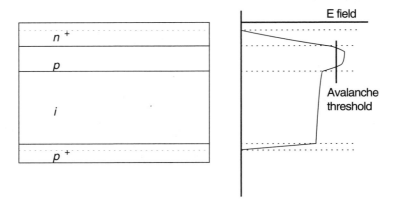

Figure 5.6   Strucutre of an avalanche photodiode (APD) (left) and resulting electric field (right).

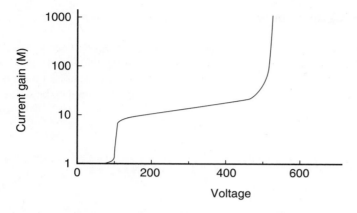

Figure 5.7 Multiplication versus reverse bias voltage.

GaAs, and InGaAs have closer ratios (5–10) thereby ensuring noisy operation in avalanche devices.

The net effect of the avalanche process is a multiplication of the current at the output terminals. The *multiplication factor M* is given by

$$M = \frac{I_M}{I},$$

(5.54)

where I_M is the output current with multiplication and I is the output current without multiplication. Since the amount of multiplication is controlled by the reverse bias (as shown in Figure 5.7), the latter current is measured under low bias conditions. It should be noted that the instantaneous amount of amplification or multiplication is a random value. The value M is the average amount of multiplication.

Responsivity of an APD

The sensitivity of an APD is also described by the *responsivity* of the detector. In a fashion similar to the pin diode, the responsivity of the avalanche photodiode is given by

$$\mathcal{R}_{APD} = \frac{\eta q \lambda}{hc} M$$

(5.55)

$$= \mathcal{R}_0 M$$

(5.56)

where $\mathcal{R}_0$ is the responsivity at unity multiplication.

5.2.3 Detector Noise

In characterizing the noise performance of an optical receiver, we will use the *signal-to-noise ratio*. The signal is the signal power delivered to a hypothetical resistor by the signal current. The noise is characterized by the noise power delivered to the same hypothetical resistor. The signal-to-noise ratio can then be written as

$$\frac{S}{N} = \frac{P_{\text{signal}}}{P_{\text{noise}}} \tag{5.57}$$

$$= \frac{<i_s^2> R}{<i_N^2> R} \tag{5.58}$$

$$= \frac{<i_s^2>}{<i_N^2>}. \tag{5.59}$$

We note that the arbitrary resistor cancels out from the numerator and the denominator, so the signal-to-noise ratio is independent of this resistor, and we need to calculate only the mean-square currents.

There are two noise mechanisms associated with photodiodes, shot noise and thermal noise.

Shot Noise

The *shot noise* can be associated with either with the quantization of charge into multiples of q or, equivalently, with the quantization of light energy into photons. The arrival of photons or, equivalently, the generation of charge carriers is characterized by Poisson statistics. The mean-square noise current $<i_N^2>$ associated with this noise source for a photodiode is given by (Yariv, 1991)

$$<i_N^2> = 2qIB, \tag{5.60}$$

where I is the average output current of the device and B is the bandwidth of the electronics accepting the noise (e.g., a preamplifier, a noise meter). The interpretation of the equation is that there is this much noise power (delivered to a 1 ohm load) in the frequency region extending from $f_0 - (B/2)$ to $f_0 + (B/2)$, where f_0 is the center frequency of the passband of the output device. Note that the AC noise power centered within the band of frequencies is dependent on the DC value of output current. Note also that, since $<i_N^2>$ is independent of the central frequency, the noise is *white noise* (i.e., the noise has a uniform frequency distribution).

The DC current out of a pin diode is made up of three components:

$$I = I_L + I_{\text{background}} + I_{\text{dark}} , \qquad (5.61)$$

where I_L is the load current out due to the incident light, $I_{\text{background}}$ is the light due to background illumination sources (assumed to be zero in fiber links), and I_{dark} is the *dark current* of the device. The dark current is the output of the device with no input illumination and is primarily due to thermal generation of charge carriers in the depletion region and surface leakage currents due to surface defects near the edges of the semiconductor.

Typically, the dark-current density is 10^{-6}–10^{-7} A/cm^2 in silicon devices, is 10^{-4}–10^{-6} A/cm^2 in InGaAs, and is 10^{-3} A/cm^2 in Ge. The high dark current of the long-wavelength detectors causes their noise characteristics to be inferior to the short-wavelength silicon-based devices.

Example: A detector has a responsivity of 0.5 A/W at a wavelength of interest and a dark current of 1 nA. Calculate the mean-square noise current and the RMS noise current due to shot noise if the noise bandwidth is 50 MHz and the incident power is 100 μW.

Solution:

$$< i_N^2 > = 2qIB \qquad (5.62)$$

$$= 2q\,(I_L + I_{\text{dark}})B \qquad (5.63)$$

$$= 2q\,(\mathcal{R}P_i + I_{\text{dark}})B \qquad (5.64)$$

$$= 2(1.6 \times 10^{-19}) \left((0.5)\,(100 \times 10^{-6}) + 10^{-9}\right)$$
$$\times (50 \times 10^{6}) \qquad (5.65)$$

$$= 8 \times 10^{-16}\ \text{A}^2 . \qquad (5.66)$$

The RMS noise current is, by definition, the square root of the mean-square noise current.

$$\sqrt{< i_N^2 >} = \sqrt{8 \times 10^{-16}} \qquad (5.67)$$

$$= 7.83 \times 10^{-8}\ \text{A} . \qquad (5.68)$$

Note: in this example the dark current turned out to be negligible compared to the signal current.

Excess Shot Noise in APDs

In the avalanche photodiode, the avalanche process contributes noise above that previously described. In this device, the mean-square noise current is (Yariv, 1991)

$$< i_N^2 > = 2qIM^2BF(M) ,$$

(5.69)

where M is the average current gain and $F(M)$ is the *excess noise factor* that is the measure of the extra noise added by the avalanche process. The function $F(M)$ depends on the detector material, the shape of the E field in the device, and relative *ionization rates* of the carriers. (The ionization rate is a measure of the ability of the hole or the electron to generate other carrier pairs in the avalanche process.) If k is the ratio of the electron ionization rate to the hole ionization rate (or the inverse, since we require $k \leq 1$), the excess noise factor, $F(M)$, can be given by (Kasper and Campbell, 1987)

$$F(M) = kM + (1 - k)\left(1 + \frac{1}{M}\right).$$

(5.70)

Typical ranges of values of k are found in Table 5.2. The lowest excess noise occurs for the lowest value of k (i.e., we want the avalanche process to be predominately caused by either holes or electrons). An approximate form of $F(M)$ is

$$F(M) \approx M^x,$$

(5.71)

where typical ranges in x are shown in Table 5.2. This excess noise factor causes some unusual effects in the application of this device, as we shall see in considering the signal-to-noise ratio. In using Equation 5.69, the current I is the DC signal current (*without gain*) plus the bulk dark current. Any surface dark currents are *not* amplified by the avalanche process and are negligible for reasonably large values of M.

Thermal Noise

The second noise source of importance in photodiodes is *thermal noise*. Any resistive load or device with an associated resistance will produce a thermal-noise mean-square current given by

$$< i_N^2 > = \frac{4kTB}{R} ,$$

(5.72)

Table 5.2: Typical range of values of k and x in the expressions for the excess noise factor $F(M)$.

Material	k	x
Silicon	0.02–0.04	0.3–0.5
Germanium	0.7–1.0	1.0
InGaAs	0.3–0.5	0.5–0.8

where k is Boltzmann's constant, T is the noise temperature of the device, B is the electronic bandwidth into which the noise is delivered, and R is the value of the resistor or input resistance. This noise is independent of the optical signal.

Having displayed the equations for the mean-square noise currents, we proceed to the consideration of the signal-to-noise ratio for the device operating into a simple load resistor.

5.2.4 Detector Signal-to-Noise Analysis

We now consider the signal-to-noise analysis of a detector loaded by a simple resistor of value R_L. We neglect, for now, the effects of any preamplifier in our effort to understand the ultimate limit placed on the system performance by the detector noise alone.

The general signal-to-noise ratio at the output of an arbitrary detector is given by

$$\frac{S}{N} = \frac{<i_s^2> M^2}{2q(I_L + I_D)M^2 F(M)B + 2qI_{surf}B + 4kTB/R_L}. \tag{5.73}$$

The term in the numerator is the mean-square signal current out of the device. The gain M has a value of 1 if the device is a pin diode. The noise terms in the denominator are as follows:

- The first term in the denominator is the amplified shot noise due to the bulk currents—the signal-dependent current (I_L) (*before amplification*) and the dark current (I_D). The excess noise is accounted for by the $F(M)$ multiplier.
- The second term of the denominator is the shot noise due to the surface leakage current I_{surf} (if any).
- The third term of the denominator is the thermal noise due to the resistance of the load resistor R_L. (It is assumed that the load resistor is considerably

smaller than the effective resistance of the reverse-biased diode and hence is the dominant thermal noise source.)

Example: Consider an avalanche photodiode operating with a gain of 50 and an unamplified bulk dark current of 10 nA. The surface dark current is 1 nA. The unamplified responsivity of the device is 0.6 A/W. The excess noise factor for the device is $M^{0.4}$ and it is operating into a 50 Ω load at a noise temperature of 300K. The noise bandwidth is 10 MHz. Calculate the signal-to-noise ratio (in dB) of this detector when irradiated with 5 nW of light.

Solution: The signal current with a gain M of 1 is

$$i_s|_{M=1} = \mathcal{R}_0 P_i \tag{5.74}$$

$$= (0.6)\,(5 \times 10^{-9}) \tag{5.75}$$

$$= 3 \times 10^{-9} \text{ A} \tag{5.76}$$

$$= I_L . \tag{5.77}$$

The excess noise factor $F(M)$ is estimated as

$$F(M) = M^{0.4} = (50)^{0.4} \tag{5.78}$$

$$= 4.78 . \tag{5.79}$$

The signal-to-noise ratio is expressed by

$$\frac{S}{N} = \frac{<i_s^2> M^2}{2q(I_L + I_D)M^2 F(M)B + 2qI_{surf}B + 4kTB/R_L} . \tag{5.80}$$

We find the first part of the shot noise as

$$N_{shot} = 2q(I_L + I_D)M^2 F(M)B \tag{5.81}$$

$$= 2(1.6 \times 10^{-19})\,(3 \times 10^{-9} + 10 \times 10^{-9})\,(50)^2$$
$$\times (10 \times 10^6) \tag{5.82}$$

$$= 4.97 \times 10^{-16} \text{ A}^2 . \tag{5.83}$$

The shot noise due to the surface leakage current is

$$N_{shot\ surface} = 2qI_{surf}B \tag{5.84}$$

$$= 2(1.6 \times 10^{-19})\,(1 \times 10^{-9})\,(10 \times 10^6) \tag{5.85}$$

$$= 2.6 \times 10^{-21} \text{ A}^2 . \tag{5.86}$$

The thermal noise is

$$N_{thermal} = \frac{4kTB}{R_L} \tag{5.87}$$

$$= \frac{4(1.38 \times 10^{-23})\,(300)\,(10 \times 10^{6})}{50} \tag{5.88}$$

$$= 3.312 \times 10^{-15} \text{ A}^2 . \tag{5.89}$$

Summing all of the noise source powers, we have

$$\sum Noise = N_{shot} + N_{shot\ surface} + N_{thermal} \tag{5.90}$$

$$= 3.81 \times 10^{-15} \text{ A}^2 . \tag{5.91}$$

Having found the total noise, we can now find the signal-to-noise ratio as

$$\frac{S}{N} = \frac{(3 \times 10^{-9})^2 (50)^2}{(3.81 \times 10^{-15})} \tag{5.92}$$

$$= 5.91 \tag{5.93}$$

$$= 7.71 \text{ dB} . \tag{5.94}$$

When using the pin photodiode, $M = F(M) = 1$, and the dominant noise sources are usually the thermal noise source. The output current of the device for a constant input (such as a pulse, when present) will be

$$i_s = I_L \tag{5.95}$$

$$= \frac{\eta q \lambda P}{hc} \tag{5.96}$$

$$= \mathcal{R}_0 P , \tag{5.97}$$

where P is the incident optical power.

When using an ADP detector, the amplification factor will make the shot noise the dominant noise source. We note that, while both the signal and noise terms increase as M^2, the excess noise factor will ensure that the noise will grow at a faster rate, $M^2 F(M)$, than the signal. A plot of the signal-to-noise ratio vs. M for this device will reveal a maximum. The optimum value of M, M_{opt}, that gives the maximum signal-to-noise ratio is found by taking the derivative of Equation 5.73 with respect to M and finding the value of M that will make this derivative equal zero. This evaluation gives

$$M_{opt}^{x+2} = \frac{2qI_{surf} + (4kT/R_L)}{xq(I_L + I_D)} \tag{5.98}$$

$$M_{opt} = \left(\frac{2qI_{surf} + (4kT/R_L)}{xq(I_L + I_D)} \right)^{\frac{1}{2+x}}. \tag{5.99}$$

Hence, operation of the APD in a low signal-to-noise environment requires adjustment of the gain to obtain optimum performance. The value of the optimum gain depends on the thermal noise, the signal level, the device dark current, and, through x, the ionization rates of the carriers. For silicon APDs, the optimum gain typically is in the range of 80–100, with improvements of 40x to 50x (16 to 17 dB) in the signal-to-noise ratio over the unamplified signal. Excessive noise in long-wavelength APDs has restricted their use. These materials suffer from nearly equal ionization rates for holes and electrons (i.e., $k \approx 1$ in Equation 5.70), compounded by relatively large dark currents.

Example: For the APD detector of the example on page 194, calculate the optimum gain that should be used.

Solution:

$$M_{opt}^{x+2} = \frac{2qI_{surf} + 4kT/R_L}{xq(I_L + I_D)} \tag{5.100}$$

$$M_{opt}^{2.4} = \frac{2(1.6 \times 10^{-19})\,(1 \times 10^{-9})}{0.4(1.6 \times 10^{-19})\,(3 \times 10^{-9} + 10 \times 10^{-9})}$$

$$+ \frac{(4)\,(1.38 \times 10^{-23})\,(300/50)}{0.4(1.6 \times 10^{-19})\,(3 \times 10^{-9} + 10 \times 10^{-9})} \tag{5.101}$$

$$= 3.98 \times 10^5, \tag{5.102}$$

$$M_{opt} = 215.4. \tag{5.103}$$

5.2.5 Linearity

The linearity of the output current vs. optical input power of optical photodiodes is excellent with typical devices exhibiting over 6 decades of linear operation. Linearity of APDs is not quite as good. Fortunately, the requirements for maximum linearity exist for analog signals with a high signal-to-noise output—a regime of operation best met by *pin* diodes, not APDs.

5.2.6 Speed of Response

The speed of response of a photodiode is limited by three factors:

1. The first factor is the *transit time* of the carriers as they drift across the depletion region. Usually the devices are designed and biased so that the carriers reach their scattering-limited velocity in the material. For silicon, the scattering-limited velocities of holes and electrons are 8.4×10^4 and 4.4×10^4 m $\cdot$ s^{-1}, respectively. For a depletion width of 10 μm, the response time caused by this effect is ≈ 0.1 ns.

2. The second factor is the *diffusion time*, which is the time that it takes carriers created in the *p* or *n* material (close to the depletion region boundary) to move by diffusion into the depletion region where the drift process takes over. A relatively small fraction of the carriers are involved in this process, but, because the diffusion process is quite slow, it can be a limiting effect on the device speed. This is especially true at low values of reverse bias, where the drift electric field is comparatively low. One way to minimize this effect (by ensuring that most of the carriers are generated in the depletion region) is to make the total depletion region much larger than $1/\alpha$ (where α is the material absorption coefficient of the material at the wavelength of interest). This guarantees that most of the light will be absorbed in this region of the device. (The increase in the size of the depletion region will increase the transit time, however.)

3. The third factor is the *RC* time constant of the device and any associated circuitry. Typically, *R* is the input resistance of the preamplifier in parallel with the load resistance, and *C* is the device capacitance. Most of the device capacitance is the junction capacitance associated with the reverse-biased diode. This capacitance is given by

$$C = \frac{\varepsilon_s A}{w}, \qquad (5.104)$$

where ε_s is the (total) permittivity of the semiconductor material, *A* is the cross-sectional area of the detecting portion of the device, and *w* is the depth of the depletion region. As seen from this equation, efforts to increase the efficiency of the device by increasing w also reduce the capacitance of the device. This increase in w also increases the transit time of the device, however, causing a tradeoff in device design. (A usual compromise is to make $w \approx 2/\alpha$.) Detectors typically have capacitance values of <1 pF. The bandwidth limitation due to this RC time constant effect is given by

$$f_{max} = \frac{1}{2\pi RC} .$$
(5.105)

To summarize, the primary mechanism limiting speed of response in a well-designed pin diode, used in a low-resistance circuit, is the transit time across the depletion region. Silicon devices have response times on the order of 1 ns or less.

In APDs, the response is typically slower because the carriers must drift into the avalanche region and then the created carriers (of the opposite type) must drift back across the depletion region, approximately doubling the total transit time. Additionally, there is a constant gain-bandwidth product constraint for an avalanche device, which results from giving the avalanche process time to occur. Typical values of this gain-bandwidth product can range up to 200 GHz.

5.2.7 Reliability

Reliability has not proved to be a major problem in photodiode operation. Based on accelerated temperature lifetime testing, the projected lifetime is approximately 10^8 hours—ample lifetime for an optical detector.

5.2.8 Temperature Sensitivity

The gain of the avalanche photodiode is quite temperature sensitive, as shown in Figure 5.8. Because the gain is also voltage dependent, it is fairly easy to incorporate a temperature-compensating feedback circuit that senses the change in gain due to a temperature change and adjusts the reverse bias in a direction to cancel the change in gain. Such circuitry is easily implemented through the use of a controllable supply voltage. Such a supply is also useful as an automatic gain-control circuit, frequently used when the avalanche diode is exposed to optical power levels with a wide dynamic range. (This gain control is necessary because the output voltage level of a high-gain APD is easily saturated with a large optical-input signal.)

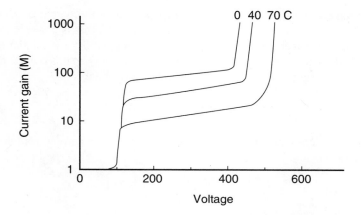

Figure 5.8 Temperature sensitivity of APD gain.

5.2.9 Analog Detector Analysis

Having considered how a simple detector (i.e., without any amplifier) responds to a digital signal, we now consider its performance for analog signals. As described in Chapter 4, an analog transmitter using intensity modulation operates with a small change in power about a dc power level. The received signal is

$$p(t) = P\left(1 + m(t)\right), \qquad (5.106)$$

where $p(t)$ is the time-varying power, P is the received DC optical power with no modulation present, and $m(t)$ is the time-varying information, normalized such that $|m(t)| < 1$. The output current from a linear receiver is given by

$$i(t) = \mathcal{R}_0 M P\left(1 + m(t)\right) \qquad (5.107)$$

$$= I_L\left(1 + m(t)\right). \qquad (5.108)$$

The mean-square signal current is the mean-square value of the AC portion of the current and is given by

$$< i_s^2 > \; = I_L^2 < m^2(t) > , \qquad (5.109)$$

where we have assumed that the time-varying portion of the output current is the only part of interest (or the only part of the output current that is passed by the AC electronics after the receiver). For the purpose of computation, a particular test signal is selected for the time-varying signal. This test signal is assumed to be a cosine modulation at a frequency ω. The equation for this test signal is

$$p(t) = P\left(1 + m\cos\left(\omega t\right)\right),\tag{5.110}$$

where m is now a constant with a value between 0 and 1. The current from the detector into the load is

$$i_L(t) = M\mathcal{R}_0 P\left(1 + m\cos\left(\omega t\right)\right).\tag{5.111}$$

The mean-square signal current is given by

$$<i_s^2> = \frac{(mM\mathcal{R}_0P)^2}{2},\tag{5.112}$$

since $<\cos^2(\omega t)> = 1/2$. (Note that we have included the gain term M^2 explicitly in this expression.) The detector noise is represented by

$$<i_N^2> = 2q(I_L + I_D)M^2F(M)B + 2qI_{surf}B + 4kTB/R_{eq}\tag{5.113}$$

from the denominator of Equation 5.73. Assuming negligible surface leakage current ($I_{surf} = 0$), the signal-to-noise ratio becomes

$$\frac{S}{N} = \frac{(1/2)\,(\mathcal{R}_0mMP)^2}{2q(\mathcal{R}_0P + I_D)M^2F(M)B + 4kTB/R_{eq}}\tag{5.114}$$

$$= \frac{(1/2)\,(I_LMm)^2}{2q(I_L + I_D)M^2F(M)B + 4kTB/R_{eq}}.\tag{5.115}$$

To simplify this expression, we now want to make various assumptions about which noise is dominant.

Photodiode Analog Response—Thermal-Noise Dominant

For the pin diode, $M = 1$. For a small optical signal and small dark current in the detector, the thermal-noise term is the dominant noise source, so the signal-to-noise ratio becomes

$$\frac{S}{N} \approx \frac{(1/2)m^2 I_L^2}{4kTB/R_{eq}} \qquad (5.116)$$

$$\approx \frac{(1/2)m^2 \mathcal{R}_0^2 P^2}{4kTB/R_{eq}} \; . \qquad (5.117)$$

Hence, for this case, we want to maximize the modulation index, to maximize the responsivity, to increase the signal power, to reduce the temperature (if possible), and to make the equivalent resistance as large as possible.

Example: Consider an analog amplitude modulation system transmitting a 15 KHz tone with an 80% modulation index. A photodiode detector with a responsivity of 0.4 A/W operates into a 50 Ω load with a noise temperature of 400K. The bulk dark current is 1 nA; the surface dark current is negligible. The system bandwidth is 50 MHz.

(a) For an incident power of 1 µW, calculate the signal-to-noise ratio assuming that the signal and dark currents are so small that the thermal noise will be dominant.

Solution: For the case where the thermal noise is dominant, we have

$$\frac{S}{N} \approx \frac{(1/2)m^2 \mathcal{R}_0^2 P^2}{4kTB/R_{eq}} \qquad (5.118)$$

$$= \frac{m^2 \mathcal{R}_0^2 P^2 R_{eq}}{8kTB} \qquad (5.119)$$

$$= \frac{\left[(0.8)\,(0.4)\,(1 \times 10^{-6})\right]^2 (50)}{(8)\,(1.38 \times 10^{-23})\,(400)\,(50 \times 10^6)} \qquad (5.120)$$

$$= 2.32 \qquad (5.121)$$

$$= 3.65 \text{ dB} \; . \qquad (5.122)$$

(b) Calculate the ratios of $<i_N^2>$ thermal to $<i_N^2>$ signal shot and $<i_N^2>$ dark shot to be certain that the thermal noise is dominant.

Solution: We find the ratio of the thermal noise to the signal-dependent shot noise to be

$$\frac{<i_N^2>_{\text{thermal}}}{<i_N^2>_{\text{signal shot}}} \quad (5.123)$$

$$= \frac{4kTB/R_L}{2q\mathcal{R}_0 P_i B} \quad (5.124)$$

$$= \frac{4(1.38 \times 10^{-23})(400)(50 \times 10^6)/50}{2(1.6 \times 10^{-19})(0.8)(1 \times 10^{-6})(50 \times 10^6)} \quad (5.125)$$

$$= \frac{4.416 \times 10^{-22}}{2.56 \times 10^{-31}} \quad (5.126)$$

$$= 1.72 \times 10^3 . \quad (5.127)$$

The ratio of the thermal noise to the shot noise due to the signal is

$$\frac{<i_N^2>_{\text{thermal}}}{<i_N^2>_{\text{dark shot}}}$$

$$= \frac{4kTB/R}{2qI_D B} \quad (5.128)$$

$$= \frac{4(1.38 \times 10^{-23})(400)(50 \times 10^6)/50}{2(1.6 \times 10^{-19})(1 \times 10^{-12})(50 \times 10^6)} \quad (5.129)$$

$$= \frac{4.42 \times 10^{-22}}{3.2 \times 10^{-28}} \quad (5.130)$$

$$= 1.38 \times 10^6 . \quad (5.131)$$

Hence, we have proven that the thermal noise, indeed, is dominant.

Photodiode Analog Response—High-Illumination Case

For a large optical signal incident on the photodiode, the shot-noise term that is dependent on the average output current will be the dominant term. For this case

$$\frac{S}{N} \approx \frac{m^2 I_L}{4qB} \quad (5.132)$$

$$\approx \frac{m^2 \mathcal{R}_0 P}{4qB} . \quad (5.133)$$

Since this latter case assumes that all thermal noise is negligible, that all device-related noise is also negligible, and that only the signal-dependent shot noise is left, this signal-to-noise is the theoretical limit, called the *quantum-limited signal-to-noise ratio*. This name results from the dominant noise term being caused by the quantization of electronic charge into integer multiples of q or, equivalently, the light being quantized into photons.

APD Analog Response

For the operation of an APD, one again finds the existence of an optimum value of M that will maximize the signal-to-noise ratio (Equation 5.99). This equation is valid for low levels of optical signal. For large values, the shot-noise term is again dominant. Since the shot noise increases faster than the signal (due to the excess noise factor), the S/N ratio of an avalanche photodiode receiver is *lower* than the S/N ratio of a pin diode receiver, and the APD should not be used in this regime of large-signal operation.

Summary for Analog Response

For low optical-power levels, an APD offers significant advantages. For small values of M, the signal can be amplified more than the noise, and the overall signal-to-noise level can be raised by increasing M (up to $M = M_{\text{opt}}$). Further increases in M have the effect of gradually decreasing the signal-to-noise ratio.

For higher optical-power levels, the advantage of APDs diminishes and, eventually, disappears (i.e., the value of M_{opt} eventually approaches a value of one for high incident power). The value of receiver power where the signal-to-noise ratio begins to decrease varies with the device and data rate and should be determined analytically or experimentally before committing to the more expensive APD.

5.2.10 Noise Equivalent Power

Another figure of merit used to describe the noise performance of a photodetector is the *noise equivalent power*. For a device with negligible dark current, the signal-to-noise ratio at the output of the pin photodiode with a constant signal of P watts incident is

$$\frac{S}{N} = \frac{(\eta q \lambda P/hc)^2}{(2\eta q^2 \lambda P/hc)B + \dfrac{4kTB}{R}}. \tag{5.134}$$

For a given value of thermal noise, there will be a value of P (the optical power) that will make the signal-to-noise ratio equal to 1. The power level that makes the $S/N = 1$ is called the *noise equivalent power* (NEP) of the receiver. Usually, in a pin detector designed for fast response times, the thermal noise is the dominant noise source, producing an NEP of

$$\text{NEP} = \frac{hc\sqrt{<i_N^2>_{\text{thermal}}}}{\eta q \lambda} . \tag{5.135}$$

Note that a value of receiver bandwidth, resistance values, and operating temperature should be specified with the NEP to allow for a meaningful comparison between detectors.

The signal-to-noise ratio for an APD can be written as

$$\frac{S}{N} = \frac{(\eta q \lambda / hc)^2 M^2 P^2}{(2 \eta q^2 \lambda / hc) P M^2 F(M) B + \dfrac{4kTB}{R}}, \tag{5.136}$$

where we see that both the signal and the shot-noise terms have the M^2 amplification and that the shot-noise term is multiplied by the excess noise-factor term $F(M)$. For the simplified expression of $F(M)$ given in Equation 5.71, Equation 5.99 estimates the optimum gain.

5.3 Receiver Sensitivity and Bit-Error Rate

A digital receiver (Figure 5.9) must take a weak optical signal and convert it into an electrical signal, decide whether the electrical output represents a logical **1** or **0** (using a comparator), and generate an electronically compatible voltage representative of the logic state. Since the output of the detector-preamplifier combination is contaminated by noise, the next step in our analysis is consideration of the value of the threshold voltage of the comparator (i.e., the *detection level*). This value will partially determine the *bit-error rate* (BER) of the receiver (Personick, 1981b; Kasper, 1988; Yariv, 1991). By removing the amplification effects of the preamplifier, the threshold voltage can be converted to an equivalent threshold current at the diode output. We are concerned whether the instantaneous current (as corrupted by noise) is above or below this threshold.

Figure 5.10 illustrates the concept. The top curve is the noise-free output from a hypothetical detector. The value of threshold is shown on this curve as a fraction k of the expected output of a logical **1**. The noise-free signal in the top curve

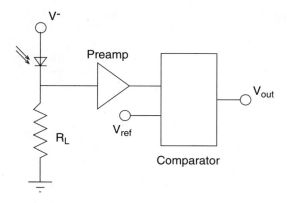

Figure 5.9 Block diagram of digital receiver.

gives the correct logical values, which represents perfect recovery of the transmitted signal. The middle curve shows a noise-corrupted received signal. For the threshold shown, the bottom curve is produced at the comparator output with one of the output bits in error. The number of errors in the recovered signal is a function of k, the threshold position (expressed as $ki(\mathbf{1})$ where $0 < k < 1$).

While various assumptions can be made about the noise statistics of the output (with increasing accuracy bringing increasing computational complexity), we will use the simplest assumption to illustrate the design process. First is the assumption that the output currents of the detector are Gaussian random variables. The mean value of the current with a logical $\mathbf{1}$ is $\overline{i(\mathbf{1})}$ and the mean value of output current for a logical $\mathbf{0}$ is $\overline{i(\mathbf{0})}$. We will assume $\overline{i(\mathbf{0})} = 0$ for simplicity, since this is usually true in a fiber link. If a logical $\mathbf{0}$ is sent, an error will be made if the noise current is positive and of amplitude

$$i_N > k\,\overline{i(\mathbf{1})}\,. \tag{5.137}$$

Similarly, if a given input is a logical $\mathbf{1}$, we would have an incorrect output if i_N (the noise current contribution) were negative and met the condition

$$i_N < -(1-k)\overline{i(\mathbf{1})}\,. \tag{5.138}$$

(The probabilities of these errors are the areas of the probability density functions as illustrated in Figure 5.11.) The probability of making either one error or the other depends on the statistics of the data stream.

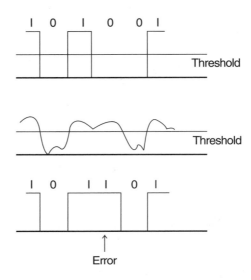

Figure 5.10 Recovery of signal. (a) Noise-free signal.
(b) Noise-corrupted signal. (c) Recovered signal with errors.

The total probability of making an error is the sum of the probability of calling a **1** a **0** (given that a **1** was sent), plus the probability of calling a **0** a **1** (given that a **0** was sent). Mathematically, this written as

$$P_e = P(0|1)P(1) + P(1|0)P(0) , \qquad (5.139)$$

where $P(0|1)$ is the probability of deciding that the output is a **0** when a **1** is sent, $P(1)$ is the probability that a **1** is sent, $P(1|0)$ is the probability of deciding that the output is a **1** when a **0** is sent, and $P(0)$ is the probability that a **0** is sent. If $P(1)$ is a and $P(0)$ is b, then the combined error probability P_e is

$$P_e = aP\left[i_N < -(1-k)\overline{i(\mathbf{1})} \right] + bP\left[i_N > k\overline{i(\mathbf{1})} \right], \qquad (5.140)$$

where $P(\,\cdot\,)$ is the probability of the event occurring.

We now make some more simplifying assumptions. For most messages we can assume that logical **1**s and **0**s are equally probable, so $a = b = 1/2$. At the same time, we choose to let the threshold factor k equal $1/2$ to set the threshold midway between the upper and lower noise-free values.

If we assume that the noise current is Gaussian distributed, the mean-square noise current $<i_N^2>$ is σ^2. The Gaussian probability distribution is

$$p(i_N)\,di_N = \frac{1}{\sqrt{2\pi}\,\sigma}\,\exp\left(-\frac{i_N^2}{2\sigma^2}\right) di_N\,, \qquad (5.141)$$

where $p(i_N)\,di_N$ is the probability of the output noise current being between the values i_N and $i_N + di_N$, and the standard deviation σ is a measure of the width of the probability distribution. As noted above, the mean-square noise current, predicted from consideration of the noise generation mechanism, is equal to the square of the standard deviation σ^2, i.e.,

$$\sigma^2 = <i_N^2>\,. \qquad (5.142)$$

From Figure 5.11, we note that the error of Equation 5.137 is the area underneath the tail of the upper probability distribution below the threshold and that the error of Equation 5.138 is the area under the tail of the lower distribution above the threshold.

Noting that $p(i_N) = p(-i_N)$ in Equation 5.141, we can compute

$$P_e = p\left(i_N > \frac{\overline{i(1)}}{2}\right) \qquad (5.143)$$

$$= \int_{\overline{i(1)}/2}^{\infty} p(i_N)\,di_N \qquad (5.144)$$

$$= \frac{1}{\sqrt{2}\,\sigma\sqrt{\pi}} \int_{\overline{i(1)}/2}^{\infty} \exp\left(-\frac{i_N^2}{2\sigma^2}\right) di_N\,. \qquad (5.145)$$

By a change of variables, this integral is put into the form

$$P_e = \frac{1}{\sqrt{\pi}} \int_{\overline{i(1)}/2\sqrt{2}\,\sigma}^{\infty} \exp\left(-\xi^2\right) d\xi\,. \qquad (5.146)$$

This last integral is now in the form of the *error function*, erf(z), defined by the integral

$$\text{erf}(z) = \frac{2}{\sqrt{\pi}} \int_{0}^{z} \exp\left(-\xi^2\right) d\xi\,. \qquad (5.147)$$

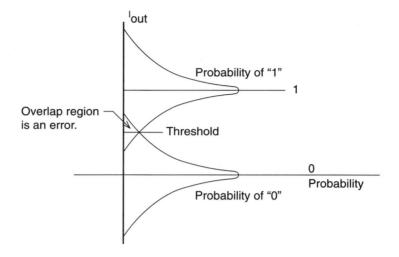

Figure 5.11 Probability density curves for a logical **1** and a logical **0**.

We can then write

$$P_e = \frac{1}{2}\left(1 - \mathrm{erf}\left(\frac{\overline{i(1)}}{2\sqrt{2}\,\sigma}\right)\right), \tag{5.148}$$

but $1 - \mathrm{erf}(z) = \mathrm{erfc}(z)$ defines the *complementary error function*, so we have the result

$$P_e = \frac{1}{2}\,\mathrm{erfc}\left(\frac{\overline{i(1)}}{2\sigma\sqrt{2}}\right) \tag{5.149}$$

$$= \frac{1}{2}\,\mathrm{erfc}\left(\frac{\overline{i(1)}}{2\sqrt{2}\sqrt{<i_N^2>}}\right) \tag{5.150}$$

$$= \frac{1}{2}\,\mathrm{erfc}\left(\frac{\sqrt{S/N}}{2\sqrt{2}}\right). \tag{5.151}$$

(Here we have noted that $\overline{i(1)}/\sqrt{<i_N^2>}$ is $\sqrt{S/N}$ for our definition of S/N.) A plot of Equation 5.151 is shown in Figure 5.12; the error probability is plotted vertically and the required $\sqrt{S/N}$ is plotted horizontally.

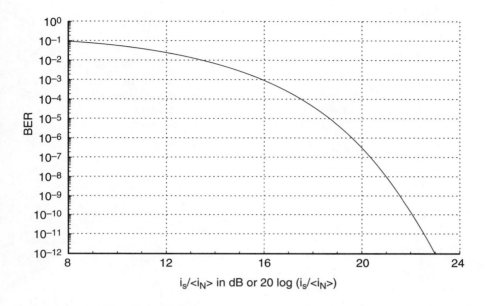

Figure 5.12 Dependence of bit-error rate (error probability) on
signal-to-noise ratio. (Here $i_s = i(\mathbf{1})$.)

To use this curve, one begins with the desired error probability and determines the required signal-to-noise ratio. For example, a BER of 10^{-9} is a probability-of-error of 10^{-9} and is frequently used in modern digital communications. From the figure, we estimate the required ratio of $i_s / \sqrt{<i_N^2>}$ as 21.5 dB or a value of 11.89. From a knowledge of the limiting noise source in the detector (i.e., thermal noise for a pin diode or shot noise for an APD), we can calculate $\sqrt{<i_N^2>}$. Knowing the required signal-to-noise ratio $\sqrt{(S/N)}$, we find the minimum RMS signal current from

$$<i_s>_{min} = \sqrt{(S/N)_{min}} \sqrt{<i_N^2>} . \tag{5.152}$$

The signal current is the product of the optical power and the responsivity $\mathcal{R}$ of the detector (i.e., $i_s = \mathcal{R}P$), giving

$$P_{min} = \frac{\sqrt{(S/N)_{min}} \sqrt{<i_N^2>}}{\mathcal{R}} . \tag{5.153}$$

This optical power must be available at the detector to achieve the desired error rate. One can then use this value to investigate possible tradeoffs between the fiber losses, connector losses, coupling efficiency, and source power levels to complete the design.

Example: Calculate the minimum power required to maintain a bit error of 10^{-6} achieved with a photodiode detector with a responsivity of 0.4 A/W. Assume that the signal-to-noise ratio is limited by thermal noise (with a 50 Ω load, a 400K noise temperature, and a 10 MHz noise bandwidth).

Solution: For BER = 10^{-6}, we find from Figure 5.12,

$$20 \log \left(\frac{<i_s>}{<i_N>} \right) \approx 19.6 \text{ dB} \tag{5.154}$$

$$\frac{<i_s>}{<i_N>} = 10^{19.6/20} \tag{5.155}$$

$$= 9.55 . \tag{5.156}$$

Since $<i_N> = \sqrt{<i_N^2>}$,

$$<i_s> = 9.55 \sqrt{<i_N^2>} \tag{5.157}$$

$$= 9.55 \sqrt{4kTB/R} \tag{5.158}$$

$$= 9.55 \sqrt{4(1.38 \times 10^{-23})(400)(10^7)/50} \tag{5.159}$$

$$= 6.35 \times 10^{-7} \text{ A} \tag{5.160}$$

$$= 635 \text{ nA} . \tag{5.161}$$

$$P_{min} = \frac{<i_s>}{\mathcal{R}_0} \tag{5.162}$$

$$= \frac{635 \times 10^{-9}}{0.4} \tag{5.163}$$

$$= 1.587 \times 10^{-6} \text{ W} \tag{5.164}$$

$$= 1.587 \text{ } \mu\text{W} . \tag{5.165}$$

Recall, however, that various assumptions have been made in our analysis. The assumption of Gaussian noise reduced the complexity of the problem, but the result produces an $i_s/\sqrt{<i_N^2>}$ value that is too low. We have also assumed that $i(0) = 0$ and that a logical **1** and logical **0** are equally probable. (This latter assumption caused us to assume that the threshold was midway between $i(0)$ and $i(1)$.) Furthermore, we have implicitly assumed a noise-free amplifier and that there is negligible pulse spreading in the received pulses.

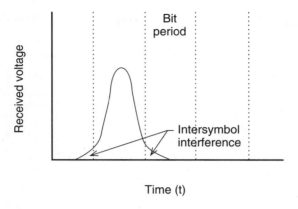

Figure 5.13 Example of intersymbol interference. (Note the
energy spilling into adjacent bit intervals.)

This last effect can lead to a form of noise termed *intersymbol interference*.
As seen in Figure 5.13, any energy that spills out of the expected pulse position is
lost to the receiver for use in detecting the pulse of interest. If large enough, this
energy might trigger a false detection in the adjacent time slot. While this intersymbol interference may be partially corrected by the use of a properly designed
equalization amplifier (at the expense of increased noise levels), the design of such a
device is beyond the scope of this text. It has been found that the intersymbol interference is negligible if the bit spacing T_b is kept larger than 4 times the RMS pulse
spreading of the fiber τ where the RMS pulsewidth is given (as in Chapter 4) by

$$\sigma_{rms} = \int t^2 h(t) \, dt - \int t \, h(t) \, dt \,, \tag{5.166}$$

where $h(t)$ is the pulse shape at the fiber output for an impulse input. If $h(t)$ is Gaussian (or approximated as Gaussian), then the RMS pulsewidth can be shown to be
numerically equal to the pulsewidth measured at the $1/e$ points.

5.3.1 Photon Arrivals and BER

An alternative method of determining the signal-to-noise ratio required to
maintain a given BER is to express the result in terms of the number of photons that
are required to be received in a single bit period when a logical **1** is received. This
number of photons is frequently called the *Q-parameter*. Assuming that the noise is
Gaussian, the Q-parameter is defined as (Keiser, 1991)

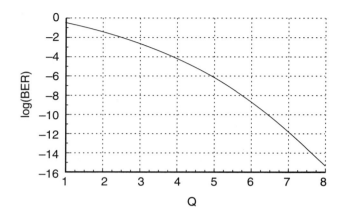

Figure 5.14 Plot of the approximation for BER vs. Q.

$$Q = \frac{i_{th} - \overline{i_0}}{\sigma_0} \tag{5.167}$$

$$= \frac{\overline{i_1} - i_{th}}{\sigma_1}, \tag{5.168}$$

where i_{th} is the detection threshold setting (controlled by the receiver designer), $\overline{i_0}$ ($\overline{i_1}$) is the current at the receiver when a logical **0** (logical **1**) is received, and σ_0 (σ_1) is the standard deviation of the Gaussian noise distribution of the current associated with the reception of a logical **0** (logical **1**). Usually, we assume that $\overline{i_0} = \sigma_0$ and that $\overline{i_0} = 0$; then (Keiser, 1991)

$$Q = \frac{\overline{i_1}}{2\sigma_1}. \tag{5.169}$$

The bit-error rate (BER) is the probability of making an error and, for a Gaussian noise assumption, is given by (Keiser, 1991)

$$BER = \frac{1}{2}\left[1 - erf\left(\frac{Q}{\sqrt{2}}\right)\right]. \tag{5.170}$$

A useful approximation to the right side of this equation is (Keiser, 1991)

$$BER \approx \frac{e^{-\frac{Q^2}{2}}}{Q\sqrt{2\pi}}. \tag{5.171}$$

Table 5.3: Values of Q-parameter for several values of BER (from approximate formula).

BER	Q
10^{-3}	3.12
10^{-4}	3.73
10^{-5}	4.28
10^{-6}	4.76
10^{-7}	5.21
10^{-8}	5.62
10^{-9}	6.00
10^{-10}	6.36
10^{-11}	6.71
10^{-12}	7.04
10^{-13}	7.35
10^{-14}	7.65

A plot of this approximation is shown in Figure 5.14. The values of Q from the approximation for several benchmark values of BER are given in Table 5.3.

5.4 Optical Receiver Noise and Sensitivity

The properties of the receiver are determined by the combination of the detector and amplifier rather than by the detector alone. In this section we will consider some typical receiver circuits and indicate some of the design considerations that must be made in applying the techniques to a fiber link. This section will follow the work in Kasper (1988) and Personick (1977, 1979, 1981b).

The combination of the detector and the preamplifier is called the receiver *front-end*. Generally there are three common implementations (although there are many designs that do not fall into these three categories).

- The first is the *low-impedance front-end* (Figure 5.15a). The detector operates into a low-impedance amplifier (usually with an input impedance of 50 Ω) through a coaxial cable. This design is popular because of the ready availability of low-impedance wideband RF amplifiers. To transfer the

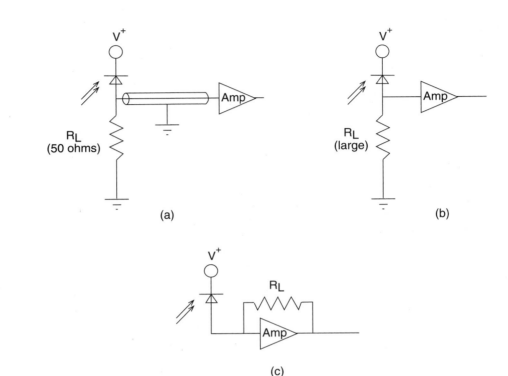

Figure 5.15 Block diagrams. (a) A low-impedance front-end.
(b) A high-impedance front-end. (c) A transimpedance front-end.

maximum power from the detector load resistor to the amplifier, the load
resistor is chosen to equal the input resistance of the amplifier (e.g., a 50 Ω
amplifier would call for a 50 Ω load resistor). This design does not provide
much sensitivity (due to the small voltage developed across the input to the
amplifier) and suffers from high noise (due to the $1/R$ dependence of the
thermal noise) from the load resistance.

■ The second design is the *high-impedance front-end* (Figure 5.15b). A larger
signal voltage (and less thermal noise) can be achieved by applying the
detector output current to a load and amplifier combination with a high
value of resistance. To achieve high input resistance, the amplifier uses an
FET and the load resistor is made large; the parallel combination is numerically equal to the load resistor R_L. Several capacitances are in parallel with

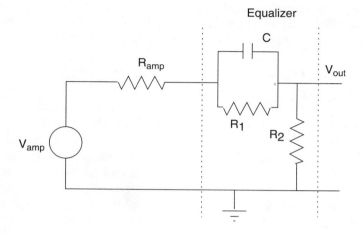

Figure 5.16 An RC equalizer for an integrating front-end.

the combined load and amplifier resistance: the detector capacitance, the
amplifier input capacitance, and parasitic capacitances. The total capacitance C_T is the sum of these individual capacitances. A current generator
driving a parallel RC circuit has the characteristics of an integrator; hence,
this type of front-end is also called an *integrating front-end*. The bandwidth
of the combination is $1/2\pi R_L C_T$. For values of R_L in the 100s of kilohms to
a few megohms and C_T values of a few picofarads or less, the bandwidth is
quite low for high data-rate systems. An equalization amplifier must be
added to compensate for this poor bandwidth.

Figure 5.16 shows one such equalizer. The preamp and detector are
represented by an equivalent frequency-dependent voltage source $V_{amp}(\omega)$
and resistance R_{amp}. The transfer function of the circuit is

$$\frac{V_{out}(\omega)}{V_{amp}(\omega)} = \frac{R_2(1 + j\omega R_1 C)}{R_1 + R_{amp} + R_2 + j\omega R_1 C(R_{amp} + R_2)}. \tag{5.172}$$

If we choose R_1 and C such that

$$\frac{1}{R_1 C} = \frac{1}{R_L C_T}, \tag{5.173}$$

the zero of the equalizer will cancel the pole of the integrating front-end and the combined bandwidth will be larger than that of the front-end alone. The bandwidth of the combined circuit, $f_{combined}$, will be

$$f_{combined} = \frac{1}{R_1 C} \frac{R_1 + R_{amp} + R_2}{R_{amp} + R_2} . \qquad (5.174)$$

Picking $R_1 >> R_{amp} + R_2$ will ensure a higher bandwidth than the integrating front-end alone.

Another disadvantage of the high-impedance front-end, besides requiring an equalizer, is that the dynamic range of the receiver is limited. (This means that the range of optical power values, from the weakest to the strongest, is limited.) This is because the low-frequency components that are integrated can quickly saturate the preamplifier output.

■ The third design is the *transimpedance front-end*. (Figure 5.15c). Here, a feedback resistor R_L connects the output and input of the preamp. Electronically, this circuit is a current-to-voltage convertor. The bandwidth of this amplifier is a factor of A (the amplifier gain) times $1/2\pi R_L C_T$ (i.e., the bandwidth is a factor of $A + 1$ larger than the unequalized high-impedance amplifier bandwidth (Kasper, 1988)). This wide bandwidth obviates the need for an equalization amplifier. Similarly, the low-frequency components are reduced by the same factor of $A + 1$, lowering the possibility of electronically saturating the amplifier and increasing the dynamic range of the front-end.

5.5 Amplifier Noise

So far, we have been working with the noise from a simple detector driving a load resistance. This analysis is useful to determine the fundamental noise limits of the detector, as the addition of an amplifying stage after the detector will only increase the noise. We now want to do some calculations to determine the signal-to-noise ratio with the added amplifier with gain G.

5.5.1 Amplifier Noise Figure

The noise added by an amplifier is a thermal noise source. Figure 5.17(a) shows a detector and amplifier combination. We show the amplifier noise as a current source across the output terminals. The total mean-square noise current from the three noise sources shown in Figure 5.17(b) is

$$<i_N^2>_{\text{total}} = G^2 <i_{\text{shot}}^2> + \frac{4kTBG^2}{R_L} + <i_N^2>_{\text{amp}} . \tag{5.175}$$

We now want to create an artificial noise source across the input which will generate the equivalent amount of noise. For reasons to be described shortly, we want to replace the $<i_N^2>_{\text{amp}}$ term with an equivalent noise source of value

$$<i_N^2>_{\text{amp}} = \frac{\dfrac{4kTBG^2}{R_L}}{F_n - 1} . \tag{5.176}$$

The quantity F_n is the *noise figure* (Meer, 1989) of the amplifier (sometimes called the *noise factor* of the amplifier). For the equality to hold, we want

$$F_n = 1 + \frac{<i_N^2>_{\text{amp}} R_L}{4kTBG^2} . \tag{5.177}$$

If we move the artificial noise generator through the amplifier, we will need to divide the mean-square noise by G^2. (Figure 5.17(c) shows the artificial noise generator shifted to the amplifier input.) We can combine the noise generator of the load resistor and the artificial noise generator representing the amplifier noise by adding their mean-square noise currents,

$$\frac{4kTB}{R_L} + \frac{4kTB(F_n - 1)}{R_L} = \frac{4kTBF_n}{R_L} . \tag{5.178}$$

Figure 5.17(d) shows the combined noise generator. From this calculation we can see that the total thermal noise of the load resistor and the amplifier can be represented by multiplying the noise from the load resistor by F_n. This increase is the noise penalty imposed by the amplifier.

The noise figure of an amplifier is usually specified in units of dB; it must be converted to a numerical value for use in the formulas of interest. A typical noise figure for a good, low-noise amplifier is 3 dB (implying a total noise that is twice the noise of the load alone). Otherwise, the value might be 6 dB or larger. The value of F_n is determined by measuring the total noise output, subtracting the shot noise and thermal noise of the load resistor, and applying Equation 5.177 to calculate F_n.

The signal-to-noise ratio for a pin-diode detector and amplifier is

$$\frac{S}{N} = \frac{G^2 \mathcal{R}^2 P^2}{2q(\mathcal{R}P + I_{\text{dark}})G^2 B + \dfrac{4kTBF_n G^2}{R_L}} . \tag{5.179}$$

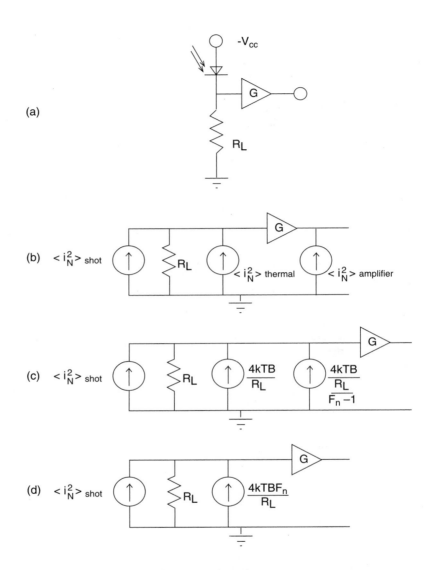

(a)

(b) $<i_N^2>$ shot

(c) $<i_N^2>$ shot

(d) $<i_N^2>$ shot

Note: all generator labels show mean-square currents

Figure 5.17 (a) A detector-amplifier receiver. (b) Noise model of receiver with detector shot-noise source, load-resistor thermal-noise source, and amplifier-output–noise source shown. (c) Noise model of receiver with the output noise model of part (b) modeled by an equivalent noise generator at the input of the amplifier. (d) Noise model with resistor noise generator and equivalent amplifier noise generator combined into a single generator.

The G^2 terms will cancel out from the numerator and denominator but have been left in the expression for completeness. Note the appearance of the F_n term in the thermal noise term in the denominator.

Considering an APD detector working with an amplifier, we have

$$\frac{S}{N} = \frac{G^2 \mathcal{R}_0^2 P^2 M^2}{2q(\mathcal{R}_0 P + I_{\text{dark}})G^2 M^2 F(m)B + 2qI_{surf}B + \dfrac{4KTBF_n G^2}{R_L}} \cdot \qquad (5.180)$$

Notice that the thermal noise does *not* have the M^2 multiplier term. Due to the excess noise factor of the avalanche process, the signal-to-noise ratio will again have a maximum value at an optimum value of multiplication factor. This optimum value M_{opt} is found by solving the equation

$$M_{\text{opt}}^{2+x} = \frac{2qI_{\text{surf}} + \dfrac{4kTF_n}{R_L}}{xq(\mathcal{R}_0 P + I_{\text{dark}})} \qquad (5.181)$$

for M_{opt}.

These expressions will allow us to calculate the signal-to-noise ratio for simple detector/amplifier receivers. Now that we can accomplish this calculation, we need some signal-to-noise benchmarks to determine if we have enough signal-to-noise ratio for our application. Each analog application has its own required minimum signal-to-noise ratio for successful transmission. For example, the North American television standard requires a 35 dB minimum signal-to-noise ratio at the receiver for successful television reception. Other applications have different standards.

5.5.2 Noise in FET Front-Ends

The implementation of the preamplifiers described earlier can use either FETs or bipolar transistors. FETs have superior noise properties compared to the bipolar transistor. The development of GaAs microwave FETs has enabled FET amplifiers to be realized for wideband high–data-rate receivers. Figure 5.18 shows a representative common-source preamp. The principal sources of noise in the amplifier are thermal noise from the FET channel resistance, the thermal noise from the load resistor R_L, the shot noise due to the FET gate leakage current, and the electronic $1/f$ noise of the FET. The mean-square noise current of this amplifier has been shown (Smith and Personick, 1982; Muoi, 1984; Kasper, 1988) to be

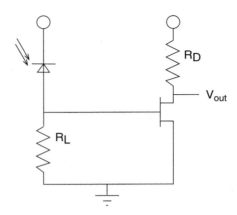

Figure 5.18 Typical common-source FET fiber receiver front-end.

$$<i_{\text{amp}}^2> \ = \ \frac{4kT}{R_L}I_2B_R + 2qI_{\text{gate}}I_2B_R + \frac{4kT\ \Gamma}{g_m}\left(2\pi C_T^2\right)f_cI_fB_R^2$$

$$+ \ \frac{4kT\ \Gamma}{g_m}\left(2\pi C_T^2\right)I_3B_R^3 \ . \tag{5.182}$$

Here B_R is the bit rate, R_L is the load resistor (or feedback resistor for a transimpedance amplifier), I_{gate} is the FET gate leakage current, g_m is the FET transconductance, C_T is the total input capacitance, f_c is the FET $1/f$-noise corner frequency, and Γ is the FET channel-noise factor. (Table 5.4 gives the ranges of typical FET parameters for three FET types.)

The parameters I_2, I_3, and I_f are the *Personick integrals* (Personick, 1973; Smith and Personick, 1982; Kasper, 1988). (Be careful not to confuse them with currents.) They depend only on the pulse shape entering and leaving the fiber and the type of coding used to encode the data. Table 5.5 gives values (Kasper, 1988) of integrals for rectangular pulses entering the fiber and a particular shape (pulses having a raised cosine spectrum) that Personick used in his original analysis. Here NRZ (non-return-to-zero) coding is the usual on-off coding that follows the data; RZ (return-to-zero) coding ensures that there is a data transition during every bit period. This RZ coding is usually used to encode the clock information on the data stream. (Coding is discussed in more detail in Chapter 7.)

The FET channel-noise factor Γ is an FET parameter that describes the noise contribution from the channel resistance and the gate-induced noise; it also includes the effects that are due to these noise sources being correlated with each other (Kasper, 1988).

Table 5.4: Typical values of FET parameters (From Kasper, B.L., "Receiver design," in *Optical Fiber Telecommunications II*, [S.E. Miller and I.P. Kaminow, eds.], pp. 689–722, New York: Academic Press, 1988.)

	GaAs MESFET	Si MOSFET	Si JFET
g_m (mS)	15–50	20–40	5–10
C_{gs} (pF)	0.2–0.5	0.5–1.0	3–6
C_{gd} (pF)	0.01–0.05	0.05–0.1	0.5–1.0
Γ	1.1–1.75	1.5–3.0	0.7
I_{gate} (nA)	1–1,000	0	0.01–0.1
f_c (MHz)	10–100	1–10	<0.1

The total capacitance C_T of the detector-preamp combination is given by

$$C_T = C_d + C_s + C_{gs} + C_{gd}, \tag{5.183}$$

where C_d is the detector capacitance, C_s is the stray capacitance, C_{gs} is the gate-to-source capacitance of the FET, and C_{gd} is the gate-to-drain capacitance of the FET. The stray capacitance is usually estimated, or a value is obtained from measured data. Integrated-circuit detector-preamp combinations have minimum stray capacitance.

The corner frequency f_c of the $1/f$ noise is an FET parameter defined as that frequency where the $1/f$ electronic noise of the device (dominant at low frequency operation) becomes equal to white thermal noise of the channel (characterized by Γ).

Table 5.5: Values of Personick integrals for RZ and NRZ (50% duty cycle) coding. Input pulses are rectangular; output pulses have raised cosine spectrums. (From Kasper, B.L., "Receiver design," in *Optical Fiber Telecommunications II*, [S.E. Miller and I.P. Kaminow, eds.], pp. 689–722, New York: Academic Press, 1988.)

	Coding	
	NRZ	RZ
I_1	0.548	0.500
I_2	0.562	0.403
I_3	0.0868	0.0361
I_f	0.184	0.0984

The first term of Equation 5.182 is due to the thermal noise of the load resistor (or the feedback resistor). It can be minimized by making the resistor large. (This reduces the receiver dynamic range, however.) The second term is due to the shot noise associated with gate leakage current; it is minimized by choosing an FET with a low value of I_{gate}. The third term is due to the $1/f$ noise of the preamp; it is minimized by choosing a device with a low amount of $1/f$ noise (as indicated by a low value of f_c). The fourth term is due to the FET channel noise; it is minimized by choosing an FET with the maximum value of g_m/C_T^2 (Kasper, 1988).

Frequently, the performance of FET preamps is expressed in terms of the FET short-circuit common-source gain-bandwidth product f_T. It can be shown that (Kasper, 1988)

$$f_T = \frac{g_m}{2\pi(C_{gs} + C_{gd})}.$$

(5.184)

We are usually interested in optimizing the receiver performance at high bit rates. In this case, a well designed receiver will be dominated by the fourth term in Equation 5.183, due to its B_R^3 dependence. In this case, the minimum noise current is (Kasper, 1988)

$$<i_{\text{amp}}^2>_{\text{min}} \approx 32\pi kT \frac{\Gamma(C_d + C_s)}{f_T} I_3 B_R^3 \quad (\text{for large } B_R).$$

(5.185)

The best results are obtained when an FET is chosen that has a maximum figure of merit, FOM_{FET},

$$\text{FOM}_{\text{FET}} = \frac{f_T}{\Gamma(C_d + C_s)}.$$

(5.186)

Note that choice of the best FET depends on the capacitance of the optical detector. If we change the detector to one with a significantly different C_d, we should also change the preamp FET.

5.5.3 Noise in Bipolar Transistor Front-Ends

Bipolar preamplifiers (Kasper, 1988) are also used in some fiber receiver front-ends. At low bit rates their noise is higher, but at high bit rates the noise is comparable to the FET noise. Figure 5.19 shows a representative common-emitter preamp using a bipolar junction transistor (BJT). The principal sources of noise in

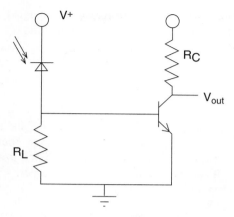

Figure 5.19 Typical common-emitter bipolar junction transistor receiver front-end.

the amplifier are thermal noise from the load resistor R_L, the shot noise due to the base and collector bias currents (I_b and I_C), and the thermal noise from the base-spreading resistance $r_{bb'}$. The mean-square noise current of this amplifier has been shown (Smith and Personick, 1982; Muoi, 1984; Kasper, 1988) to be

$$<i^2_{\text{amp}}> = \frac{4kT}{R_L}I_2B_R + 2qI_bI_2B_R + \frac{2qI_c}{g_m^2}\left(2\pi C_T\right)^2 I_3B_R^3$$
$$+ 4kTr_{bb'}\left[2\pi\left(C_d+C_s\right)\right]^2 I_3B_R^3 . \qquad (5.187)$$

The transconductance depends on the collector bias current as

$$g_m = \frac{I_c}{V_T}, \qquad (5.188)$$

where

$$V_T = \frac{kT}{q} . \qquad (5.189)$$

The total capacitance for the BJT front-end is

$$C_T = C_d + C_s + C_{b'e} + C_{b'c} , \qquad (5.190)$$

where $C_{b'e}$ and $C_{b'c}$ are capacitances from the small-signal hybrid-pi transistor model. The base-emitter capacitance $C_{b'e}$ is a function of the collector bias current I_c. It can be broken up into two components,

$$C_{b'e} = C_{je} + \frac{I_c}{2\pi V_T f_T}, \tag{5.191}$$

where C_{je} is the current-independent junction capacitance and the second term is a diffusion capacitance. In this second term, f_T is the short-circuit common-emitter gain-bandwidth product of the transistor (measured at high I_c to ensure that the diffusion term dominates). It can be shown (Muoi, 1984; Kasper, 1988) that there exists an optimum collector bias current given by

$$I_{c\ opt} = 2\pi C_0 f_T V_T \psi(B_R), \tag{5.192}$$

where C_0 is the total capacitance at zero bias

$$C_0 = C_d + C_s + C_{b'c} + C_{je}, \tag{5.193}$$

and $\psi(B_R)$ is given by

$$\psi(B_R) = \frac{1}{\sqrt{1 + \dfrac{I_2 f_T^2}{\beta I_3 B_R^3}}}. \tag{5.194}$$

with β being the transistor current gain,

$$\beta = \frac{I_c}{I_b}. \tag{5.195}$$

The total capacitance C_T can be written in terms of $\psi(B_R)$ as

$$C_T = C_0 \left[1 + \psi(B_R) \right]. \tag{5.196}$$

The mean-square amplifier-noise current at the optimum bias current can be written as (Kasper, 1988)

$$< i_{amp}^2 > |_{\text{optimum } I_c}$$

$$= \frac{4kT}{R_L} I_2 B_R + \frac{4\pi kT C_0 f_T}{\beta} \psi(B_R) I_2 B_R + \frac{4\pi kT C_0}{f_T} \frac{[1 + \psi(B_R)]^2}{\psi(B_R)} I_3 B_R^3$$

$$+ 4kT r_{bb'} \left[2\pi \left(C_d + C_s \right) \right]^2 I_3 B_R^3 . \tag{5.197}$$

Again, we are usually concerned with design of a high-frequency receiver front-end. The bipolar transistor figure of merit FOM$_{\text{BJT}}$ for this region of operation is (Kasper, 1988)

$$\text{FOM}_{\text{BJT}} = \frac{2f_T}{C_0 + \pi f_T r_{bb'} (C_d + C_s)} \quad \text{(for high } B_R) \tag{5.198}$$

$$\approx \frac{2f_T}{C_0} \quad \text{(for small } r_{bb'} \text{ and high } B_R) . \tag{5.199}$$

5.5.4 Comparison of Noise in FET and BJT Front-Ends

Figure 5.20 shows a comparison of the calculated noise for three FET front-ends and one BJT front-end as a function of bit rate (Kasper, 1988). (See problems at end of chapter.) The FET designs use the data in Table 5.6. The sum of detector and stray capacitance $(C_d + C_s)$ is assumed to be 0.2 pF, and the load resistance R_L is assumed to be very large (to make its thermal noise negligible).

The noise for the silicon bipolar transistor is modeled from a device with the parameters of Table 5.7. The optimum bias current is used to calculate the noise until the calculated optimum bias current falls below 0.1 mA. (The gain β starts to fall in value for bias currents below this value.) When the optimum bias current is computed to be below 0.1 mA, a bias current of 0.1 mA is assumed for calculating the noise current.

Figure 5.20 shows that at low bit rates the bipolar transistor front-end is inferior to the FET front-end. At high bit rates this inferiority disappears. Among the FET front-ends, the silicon MOSFET is slightly advantageous at low bit rates, with the GaAs MESFET having slightly superior noise properties at high bit rates. (Silicon JFETs lose gain at bit rates above about 200 Mb $\cdot$ s^{-1} because of their relatively low gain-bandwidth product. These devices are, therefore, not suitable for high–bit-rate designs.)

Table 5.6: Assumed FET parameters for noise and sensitivity calculations. (From Kasper, B.L., "Receiver design," in *Optical Fiber Telecommunications II*, [S.E. Miller and I.P. Kaminow, eds.], pp. 689–722, New York: Academic Press, 1988.)

	GaAs MESFET	**Si MOSFET**	**Si JFET**
g_m (mS)	40	30	6
C_{gs} (pF)	0.38	0.8	4.0
C_{gd} (pF)	0.02	0.1	0.8
Γ	1.1	2.0	0.7
I_{gate} (nA)	2.0	0	0.05

Table 5.7: Assumed BJT parameters for noise and sensitivity calculations. (From Kasper, B.L., "Receiver design," in *Optical Fiber Telecommunications II*, [S.E. Miller and I.P. Kaminow, eds.], pp. 689–722, New York: Academic Press, 1988.)

Parameter	**Value**
β	100
$r_{bb'}$	20 ohms
$C_{b'c}$	0.2 pF
C_{je}	0.8 pF
f_T	10 GHz

5.5.5 Sensitivity of Front-Ends

We now want to consider a calculation of the optical power required on the receiver to achieve a desired bit-error rate in the presence of both detector noise and amplifier noise (Kasper, 1988). We first consider the case of a pin-diode detector and then the more complicated case of an APD detector.

PIN-Diode Sensitivity

The sensitivity of current pin-diode receivers is more than 20 dB from the theoretical minimum of the quantum limit. Hence, we neglect the signal-related shot noise. The total mean-square noise current is the sum of the amplifier mean-square noise current and the mean-square shot-noise current due to the dark current of the amplifier

$$< i_N^2 >_{\text{Total}} = < i_{\text{amp}}^2 > + 2qI_dI_2B_R . \tag{5.200}$$

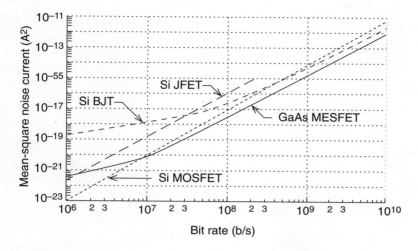

Figure 5.20 Comparison of mean-square noise current for typical FET front-ends and a BJT front-end. (Assumed device values are found in Tables 5.6 and 5.7.)

Here, the amplifier mean-square noise current is determined as in the prior sections, depending on the type of preamplifier used. Both expressions on the right side of this equation depend on the bit rate B_R. The required signal-to-noise ratio, S/N, for the desired bit-error rate is determined from Figure 5.12. The average optical power required to achieve this signal-to-noise ratio for the pin-diode receiver is (Kasper, 1988)

$$P = \frac{hc}{q\lambda} \frac{S}{N} \sqrt{<i_N^2>_{\text{Total}}} \, . \tag{5.201}$$

The required optical power can be calculated and plotted as a function of the bit rate B, once the pin-diode parameters and the amplifier type and parameters are known. (See the problems at the end of this chapter.)

APD Sensitivity

The sensitivity of an APD receiver is made more difficult to analyze because the gain of the APD is an additional variable. Also the APD produces excess noise which makes the model more difficult. In fact, there is an optimum value of diode gain M_{opt} which gives the best sensitivity. (This value of optimum gain depends on the device parameters, the preamplifier noise, and the bit rate.) At the optimum gain $M = M_{opt}$, the APD noise is approximately (but not precisely) equal to the preamplifier noise (Kasper, 1988). The shot noise due to the dark current in an APD consists of two parts (Muoi, 1984), one part due to the unamplified surface dark current I_{surf} and the other part due to the multiplied bulk dark current I_{bulk}, i.e,

$$< i^2 >_{dark} = 2qI_{surf}I_2B_R + 2qI_{bulk}M^2F(M)I_2B_R . \qquad (5.202)$$

Here $F(M)$ is the excess noise factor, as described earlier in this chapter. The surface dark current leaks around the edges of the device and does not engage in the avalanche process. The receiver sensitivity for an APD receiver is (Kasper, 1988; Muoi, 1984)

$$P = \left(\frac{hc}{q\lambda}\right)Q\left[QqB_RI_1F(M) + \sqrt{\frac{< i^2 >_{Total}}{M^2} + 2qI_{bulk}F(M)B_RI_2}\right], \qquad (5.203)$$

where Q is the signal-to-noise ratio needed to achieve the required bit-error rate, I_1 and I_2 are Personick integrals described earlier, and

$$< i^2 >_{Total} = < i_{amp}^2 > + 2qI_{surf}I_2B_R . \qquad (5.204)$$

If the multiplied part of the dark current I_{bulk} is small enough that its noise is negligible (not a very good assumption for long-wavelength detectors), then the sensitivity is

$$P = \frac{hcQ}{q\lambda}\left(\frac{\sqrt{< i^2 >_{Total}}}{M} + qQB_RI_1F(M)\right) \quad \text{(for noise due } I_{bulk} \text{ negligible).} \qquad (5.205)$$

For this same case, the optimum gain is (Muoi, 1984; Kasper, 1988)

$$(5.206)$$
$$M_{opt} = \frac{1}{\sqrt{k}}\sqrt{\frac{\sqrt{< i^2 >_{Total}}}{qI_1B_RQ} - k + 1} \quad \text{(for noise due } I_{bulk} \text{ negligible)} ,$$

where k is the ionization ratio of the avalanche process, described earlier.

If the noise contribution from I_{bulk} is *not* negligible, M_{opt} is smaller than the value predicted by Equation 5.206. The value of M_{opt} must be found graphically or numerically at each value of the bit rate by finding the value that minimizes the sensitivity of the receiver. Then the total noise and sensitivity can be calculated as a function of bit rate. A more detailed discussion can be found in Muoi (1984).

5.5.6 Extinction Ratio Effects

The extinction ratio r (the ratio of the power transmitted for a logical **0** to the power transmitted for a logical **1**) indicates whether the optical source is turned off all the way when a logical **0** is transmitted. Not turning off the source causes a reduction in sensitivity (call a *sensitivity penalty*) (Muoi, 1984; Kasper, 1988). This penalty is incurred because there will be shot noise associated with the reception of a **0** and because not all of the received optical power is being modulated. For a pin-diode receiver with a nonzero extinction ratio, the power required to achieve a desired BER is multiplied by a factor of $(1 + r)/(1 - r)$. For the APD receiver, the extinction ratio affects the optimum APD gain (which must be found numerically) and reduces the required power in a complicated fashion. (See Muoi, 1984 and Kasper, 1988 for more details.)

5.6 Eye Pattern Analysis

The performance of a high–data-rate link can be measured from the *eye pattern* of the system (Keiser, 1991). The eye-pattern measurement (Keiser, 1991) is a time-domain measurement made with the simple setup shown in Figure 5.21. The eye pattern observed on the scope provides several pieces of meaningful information about the performance of digital data links. The pseudorandom generator produces a data stream with a pattern length and bit period controlled by the operator. Within the pseudorandom data stream, the bits are produced in a highly varied fashion, but the pattern is still predictable by the user (although this prediction is not worth the

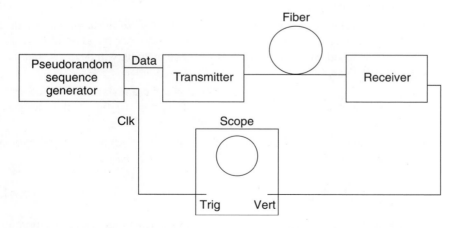

Figure 5.21 Experimental setup for observing the eye pattern of a fiber optic digital link.

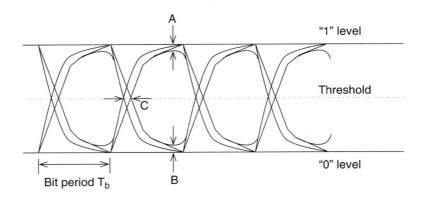

Figure 5.22 Representative eye pattern.

effort for long data streams). At the end of the data pattern, the pattern repeats itself. The received data is put into the vertical trace of the oscilloscope, while the data clock is used to trigger the oscilloscope. The resulting eye pattern (so-called for its resemblance to the human eye [see Figure 5.22]) is the superposition of output pulses from many data pulses. This pattern contains much easily observed information about the digital transmission characteristics of the optical link (see Figure 5.22).

- The horizontal width of the eye opening gives the optimum *sampling time interval* for the received signal to be sampled without error from intersymbol interference. The optimum sampling time is at the position of maximum eye opening.

- The vertical height of the eye opening is a measure of the *amplitude distortion* of the signal. As the upper limit of the frequency response of the system is reached, the vertical height of the eye opening will decrease and the eye will close.

- The spacing of "A" on the figure indicates the amount of noise present when a logical **1** is sent; the spacing of "B" on the figure indicates the amount of noise present when a logical **0** is sent.

- The width of the threshold crossing ("C" on the figure) determines the *timing jitter* (or *edge jitter*) of the system. The *jitter* is defined as

$$\text{Timing jitter (in \%)} = \frac{\Delta T}{T_b} \times 100\% ,\qquad (5.207)$$

where T_b is the bit spacing in the data stream. This jitter is an indication of the timing accuracy of the received pulse as modified by the receiving circuitry. (Meeting jitter specifications is receiving renewed attention in fiber links. To help in characterizing jitter performance, new instruments are becoming available to measure the jitter in a system.)

■ If a long string of logical **1**s and **0**s is included in the data stream, the 10 percent to 90 percent rise (and fall) times can be measured from the rise (and fall) times of the eye.

5.7 Summary

Table 5.8 summarizes the properties of pin diodes and APDs by presenting the properties of representative devices. (Individual devices can exhibit superior performance in some categories.) The silicon devices represent mature technology and operate close to theoretical limits in the short-wavelength region. At long wavelengths, the InGaAs detector exhibits the best characteristics, while germanium-based devices have fundamental difficulties with noise performance, construction of avalanche devices, and high dark current levels.

In the design of receivers, we have seen that the noise contributions of the preamplifier are important and that much design effort goes into the design of this amplifying stage. We have seen that, because of the capacitive nature of the detector, high-impedance preamplifiers, while providing the best sensitivity, integrate the current produced at the detector, necessitating the incorporation of equalization amplifiers. A frequently used alternative receiver is the transimpedance amplifier, which sacrifices some noise performance for increased dynamic range and simplicity of design and operation.

Table 5.8: Performance of representative detectors.

Characteristic	Pin Diodes			APDs	
	Silicon	Germanium	InGaAs	Silicon	Germanium
λ (μm)	0.4–1.1	0.5–1.8	1.0–1.5	0.4–1.1	0.5–1.65
Quantum efficiency	80%	50%	70%	80%	75%
Rise time (ns)	0.01	0.3	0.1	0.5	0.25
Bias voltage	15	6	10	170	40
Responsivity (A/W)	0.5	0.7	0.4	0.7	0.6
Gain	1	1	1	80-150	80-150

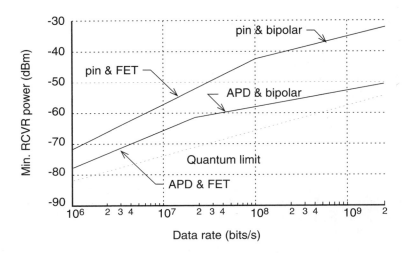

Figure 5.23 Representative sensitivities required for 10^{-9} BER
for typical receivers (for silicon detectors and transistors).

Figure 5.23 illustrates representative receiver sensitivities required for operation of a digital data link with a BER of 10^{-9}. We note that increased sensitivity is required at higher data rates and that silicon FET receivers are good up to approximately 70 Mb $\cdot$ s^{-1}. The APDs offer about 10 dB of increased sensitivity with disadvantages in requiring a high-voltage bias, increased cost, and the requirement for temperature compensation. We also note that the best detector/preamplifier combinations are approximately 10 dB from achieving quantum-limited detection. This 10 dB deficit indicates the degree of improvement in operation that might be expected with improved devices and preamplifiers. As a benchmark, we note that the pin diode requires –42 dBm at 100 Mb $\cdot$ s^{-1} while the APD requires –58 dBm.

5.8 Problems

1. (a) What are the energies (in joules and in eV) of photons having wavelengths of 820 nm and 1.3 μm?

   (b) What are the values of the free-space propagation constants for these wavelengths?

2. Calculate the cutoff wavelength of silicon ($E_g = 1.1$ eV) and germanium ($E_g = 0.785$ eV).

3. Consider the absorption coefficient for silicon as shown in Figure 5.3. Ignoring surface reflections, use a computer to plot the fraction of the incident power that is absorbed for depletion layer widths of 1, 5, 10, 20, and 50 μm over the wavelength range of 0.6 to 1.0 μm. Assume that $d = 20$ μm.

4. An avalanche photodiode has the following parameters:

   ☐ dark current = 1 nA

   ☐ surface leakage current = 1 nA

   ☐ quantum efficiency = 0.85

   ☐ gain = 100

   ☐ excess noise factor = $M^{1/2}$

   ☐ load resistance = $10^4 \Omega$.

   Consider a sinusoidal modulation on an 850 nm carrier with a modulation index m of 0.85. The average power level is –50 dBm and the detector operates at room temperature. Calculate the following S/N (in dB) if the bandwidth is 10 kHz and . . .

   (a) . . . if the signal-dependent shot noise is the dominant term. (The mean-square signal current is given by $< i_s^2 > = (mM\mathcal{R}_0 P)^2 / 2$.)

   (b) . . . if the shot noise due to the dark current is the dominant term.

   (c) . . . if the shot noise due to the surface current is the dominant noise term.

   (d) . . . if the thermal noise is the dominant noise term.

   (e) Which is really the dominant noise source? What will be the actual S/N?

5. Consider the avalanche device of the preceding problem.

   (a) Use a computer to plot the S/N vs. M for values of M ranging from 1 to 100.

   (b) Find the approximate value of M_{opt} and S/N at $M = M_{opt}$ from your graph.

   (c) Calculate the values of M_{opt} and S/N at $M = M_{opt}$ from formulas and compare with your graphical results of the previous question.

6. Consider a silicon avalanche photodiode with parameters as given below, operating in a link with no intersymbol interference present.

   ☐ $F(M) = M^{0.4}$

   ☐ Responsivity (at $M = 1$) = 0.3 A/W

   ☐ Surface dark current = 1 μA

   ☐ Temperature = 300 K

☐ $R_L = 1 \, k\,\Omega$

☐   Bulk dark current $= 1 \, nA$

☐   Bandwidth of receiver $= 10 \, MHz$.

    (a) Calculate the DC optical power that must be incident on the detector to make the optimum gain of this amplifier have a value of 80.

    (b) With a value of gain of 80, calculate the ratio (in dB) of the mean-square noise current due to the shot noise caused by the bulk dark current to the mean-square noise current due to the thermal noise.

7. Consider a silicon photodiode operating at 850 nm ($\alpha = 10^3 \, cm^{-1}$).

    (a) Calculate the area of the device if the capacitance is to be kept equal or less than 2 pF. The relative permittivity of silicon is 11.7.

    (b) Calculate the maximum bandwidth of the detector when operating into 50 Ω load.

8. Consider a pin diode with the following properties at 920 nm:

☐   Responsivity $= 0.5 \, A/W$

☐   Dark current $= 1.0 \, nA$

☐   Surface dark current is negligible

☐   Operating temperature $= 300 \, K$.

This diode is irradiated with a constant 80 nW of optical power (at 920 nm). Find the signal-to-noise ratio *(in dB)* of the detector if it is operated into an equivalent load of 10 K Ω and a (noisefree) preamp with a bandwidth of 1 MHz.

9. Show that the relation between the Q-parameter and S/N is

$$Q = \sqrt{S/N}/2 \,.$$

10. Using an iterative approach, show that $Q \approx 6.00$ for a BER of 10^{-9}.

11. Consider three FET front-ends having the parameters found in Table 5.6. If $C_s + C_d = 0.2 \, pF$ and $R_L = \infty$, plot (using a computer) the mean-square noise current of the front-end vs. the bit rate B for values of B falling between 1 Mb $\cdot$ s^{-1} and 10 Gb $\cdot$ s^{-1}. (Plot the graph on a log-log scale.)

12. Consider the bipolar transistor with parameter values given in Table 5.7. It is to be used with a silicon pin detector with a device capacitance of of 0.2 pF and a dark current of 1 nA. The stray capacitance is assumed to be 0.

    (a) Using a computer, calculate and plot the optimum bias current I_c as a function of bit rate B_R for values of B_R falling between 1 Mb $\cdot$ s^{-1} and 10 Gb $\cdot$ s^{-1}. (Plot the B_R axis on a logarithmic scale.)

    (b) For what values of B_R is the optimum bias current smaller than 0.1 mA?

(c) For the range of values of B_R such that the optimum bias is greater than 0.1 mA, plot (using a computer) the mean-square noise current of the BJT front-end on a log-log scale, using the optimum bias current.

(d) For the range of values of B_R such that the optimum bias is less than 0.1 mA, plot (using a computer, on the same figure as the previous part of this problem) the mean-square noise current of the BJT front-end, using a bias current of 0.1 mA. (Your result should be the same as the curve labeled "Si bipolar" in Figure 5.20.)

(e) Plot the minimum power required to achieve a BER of 10^{-9} with thin pin diode and BJT transistor as function of bit rate B.

13. Consider an InGaAs APD with the properties listed below, operating with a GaAs FET preamp with the properties listed in Table 5.6 at a bit rate of 1 Gb · s^{-1}. Assume that the stray capacitance C_s is zero.

Parameter	Value
C_d	0.2 pF
I_{surf}	0 nA
I_{dark}	1 nA
k	0.3

(a) Using a computer, plot the receiver sensitivity (i.e., the required optical power in dBm) for a BER of 10^{-9} ($Q = 6$) as a function of APD gain M for values of M from 1 to 50.

(b) From your plot, estimate the optimum value of M.

(c) From your plot, estimate the sensitivity of the receiver at the optimum gain.

References

Bowers, J. E. and C. A. Burrus, Jr., "Ultrawide-band long-wavelength p-i-n photodetectors," *J. Lightwave Technology*, vol. LT-5, no. 10, pp. 1339–1350, 1987.

Brain, M. and T. Lee, "Optical receivers for lightwave communication systems," *J. Lightwave Technology*, vol. LT-3, no. 6, pp. 1281–1300, 1985.

236

Forrest, S., "Optical detectors for lightwave communication," in *Optical Fiber Telecommunications II*, (S. E. Miller and I. P. Kaminow, eds.), pp. 569–599, New York: Academic Press, 1988.

Kasper, B. L. and J. C. Campbell, "Multigigabit-per-second avalanche photodiode lightwave receivers," *J. Lightwave Technology*, vol. LT-5, no. 10, pp. 1351–1364, 1987.

Kasper, B. L., "Receiver design," in *Optical Fiber Telecommunications II*, (S. E. Miller and I. P. Kaminow, eds.), pp. 689–722, New York: Academic Press, 1988.

Keiser, G., *Optical Fiber Communications,* 2nd Edition. New York: McGraw-Hill, 1991.

Lee, T. P., and T. Li, "Photodetectors," in *Optical Fiber Telecommunications*, (S. E. Miller and A. G. Chynoweth, eds.), pp. 593–626, New York: Academic Press, 1979.

Meer, D. E., "Noise figures," *IEEE Trans. on Education*, vol. 32, no. 2, pp. 66–72, 1989.

Muoi, T. V., "Receiver design for high-speed optical-fiber systems," *J. Lightwave Technology*, vol. LT-2, no. 3, pp. 243–267, 1984.

Personick, S., "Receiver design for digital fiber optic communication systems," *Bell System Technical Journal*, vol. 52, pp. 843–874, 1973.

Personick, S., "Receiver design for optical fiber systems," *Proc. IEEE*, vol. 65, no. 12, pp. 1670–1678, 1977.

Personick, S. D., "Receiver design," in *Optical Fiber Telecommunications*, (S. E. Miller and A. G. Chynoweth, eds.), pp. 627–651, New York: Academic Press, 1979.

Personick, S., "Photodetectors for fiber systems," in *Fundamentals of Optical Fiber Communications*, (M. F. Barnoski, ed.), pp. 257–293, New York: Academic Press, 1981a.

Personick, S., "Design of receivers and transmitters for fiber optic systems," in *Fundamentals of Optical Fiber Communications,* 2nd edition, (M. Barnoski, ed.), pp. 295–328, New York: Academic Press, 1981b.

Plumb, R., "Detectors for fibre-optic communication systems," in *Optical Fibre Communication Systems*, (C. Sandbank, ed.), pp. 184–205, New York: Wiley, 1980.

Saleh, B. E. and M. C. Teich, *Fundamentals of Photonics*. New York: John Wiley and Sons, 1991.

Smith, R., "Photodetectors for fiber transmission systems," *Proc. IEEE*, vol. 68, no. 10, pp. 1247–1252, 1980.

Smith, R. and S. Personick, "Receiver design for optical fiber communication systems," in *Topics in Applied Physics, Vol. 39*, (H. Kressel, ed.), pp. 89–160, Berlin: Springer-Verlag, 1982.

Yariv, A., *Optical Electronics,* 4th Edition. New York: Holt Rinehart and Winston, 1991.

237

Chapter 6

Splices, Connectors, and Couplers

6.1 Introduction

Since fibers are available in lengths typically ranging up to a few kilometers, the fiber–optic-system user must have means of interconnecting or joining lengths of fibers in a way that offers low *insertion loss* (i.e., the additional loss introduced by the connection), high strength, and simplicity of installation. The same properties are also required for the repair or expansion of a fiber-optic link. A *splice* (Mettler and Miller, 1988; Morra and Vezzoni, 1990) is a permanent joining of fibers; a *connector* allows a demountable connection between fibers or between a fiber and a source or detector.

Another need is to split the power in a fiber into two or more fibers or to combine the power from several fibers into a single fiber. Devices that perform this operation are *couplers,* and they, too, have an insertion loss. In this chapter we will also describe the various couplers available.

6.2 Joining Losses

The causes of loss in a fiber connection or splice have been subdivided into two areas, those that depend on the properties of the fibers being joined (the *intrinsic* sources of loss) and those due to factors external to the fiber parameters (the *extrinsic* sources of loss), such as fiber misalignment. We use the *coupling efficiency* η to describe the fraction of the incident power that is transmitted through a joint. It is given by

$$\eta = \frac{P_{out}}{P_{in}}, \tag{6.1}$$

where P_{in} is the optical power in the fiber on the input side and P_{out} is the optical power in the fiber at the output side of the connector. In general, the coupling efficiency is *not* the same in both directions. The *joint loss* L_j is the dB equivalent of η, given by

$$L_j = -10 \log \left(\frac{P_{out}}{P_{in}} \right). \tag{6.2}$$

(Since the output power will be less than or equal to the input power, the joint loss will be a positive value.)

6.2.1 Effects of Fiber Parameters I: Multimode Fibers

The optical power coupled from one fiber to another is limited by the lesser of the number of modes in each. Optimum coupling occurs (in either direction) when the number of modes is matched. The number of modes N in a fiber is (Keiser, 1991)

$$N = k^2 \int_0^a NA^2(r) \, r \, dr, \tag{6.3}$$

where $k = 2\pi/\lambda$, a is the radius of the fiber, and $NA(r)$ is the generalized numerical aperture for both step-index and graded-index fibers. This equation reduces to

$$N = k^2 \, NA^2(0) \int_0^a \left[1 - (r/a)^g \right] r \, dr. \tag{6.4}$$

Hence we see the factors of interest include fiber radius a, axial numerical aperture $NA(0)$, and index gradient g. The usual approach taken is to isolate the effects as if they were independent and, then, add the dB losses to provide an estimate of the overall losses.

Additionally, the losses depend on the *power distribution* among the modes of the fiber. Computations usually assume that the power is uniformly distributed over all of the modes of the fiber; in reality, the power may be unevenly distributed due to the launch conditions or due to the modal effects in very long fibers that leave most of the power in the lower–order modes of the fiber. Such power distribution effects also play a role in the extrinsic factors of loss. For example, the losses encountered due to an axial misalignment of the fibers depend on the distribution of power in the modes. Light that is concentrated in the lower modes will be more sensitive to axial displacements than fibers that have the light uniformly distributed. Because of the computational difficulties of including these effects, the approximation is to assume uniform illumination in the fiber that excites all modes uniformly.

We now want to consider the effects of joining mismatched fibers.

- Numerical aperture effects—The coupling efficiency depends on the numerical apertures of the fibers as

$$
\eta_{NA} = \begin{cases} \left(\dfrac{NA_r(0)}{NA_e(0)}\right)^2 & \text{if } NA_r(0) < NA_e(0) \\ 1 & \text{if } NA_r(0) \geq NA_e(0) \end{cases}, \tag{6.5}
$$

where $NA_r(0)$ and $NA_e(0)$ are the axial numerical apertures of the receiving and emitting fibers, respectively. The formula assumes equal radii and index profiles of both fibers.

- Fiber radius effects—The coupling efficiency for fibers of differing core radii is given by

$$
\eta_r = \begin{cases} \left(\dfrac{a_r}{a_e}\right)^2 & \text{for } a_r < a_e \\ 1 & \text{for } a_r \geq a_e \end{cases}, \tag{6.6}
$$

where a_r and a_e are the radii of the core of the receiving and emitting fibers, respectively. The fiber numerical apertures and axial index profiles are assumed equal.

- Index profile effects—The coupling efficiency due to differences in the index of refraction profile is given by

$$
\eta_g = \begin{cases} \dfrac{g_r(g_e + 2)}{g_e(g_r + 2)} & \text{for } g_r < g_e \\ 1 & \text{for } g_r \geq g_e \end{cases}, \tag{6.7}
$$

where g_r and g_e are the refractive index profile values of the receiving and emitting fibers, respectively. Figure 6.1 shows a plot of this equation.

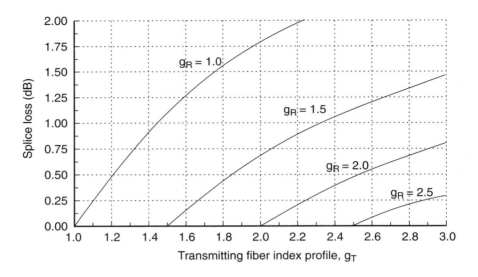

Figure 6.1 Typical calculated losses due to refractive index profile mismatches.

■ Combined effects—Having found the individual intrinsic coupling efficiencies, we estimate the total intrinsic coupling efficiency η_{total} as

$$\eta_{total} \approx \eta_{NA}\, \eta_r\, \eta_g .\tag{6.8}$$

To express the combined losses in dB we find

$$L_{total}(gdB) = -10 \log \eta_{total}\tag{6.9}$$

$$\approx -10 \log (\eta_{NA}) - 10 \log (\eta_r) - 10 \log (\eta_g)\tag{6.10}$$

$$\approx L_{NA} + L_r + L_g .\tag{6.11}$$

Example: Calculate the coupling efficiency that can be expected in coupling a 50/125 SI (emitting) fiber with an NA of 0.15 to a 62.5/125 GI ($g = 2$) receiving fiber with an NA = 0.20.

Solution: Since $NA_r > NA_e$,

$$\eta_{NA} = 1 .\tag{6.12}$$

Since $a_e < a_r$,

$$\eta_r = 1 . \tag{6.13}$$

Since $g_e(= \infty) > g_r$,

$$\eta_g = \frac{g_r(g_e + 2)}{g_e(g_r + 2)} \tag{6.14}$$

$$= \frac{g_r\left(1 + \dfrac{2}{g_e}\right)}{(g_r + 2)} \tag{6.15}$$

$$= \frac{2\left(1 + \dfrac{2}{\infty}\right)}{(2 + 2)} \tag{6.16}$$

$$= 0.5 . \tag{6.17}$$

The total efficiency is the product of the individual contributions,

$$\eta = \eta_{NA}\eta_r\eta_g \tag{6.18}$$

$$= (1)\,(1)\,(0.5) \tag{6.19}$$

$$= 0.5 . \tag{6.20}$$

The total losses are $-10 \log (0.5) = 3$ dB.

■ **Effects of other fiber parameters**—There are a variety of loss mechanisms that are functions of the fiber fabrication process and its quality control. These include

☐ ellipticity of the core,
☐ variations in the index of refraction profile,
☐ concentricity of the core within the cladding,
☐ variation in the core diameter, and
☐ other factors that depend on fabrication tolerances.

Since these factors are determined by the manufacturing process, the user has little control over them, except to establish specifications and inspection procedures.

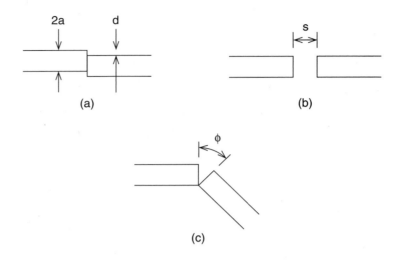

Figure 6.2 Types of misalignments. (a) Lateral displacement.
(b) Longitudinal separation. (c) Angular misalignment.

Of these factors of joint loss, the differences in core radii and numerical aperture have the dominant effects, while mismatches in the core ellipticity and the refractive index profile play lesser roles (for the same values of fractional mismatch).

6.2.2 Misalignment Effects

The effects just discussed are properties of the fiber. We now turn our consideration to factors that are under the control of the connector designer (and to some degree, the user of the connectors) (Cook and Runge, 1979; Gloge et al., 1979; Dalgleish, 1980 and 1981). These effects are primarily due to the misalignment of the fibers, and they determine the mechanical tolerances required of the connector or splicer to meet a given loss allocation.

- Lateral displacement effects—The misalignment to which a connection is most sensitive is *lateral displacement*, shown in Figure 6.2(a). Mechanical misalignments of the fibers cause losses because the areas of the fiber cores do not overlap sufficiently. In the analysis of misalignments the usual assumptions are that the fibers have equal radii, index profiles, and numerical apertures to isolate the effects of the misalignment. The fiber is also assumed to have uniform power distribution across its area. For a step-index fiber, the coupling efficiency depends on the lateral misalignment distance d as (Keiser, 1991)

$$\eta_{SI} = \frac{2}{\pi} \cos^{-1}(d/a) - \frac{d}{\pi a} \sqrt{1 - (d/2a)^2} \,. \tag{6.21}$$

This result is derived from a calculation of the overlapping areas of two equal circles with centers separated by a distance, d. A similar calculation for a graded-index fiber is complicated by the radial dependence of the numerical aperture. The result of the calculation (Keiser, 1991) based on a parabolic index ($g = 2$) predicts a coupling efficiency of

$$\eta_{GI} = \frac{2}{\pi} \left[\cos^{-1}(d/2a) - \sqrt{1 - (d/2a)^2} \, (d/6a) \left(5 - (d^2/2a^2) \right) \right]. \tag{6.22}$$

For $d/a < 0.4$ we can use the approximation (Keiser, 1991)

$$\eta_{GI} \approx 1 - \frac{8d}{3\pi a} \,. \tag{6.23}$$

A different form for the loss in a graded-index fiber that is valid for any value of g is found in Marcuse et al. (1979) as

$$\eta_{GI} \approx 1 - \left(\frac{2d}{\pi a} \right) \left(\frac{g+2}{g+1} \right). \tag{6.24}$$

■ Longitudinal displacement effects—The effects caused by a pure *longitudinal displacement* (Figure 6.2[b]) are because some of the light has spread beyond the area of the receiving fiber, as shown in Figure 6.3. For the step fiber, we have (Keiser, 1991)

$$\eta_{SI} = \left(\frac{1}{1 + (s/a) \tan \theta_c} \right)^2, \tag{6.25}$$

where s is the separation distance between the fiber ends and θ_c is the critical angle of acceptance ($= \sin^{-1} NA$).

Marcuse et al. (1979) gives a formula for this coupling efficiency for a step-index fiber as

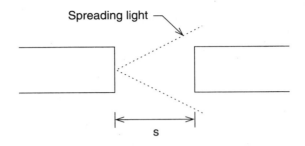

Figure 6.3 Spreading of light in longitudinally separated fibers.

$$\eta_{SI} \approx 1 - \frac{s\sqrt{2\Delta}}{4a} \,. \tag{6.26}$$

No similar formula for the losses due to longitudinal displacement seems to be available for a graded-index fiber.

Example: Calculate the coupling efficiency for a 50/125 SI fiber if the longitudinal displacement is 10% (i.e., $s/a = 10\%$) and NA = 0.2.

Solution: The critical angle is found by

$$\theta_c = \sin^{-1}(NA) \tag{6.27}$$
$$= \sin^{-1}(0.2) \tag{6.28}$$
$$= 11.54° \,. \tag{6.29}$$

The coupling efficienty is found as

$$\eta_{SI} = \left(\frac{a}{a + s \tan \theta_c} \right)^2 \tag{6.30}$$

$$= \left[\frac{25 \times 10^{-6}}{(25 \times 10^{-6}) + (25 \times 10^{-7}) (\tan 11.54°)} \right]^2 \tag{6.31}$$

$$= 0.980 \,. \tag{6.32}$$

- Angular misalignment effects—The effects of *angular misalignment* are shown in Figure 6.2(c). Marcuse et al. (1979) give the loss due to angular misalignment φ in a fiber with an arbitrary index profile g as

$$\eta \approx \frac{\sin \varphi}{\sqrt{2\pi\Delta}} \frac{\Gamma\left(\dfrac{2}{g+2}\right)}{\dfrac{2}{g+\dfrac{3}{2}}}. \tag{6.33}$$

where $\Gamma(x)$ is the Gamma function (found in books of advanced math tables).

- Combined misalignment effects—Again, we combine the misalignment effects by multiplying the individual coupling efficiencies or adding the coupling losses in dB.

6.3 Joint Losses: Single-Mode Fiber

The mode field diameter (MFD) of a single-mode fiber is physically useful because it relates to the performance of the fiber in an optical system. Specifically, the MFD determines the sensitivity of the joint losses at a connector or splice to misalignment. The MFD will also determine the sensitivity of the fiber to excess losses due to microbends and macrobends. In addition, the cutoff wavelength and the dispersion properties of the fiber can be inferred from a spectral measurement of the MFD. The following sections discuss some of these dependencies.

Joint losses occur at splices and connectors. Since the power transmitted through a joint depends on the fields supported on each side, the shape and diameter of the fields are expected to be key parameters in determining the losses at the joint.

Nemota and Makimoto (1979) give an expression for the coupling loss (in dB) for two single-mode fibers with unequal mode-field diameters between the transmitting fiber and the receiving fiber (W_1 and W_2, respectively) with a lateral offset d, a longitudinal offset s, and an angular offset φ. This single-mode loss, based on a Gaussian-wave assumption, is

$$L = -10 \log \left[\frac{16n_1^2 n_3^2}{(n_1 + n_3)^4} \frac{4\sigma}{q} e^{-\frac{\rho u}{q}} \right], \tag{6.34}$$

where λ is the source wavelength, n_1 is the refractive index of the fiber cores (assumed to be the same on both sides of the joint), n_3 is the refractive index of the medium between the fibers (if applicable),

$$\sigma = \left(\frac{W_2}{W_1}\right)^2 , \tag{6.35}$$

$$k = \frac{2\pi n_3}{\lambda} , \tag{6.36}$$

$$\rho = (kW_1)^2 , \tag{6.37}$$

$$F = \frac{d}{kW_1^2} \quad \text{(a lateral offset parameter)} , \tag{6.38}$$

$$G = \frac{s}{kW_1^2} \quad \text{(a longitudinal offset parameter)} , \tag{6.39}$$

$$q = G^2 + (\sigma + 1)^2 , \text{ and} \tag{6.40}$$

$$u = (\sigma + 1)F^2 + 2\sigma FG \sin \varphi + \sigma(gG^2 + \sigma + 1) \sin^2 \varphi . \tag{6.41}$$

Equation 6.34 can be used to predict the losses expected due to each of the extrinsic alignment parameters or for any given combination of the parameters.

6.4 Reflection Losses

The coupling efficiency at a perpendicular interface is given by the *Fresnel reflection loss* formula,

$$\eta_{\text{reflection}} = 1 - \left[\frac{n_1 - n_2}{n_1 + n_2}\right]^2 , \tag{6.42}$$

where n_1 is the index of one medium and n_2 is the index of the other. The reflection losses (in dB) are found by calculating $-10 \log (\eta)$. The reflection losses are the same regardless of the direction of travel of the light. The losses at an air-glass inter-face ($n_{\text{glass}} = 1.5$, $n_{\text{air}} = 1.0$) in a connector can add 0.2 dB loss for each fiber face. These losses can be eliminated by the application of an index-matching gel to the connector or through the use of an index-matching epoxy in making a permanent joint.

6.5 Fiber End Preparation

All expressions for coupling efficiency or loss assume that the end of the fiber is a perfect transmitter. Any pits or imperfections in the end will scatter light into higher angles, many of them lying outside of the numerical aperture of the receiving fiber. Similarly, the end faces must be parallel to ensure uniform illumination or reception. End preparation techniques have evolved to ensure a smooth perpendicular end of the fiber for splicing or connecting. (Recent technqiues have been developed to cut the fiber ends at an angle to minimize back reflections.)

- The *grinding and polishing technique* uses progressively finer abrasives to smooth and polish the ends. The perpendicularity of the fiber end is assured by the use of a mechanical fiber holder that aligns and holds the fiber perpendicular to the polishing surface. This technique is labor- and time-intensive.

- The *score-and-break technique* uses traditional glass-cutting methods to provide a uniform surface. A fiber is placed under mild tension as in Figure 6.4 and is scratched with a scoring blade. The scratch creates a stress concentration at the tip of the scratch. The tension is then increased and the crack tip propagates across the fiber. If the curvature of the fiber and the application of the tension are carefully controlled, the crack will propagate perpendicular to the fiber axis and create a clean, smooth break. Improper control of the crack propagation can cause the formation of a *lip* or a *hackle* on the fiber end, as shown in Figure 6.5. Microscope inspection of the fiber end is necessary to detect such problems that would require a new scoring. A number of tools have been commercially developed to implement this technique, which has the advantage of taking very little time for an experienced user.

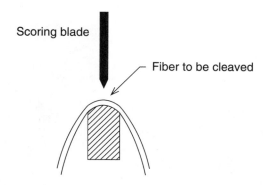

Scoring blade

Fiber to be cleaved

Figure 6.4 Score-and-break technique of cutting fiber with smooth ends.

Figure 6.5 Improper fiber ends.

6.6 Splices

Three types of (Gloge et al., 1979; Barnoski, 1981; Kato et al., 1982; Mettler and Miller, 1988) splices have proved popular in fiber technology:

■ Perhaps the most popular splice is the *fusion splicing technique* (Kato et al., 1982), illustrated in Figure 6.6. The fusion splicer uses micro-manipulators to bring the prepared ends of the fiber into close alignment. The ends are then heated, typically with an electric arc, until they grow molten and fuse together. As the joint cools, the surface tension at slightly misaligned fibers will pull the fibers into alignment. In these splicers the arc voltage is kept low, until the fiber ends are rounded, thereby avoiding bubble formation, and then increased to complete the fusing process. Losses caused by splices made with this technique are typically a few tenths of a dB, with the primary problem being reduced fiber strength in the region near the joint. This reduced strength (about 60% of the strength before making the joint) is caused by the development of microcracks on the fiber surface in the region that has been stripped of its buffering and by chemical changes in the glass

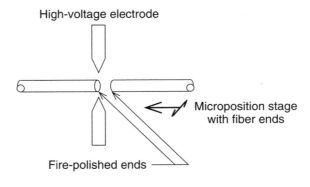

Figure 6.6 Fusion splicing technique.

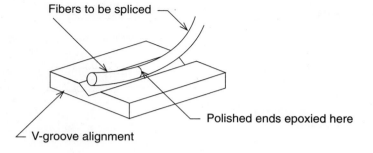

Fibers to be spliced

Polished ends epoxied here

V-groove alignment

Figure 6.7 V-groove splicing technique.

due to the heating. This reduction of strength is countered by the use of a high-strength wrapping placed around the spliced region, which protects the fiber and provides stress relief for the splice region.

■ The *V-groove splice* (Figure 6.7) uses a V-shaped groove as an alignment aid to bring the two fibers into mechanical alignment, before applying an adhesive, such as epoxy, or being fastened in place with a cover plate. The channel can be made in plastic, silicon, ceramic, or metals. A variation on this technique, called the *loose-tube splice*, uses the corner of a rectangular tube as the alignment aid, as shown in Figure 6.8. Since this technique uses the outside surface of the fiber as a reference surface, it is susceptible to losses due to variations in core ellipticity, core concentricity, and core size. The fiber ends require preparation before splicing. Losses as low as several

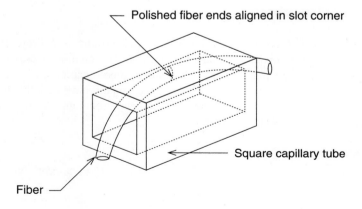

Polished fiber ends aligned in slot corner

Square capillary tube

Fiber

Figure 6.8 Loose-tube splice.

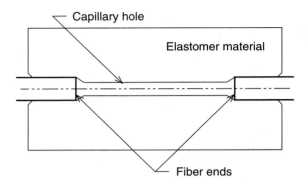

Figure 6.9 Splice using elastic material. (The fibers are shown only partially inserted.)

hundredths of a dB have been reported with these techniques. Note, however, that splicing fibers of unequal diameters by this technique will result in unacceptable misalignments.

- Another type of splice, shown in Figure 6.9, uses an *elastic material* to bring the fibers into alignment. With a central circular hole slightly smaller than the fiber diameter, the restoring forces will center the fiber (again with respect to the outside surface). The advantages of this technique are that a wide range of fiber diameters can be inserted into the device and fibers of unequal diameters can be aligned.

6.7 Connectors

Optical-fiber connectors (Dalgleish, 1981; Young and Frey, 1988) are designed to allow disconnection and reconnection. The goal is to provide a low-insertion-loss connector that will provide reliable reproducible connections. Most connector designs incorporate the fiber into a precision alignment aid that then plugs into a receptacle in the connecting piece. Various environmental factors can make the design increasingly difficult, including

- dust levels,
- pressure differentials across the connector (as in aircraft bulkheads or underwater connectors),

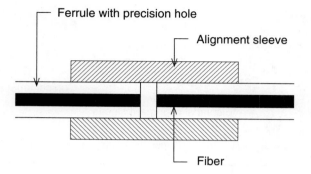

Figure 6.10 Connector made with alignment hole in a ferrule.
The sleeve aligns the ferrules that contain the fibers.

- presence of water vapor and water, both around the connector and within the fiber cable (as in underwater connectors), and

- connector mating requirements (e.g., is a male-female arrangement satisfactory or is a hermaphrodite connector required?).

The different types of connectors are generally divided according to the alignment aid that is used. In application, however, the user is concerned with the fiber size accepted by the connector, the connector type (i.e., SMA, BNC, or other standard and nonstandard fastener types), and the losses of the connector.

Different technologies are used to align the fibers.

- One technique uses a precision-drilled hole in a cylinder, called a *ferrule*. The ferrule then fits an *alignment sleeve* to bring the fiber ends into alignment, as shown in Figure 6.10. The fiber is stripped of its protective jacket and buffering and inserted into the housing and out through the hole in the ferrule. After epoxying the fiber into the ferrule, the end is prepared by the grind-and-polish technique. The main problem is to carefully center the hole in the alignment sleeve and to make the hole slightly larger than the outside diameter of the fiber. Ferrules are commonly made of aluminum (for low-cost connectors), stainless steel, or ceramics.

- A similar technique developed by Bell Laboratories, the *biconic connector*, injection molds the alignment element (Figure 6.11). This element, with the shape of a biconical taper, is designed to mate with the housing so that the fiber is centered.

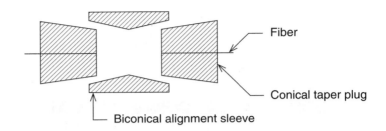

Fiber

Conical taper plug

Biconical alignment sleeve

Figure 6.11 Biconical taper connectors (cross-section).

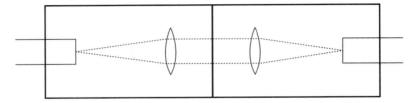

Figure 6.12 An expanded-beam connector.

■ Another approach is the *expanded beam connector*, as shown in Figure 6.12. A microlens is inserted at the fiber's end to collimate the beam. The receiving fiber has a similar collimator to receive the beam and focus it on the core of the fiber. The collimation reduces the requirements on the lateral and longitudinal alignments required, at the penalty of increasing the angular alignment required. (This is a desirable trade-off, since the angular tolerance is more easily controlled.) The lenses are either gradient-index lenses of the proper length to form a collimator or some other structure that will receive the fiber and hold it in alignment.

6.8 Commercial Connectors

Several connector types have become classic connectors that are widely used in the field. For patent and proprietary rights reasons, several manufacturers have invented their own type of connectors that are used in their systems. Frequently these connector designs are "second-sourced" or cross-licensed with other connector companies, expanding their impact. Reliability issues (Young, 1991) require that special attention be paid to the materials used and the method of attaching the con-

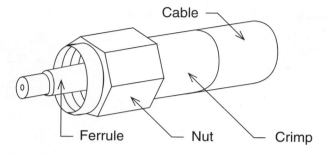

Figure 6.13 An SMA connector.

nectors to the fiber cable. Typical insertion losses for connectors range from several tenths of a dB to a few dB in value.

- The *SMA connector* (Figure 6.13) was borrowed from the RF field and has proved to be a popular connector for multimode fibers. A single-mode version is also available. The fiber is inserted into a ferrule that matches the outside diameter of the fiber and is epoxied in place. The fiber tip is polished (along with the end of the mounting barrel).

- The *biconic connector* (Figure 6.14) was developed by AT&T (American Telephone and Telegraph) and is in wide use in their single-mode systems. It uses a molded, ground plastic or ceramic plug to achieve the close tolerances required for a single-mode connector. It is available in both a single-mode version and a multimode version.

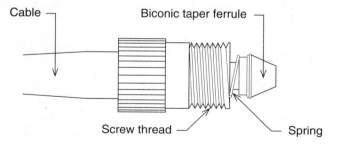

Figure 6.14 A biconic connector.

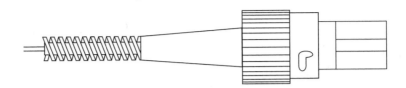

Figure 6.15 An ST connector.

- The *ST connector*, developed by AT&T, is also widely used in single-mode systems. (ST is a registered trademark.) It is available for multimode systems as well. The connector, shown in Figure 6.15, features a spring-loaded bayonet clip. It requires no polishing (i.e., the fiber is cleaved before insertion) and is fairly easy to terminate.

- The *FC (or "D3") connector* (Figure 6.16) is an NTT (Nippon Telephone and Telegraph) design for single-mode fibers. The D3 connector is an NEC (Nippon Electronics Corporation) clone of the FC connector. It, too, is a spring-loaded connector with a screw-on nut. It uses a metal ferrule to align the fiber.

- The *FC/PC connector* is an offshoot of the FC connector. It uses a pure ceramic ferrule for increased alignment accuracy over the metal/ceramic one used in the FC connector. It is primarily used for long-haul telecommunications applications, where the increased performance and cost are justified.

- The *D4 connector* was designed by NEC. It is similar to the D3 connector, but smaller. Figure 6.17 shows a cross-section of this type of connector.

- The *SC connector* (Figure 6.18) is a newer plastic-case connector that uses a push-pull configuration. It has a ceramic ferrule in the center for fiber alignment.

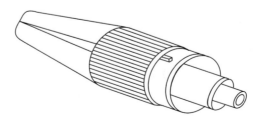

Figure 6.16 An FC connector.

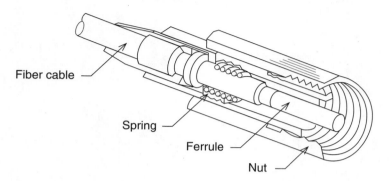

Figure 6.17 Cross-section of a D4 connector.

■ The *FDDI connector* (Figure 6.19) is a dual-fiber connector, designed for use in FDDI data links. (The FDDI network is discussed in detail in Chapter 8.)

6.9 Splice/Connector Loss Measurement

We have mentioned several times that the losses associated with a splice or connector depend on many variables—especially the optical-power launch conditions, the type of source used, and the characteristics of the fiber on either side of the joint (Kapron, 1990). The experimental setup shown in Figure 6.20 is used to measure the losses in a fiber splice or connector. We measure the power at the input and output of the connector (P_1 and P_2) and define the *insertion loss L_s* as

$$L_s = -10 \log \left(\frac{P_2}{P_1} \right). \tag{6.43}$$

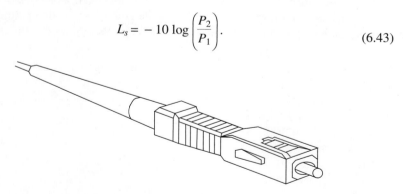

Figure 6.18 An SC connector.

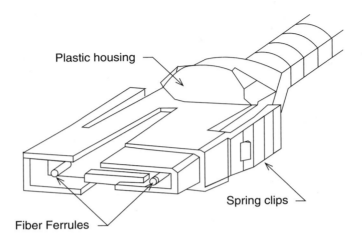

Figure 6.19 An FDDI connector.

With a short length of fiber, we find that the loss is strongly dependent on the launching condition of the source as reflected by the numerical aperture of the source. With a sufficient length of the fiber, however, we find that the connector has been isolated from the effects of the source launching conditions due to the mode conversion effects of the fiber. Hence we require that an *equilibrium mode distribution* be established in the illuminating fiber. Two techniques have evolved for this.

- One technique is to include a long length of fiber as a "pigtail" to allow the equilibrium pattern to be established before the light reaches the connector.
- The other technique is to include an *equilibrium mode simulator,* which usually is a short length of fiber undergoing a serpentine path that allows ample opportunity for mode mixing. A frequently used equilibrium mode

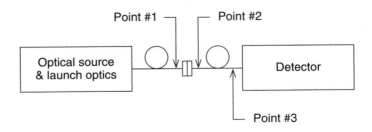

Figure 6.20 Experimental setup for measuring splice losses.

simulator is a 1 meter (or so) length of fiber wrapped around a 1.6 cm diameter spool (often called a "mandrel").

Splices, in turn, can have an effect on the measured losses of a length of multimode fiber that follows a splice due to the splice's effect on the mode distribution (i.e., the splice will redistribute the power in the modes of the fiber). Observation of this effect involves measurements of power (P_1 and P_3) in the setup of Figure 6.20. The measured loss ($-10 \log (P_3/P_1)$) is frequently greater than the sum of the insertion loss and the fiber loss. Since this additional loss is due to the splice rather than the fiber, the loss is properly added to the splice loss rather than the fiber loss. The total loss of the splice or connector is

$$L_{s\,total} = -10 \log \left(\frac{P_3}{P_1}\right) - \alpha L,\qquad (6.44)$$

where αL is the product of the fiber loss coefficient and the fiber length. The excess loss $\delta L_{s\,excess}$ of the splice is the difference between $L_{s\,total}$ and the insertion loss L_s of the splice,

$$\delta L_{s\,excess} = L_{s\,total} - L_s.\qquad (6.45)$$

For a splice with reasonable extrinsic losses (e.g., good alignment) between identical fibers, $L_{s\,total}$ is usually greater than L_s by a factor of two or so. The conclusion is that, while the insertion loss L_s is a well defined local loss quantity, the true total splice loss $L_{s\,total}$ is highly variable and depends on not only the splice but also the length, mode mixing properties, and modal attenuation of the receiving fiber. Usually L_s is the parameter quoted for a splice as measured with a long transmitting fiber and a short receiving fiber. This value, however, can be only partially representative of the splice's effect in a system.

Example:

(a) We want to measure a connector's loss. The power in the fiber at the connector input is 100 µW; the output power immediately after the connector is 83.2 µW. Calculate the insertion loss of the connector.

Solution:

$$L_s = -10 \log (P_{out}/P)\qquad (6.46)$$
$$= -10 \log (83.2/100)\qquad (6.47)$$
$$= 0.8 \text{ dB}.\qquad (6.48)$$

(b) When 1.8 km of 1.5 dB/km fiber is added after the connector, the measured output power at the end of the added fiber is 35.5 µW. Calculate the excess loss of the connector.

$$\delta L_{s\,excess} = -10 \log\left(P_{out}/P\right) - L_s - \alpha L \qquad (6.49)$$

$$= -10 \log\left(\frac{35.5}{100}\right) - (0.8) - (1.5)\,(1.8) \qquad (6.50)$$

$$= +4.50 - 0.8 - 2.7 \qquad (6.51)$$

$$= 1.0 \text{ dB} . \qquad (6.52)$$

6.10 Couplers

In fiber-optic systems there can be applications where it is desirable to combine separate optical signals or to divide the optical signal. Such *multiplexing* and *demultiplexing* tasks are handled by *optical couplers* (Barnoski, 1981). Usually it is desired that each output share the signal equally, although it is possible, in some designs, to weight the *coupling fraction* between a source line and an output line. If one input equally feeds N output lines, then the *power splitting factor* $L_{pwr\,split}$ (in dB) is

$$L_{pwr\,split} = -10 \log\left(\frac{1}{N}\right) \qquad (6.53)$$

$$= 10 \log\left(N\right) . \qquad (6.54)$$

This loss is expected and is unavoidable. Any *extra* losses, however, are included in the *insertion loss* L_{insert} of the device,

$$L_{insert} = -10 \log\left(\frac{\sum_{j=1}^{N} P_j}{P_i}\right), \qquad (6.55)$$

where P_i is the power into the i-th terminal and the summation is of all of the power out of the other ports of the device. Of more utility is a splitting matrix that provides the measured or specified loss L_{ij} for an output j when the splitter is excited at input i.

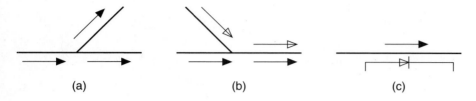

Figure 6.21 Examples of optical couplers. (a) A splitter
or tee coupler. (b) A combiner. (c) A monitor.

6.10.1 Coupler Descriptions

Since couplers can be used to do more than just split and combine light,
specific terms are used to describe their functions. It is instructive to consider the
terminology applied to the various coupling functions as they illustrate potential
applications of the devices.

- A *splitter* (Figure 6.21[a]) splits or divides an input signal into two or more
  output channels. The fraction of the input optical power that is delivered to
  any output arm is under the designer's control. For a two-port splitter, standard designs are 50:50 (i.e., 50% of the input light is found in each output
  port) and 90:10. Other ratios are available by custom order.

- A *polarizing splitter* also splits the signals into two output channels, but the
  polarizations in each output are orthogonal. These devices work only for
  single-mode fibers.

- A *combiner* (Figure 6.21[b]) combines two or more input channels into one
  output channel. Many (but not all) passive devices are reciprocal, so a splitter can sometimes be used as a combiner as well.

- A *monitor* (Figure 6.21[c]) is a splitter that couples very little light (e.g.,
  1%) into the monitor port. A detector is placed at the output of the monitor
  port to measure the light output. Knowing the coupling ratio, we can establish the light level in the fiber.

- A *directional coupler* (Figure 6.22[a]) is a nonreciprocal device that isolates one input from the other.

- A *multiplexer* (actually a *wavelength multiplexer*) is a combiner that joins
  two or more source signals of differing wavelengths. A *demultiplexer* (actually a *wavelength demultiplexer*) splits the signals at the receiving end

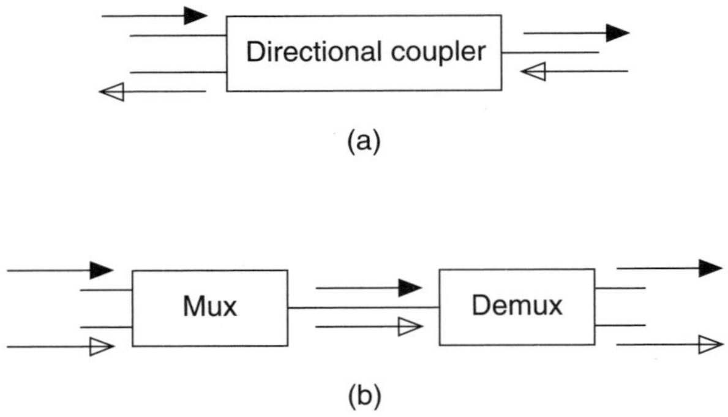

Directional coupler

(a)

Mux   Demux

(b)

Figure 6.22 More examples of optical couplers.
(a) A directional coupler. (b) A multiplexer and demultiplexer.

according to wavelength. Figure 6.22(b) illustrates their application in a unidirectional wavelength-division multiplexing (WDM) system. (Wavelength-division multiplexing is considered in more detail in Chapter 10.)

■ A *circulator* is a three-terminal device that allows light entering at one terminal (terminal #1) to be coupled out of another terminal (terminal #2) with (ideally) 100% efficiency. Light entering terminal #2 is coupled out of the third terminal (#3) with (ideally) 100% efficiency. These devices would allow a bidirectional link to work without incurring the 25% coupling efficiency of using a 2×2 splitter at each end.

6.10.2 Coupler Fabrication

Three different technologies have developed in the fabrication of these coupling elements.

1. The most popular technology is the *biconical taper coupler* (or *fused coupler*). If the claddings of two (or more) fibers are partially removed and the fibers are placed in close proximity over some length, then some light will couple from one fiber into the other(s). The fraction that couples can be controlled by the thickness of the remaining cladding and the length of the region where the fibers are in proximity. As shown in Figure 6.23, this type of coupler can be made by taking a group of fibers with the claddings exposed (i.e., with no protective jacketing), applying tension, and heating

the junction. The coupling fraction is controlled by the amount of tension and the time of heating. Surprisingly, equal coupling can be achieved for all fibers with low insertion loss. In excess of 100 fibers have been formed into a star coupler by this technique.

2. The second technique uses mode-mixing rods (Figure 6.24) as the coupling mechanism. A *mode-mixing rod* is glass rod of a few millimeters diameter with sufficient length to allow the light from all input locations to fully expand to (ideally) uniformly illuminate the end of the rod. All output fibers are uniformly excited. In addition to this transmissive configuration, the rod can be cut in half, a mirrored reflecting surface can be applied, and the output fibers can be moved to the input side of the device to make a reflective system.

3. The third technology uses mirrors with limited apertures placed in strategic locations to intercept a fraction of the illumination beam emitted from the source fiber. By properly shaping the curvature of the mirror and locating the output fibers properly, the output fibers will capture the focussed light, which is a fraction of the incident light, thereby performing the coupling function.

6.10.3 Typical Coupler Specifications

The losses of a coupler consist of the desired *splitting loss* and the undesired *excess loss* that exists beyond the splitting loss. Typical excess losses are on the order of 1 dB. Most coupler losses are specified in a matrix that relates each input fiber to each output fiber.

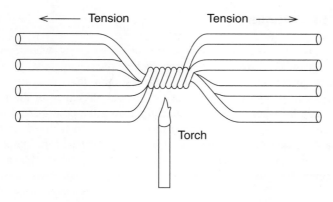

Figure 6.23 Fabrication of a biconical taper coupler.

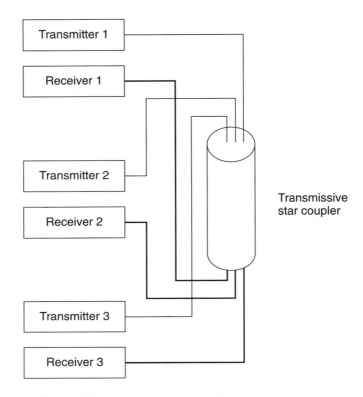

Figure 6.24 A star coupler made with a mode-mixing rod.

6.11 Summary

We have seen that the losses incurred at a fiber joint—whether a splice or a connector—depend on many factors. These include the fiber geometry (core ellipticity, core-cladding concentricity, area mismatches, etc.), the waveguide characteristics of the fiber (NA, index profile), the mechanical alignment of the fiber (lateral and longitudinal displacement, angular misalignment), the power distribution within the fiber (due to excitation conditions or mode conversion within long fibers), and the fiber end-face quality (scratches, presence of lips or hackles). Despite the large number of parameters, commercial manufactures have proved quite successful in producing connectors and splicing techniques with acceptable low losses for most applications.

With fiber-optic couplers, we are able to combine and separate the light in a fiber. The primary parameters are the excess insertion loss and the splitting loss of the coupler. As will be described further in Chapter 10, couplers can also be made to be wavelength sensitive.

6.12 Problems

1. Consider two fibers with the properties given in the table below. Assuming perfect alignment, calculate the splicing losses for. . .

   (a) . . . light going from fiber #1 into fiber #2, and

   (b) . . . light going from fiber #2 into fiber #1.

	Fiber 1	Fiber 2
n_1	1.45	1.48
Δ	1.5%	1.2%
a	50 μm	30 μm
g	1.80	2.00

2. Consider two 50/125 μm SI fibers. Calculate the coupling coefficient if the fibers are laterally misaligned by 5% (i.e., $d/a = 0.05$).

3. Consider a connector that joins two single-mode fibers with the same mode-field *radius* of 4 μm at 1300 nm. The core index of each fiber is 1.47; the index of the medium between the fiber ends is $n = 1$.

   (a) Using a computer, plot the connector loss (in dB) as a function of lateral displacement d for $0 \leq d \leq 8$ μm. Assume that the longitudinal offset and the angular misalignment are zero.

   (b) Using a computer, plot the connector loss (in dB) as a function of longitudinal displacement s for $0 \leq s \leq 8$ μm. Assume that the lateral offset and the angular misalignment are zero.

   (c) Using a computer, plot the connector loss (in dB) as a function of angular misalignment θ for $0 \leq \varphi \leq 1.0$ degrees. Assume that the lateral and longitudinal offsets are zero.

4. (a) Sketch a bidirectional link that uses an ideal 2×2 coupler at each end. Show that only 25% of the transmitter power from one end will reach the receiver at the other end.

   (b) Sketch a bidirectional link that uses an ideal circulator at each end. Show that ideally 100% of the transmitter power will reach the receiver.

References

Barnoski, M. K., ed., *Fundamentals of Optical Fiber Communications,* 2nd Edition. New York: Academic Press, 1981.

Cook, J. and P. K. Runge, "Optical fiber connectors," in *Optical Fiber Tele-communications*, (S. E. Miller and A. G. Chynoweth, eds.), pp. 483–497, New York: Academic Press, 1979.

Dalgleish, J., "Splices, connectors, and power couplers for field and office use," *Proc. IEEE*, vol. 68, no. 10, pp. 1226–1232, 1980.

Dalgleish, J., "A review of optical fiber connection technology," in *Optical Fiber Technology II*, (C. Kao, ed.), pp. 206–212, New York: IEEE Press, 1981.

Gloge, D., A. H. Cherin, C. M. Miller, and P. W. Smith, "Fiber splicing," in *Optical Fiber Telecommunications*, (S. E. Miller and A. G. Chynoweth, eds.), pp. 455–482, New York: Academic Press, 1979.

Kapron, F. P., "Fiber-optic test methods," in *Fiber Optics Handbook for Engineers and Scientists*, (F. C. Allard, ed.), pp. 4.1–4.54, New York: McGraw-Hill, 1990.

Kato, Y., S. Sekai, N. Shibata, S. Tachigama, and Y. Toda, "Arc-fusion splicing of single-mode fibers, 2: A practical splicing machine," *Applied Optics*, vol. 21, no. 11, pp. 1916–1921, 1982.

Keiser, G., *Optical Fiber Communications,* 2nd Edition. New York: McGraw-Hill, 1991.

Marcuse, D., D. Gloge, and E. A. Marcatili, "Guiding properties of fibers," in *Optical Fiber Telecommunications*, (S. E. Miller and A. G. Chynoweth, eds.), pp. 37–100, New York: Academic Press, 1979.

Mettler, S. C. and C. M. Miller, "Optical fiber splicing," in *Optical Fiber Telecommunications II*, (S. E. Miller and I. P. Kaminow, eds.), pp. 263–300, New York: Academic Press, 1988.

Morra, P. and E. Vezzoni, "Fiber-optic splices, connectors, and couplers," in *Fiber Optics Handbook for Engineers and Scientists*, (F. C. Allard, ed.), pp. 3.1–3.86, New York: McGraw-Hill, 1990.

Nemota, S. and T. Makimoto, "Analysis of splice loss in single-mode fibers using a Gaussian field approximation," *Optical Quantum Electronics*, vol. 11, no. 5, pp. 447–457, 1979.

Young, W. and D. Frey, "Fiber connectors," in *Optical Fiber Telecommunications II*, (S. E. Miller and I. P. Kaminow, eds.), pp. 301–325, New York: Academic Press, 1988.

Young, W. C., "Introduction to reliability-related problems in optical fiber connectors," *Optical Engineering*, vol. 30, no. 6, pp. 821–823, 1991.

Part Three

Link Design

Chapter 7

Optical-Link Design

7.1 Introduction

Now that we have considered the building blocks of an optical link, we need a procedure to design a usable optical link to meet desired specifications. The procedure described in this chapter is iterative; that is, certain assumptions are made and the design is carried out based on those assumptions. The design is not finished at that point, however, as the designer must verify that it meets the objectives and represents an economical, as well as a technical, solution. If not, another pass through the design procedure is required. In particular, the assumptions need to be inspected to determine if changes might provide a simpler or cheaper alternative.

Techniques for optically amplifying the signal power to increase the length of the link are also discussed in this chapter. In particular, erbium-doped fiber amplifiers offer great capability.

7.2 Source Selection

The starting point for a link design is choosing the operating wavelength, the type of source (i.e., laser or LED), and whether a single-mode or multimode fiber is required. In a link design, one usually knows (or estimates) the data rate required to meet the objectives. From this data rate and an estimate of the distance, one chooses the wavelength, the type of source, and the fiber type.

Table 7.1: Source data rate–distance performance limits.

	Short λ	**Long** λ
LED	$< 150\ \text{Mb} \cdot \text{s}^{-1} \cdot \text{km}$	$< 1.5\ \text{Gb} \cdot \text{s}^{-1} \cdot \text{km}$
Laser	$< 2.5\ \text{Gb} \cdot \text{s}^{-1} \cdot \text{km}$	$< 25\ \text{Gb} \cdot \text{s}^{-1} \cdot \text{km}$

We begin knowing that a silica-based fiber operating with an LED source in the 800 to 900 nm region has a data rate–distance product of about $150\ \text{Mb} \cdot \text{s}^{-1} \cdot \text{km}$ (due to the spectral dispersion). The same fiber operating with a laser source in the same region of the spectrum has a product of approximately $2.5\ \text{Gb} \cdot \text{s}^{-1} \cdot \text{km}$. In the region near 1300 nm, an LED can achieve a product of $1.5\ \text{Gb} \cdot \text{s}^{-1} \cdot \text{km}$ and a laser can achieve products in excess of $25\ \text{Gb} \cdot \text{s}^{-1} \cdot \text{km}$. These benchmarks are summarized in Table 7.1. From these products, the decision is tentatively made whether to work with the lower-cost, short-wavelength sources, or, if higher performance is required, to use the long-wavelength sources. A tentative choice of whether to use a laser or LED source can be made at this time as well, but this decision will be refined later.

The choice of fiber type involves the decision to use either multimode or single-mode fiber, and, if multimode, whether to use graded-index or step-index profiles. This choice is dependent on the allowable dispersion and the difficulty in coupling the optical power into the fiber. If an LED is chosen, then the obvious choice of fiber is a multimode fiber, because the coupling losses into a single-mode fiber are too severe. (A long-wavelength LED can sometimes be used with a single-mode fiber for short-distance links.) For a laser source, either a multimode or single-mode fiber can be used. The choice depends on the required data rate, as losses in both types of fiber can be made quite low.

7.3 Power Budget

With the choice of the wavelength region and a tentative choice of fiber type made, one then proceeds to compute the power levels required at various locations in the circuit, as follows. From the data rate desired, combined with the desired bit-error rate (BER), we assume either a pin-diode detector or an APD and find the required detector power from sensitivity curves, such as Figure 5.23. The required receiver power can also be calculated from models, such as those used to derive expressions for the signal-to-noise ratio of the receiver in concert with calculations (or plots) that relate the error rate to the signal-to-noise ratio. As an initial estimate, we would probably begin by assuming a pin detector for its lower cost and simplicity of circuitry, unless we suspect that the application in mind is going to require an APD. The choice of preamplifier (i.e., whether it is a low-impedance receiver, an integrating front-end receiver, or a transimpedance receiver) depends on the data rate and the amount of noise that can be tolerated. From this, we obtain the receiver power P_R necessary to achieve the required performance.

Usually, we then choose a tentative source, based on considerations from Section 7.2. With this choice of source, we know the power P_T available to be coupled into the fiber. The ratio of P_T/P_R, expressed in dB, is the amount of acceptable loss that can be incurred and still meet the specifications. This is expressed by the following equation,

$$\text{Losses(dB)} + l_M = 10 \log \left(\frac{P_T}{P_R} \right), \tag{7.1}$$

where l_M is the *system margin*.

These losses can be allocated in any desired fashion by the system designer. Generally, the probable losses will be as follows:

- The source-to-fiber coupling loss l_T (dB) at the transmitter.
- The connector insertion loss l_C for each pair of connectors (or the splice insertion loss l_S for splices). If there are n connectors or splices, then the total losses will be nl_C (or nl_S), assuming that all losses are equal.
- The fiber-to-receiver loss l_R. In a well-designed link with a fairly sizable detector, this loss is usually negligible. Often the detector size is kept to a minimum to allow fast response speeds (since decreased size corresponds to decreased capacitance of the detector).
- Allowance for device aging effects (especially for the reduction in laser power over time) and future splicing requirements l_A.
- Fiber losses, expressed as the product of the losses per kilometer, α, times the link length L.

Equation 7.1 can then be written as

$$10 \log \left(\frac{P_T}{P_R} \right) = \alpha L + l_T + nl_S + l_R + l_A + l_M . \tag{7.2}$$

After we solve Equation 7.2 for the system margin, we find

$$l_M = P_T(\text{dBm}) - P_R(\text{dBm}) - \alpha_0 L - l_T - nl_s - l_R - l_A . \tag{7.3}$$

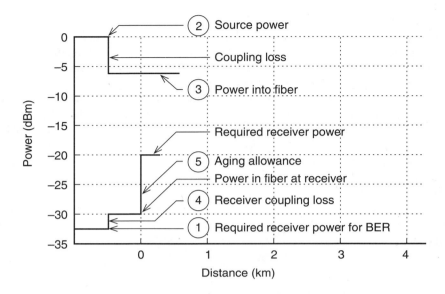

Figure 7.1 Graphical analysis of power budget. (The numbers refer to the steps of the design procedure described in the text.)

(We can also use dBμ, if desired.) A positive system margin ensures proper operation of the circuit; a negative value indicates that insufficient power will reach the detector to achieve the BER.

To illustrate the design procedure, we will use a graphical example. (Alternatively, equations can be solved by substituting values.) One uses a linear graph that plots power (in dBm or dB μ) for the vertical axis.

1. We begin by entering the required receiver power to achieve the specified BER as a horizontal line, as shown on the bottom of Figure 7.1 (labeled "1").

2. The power from the source is then plotted as a horizontal line at the appropriate power level as indicated (labeled "2").

3. The transmitter coupling loss is plotted as a vertical decrease in power level, providing the optical power in the fiber (labeled "3"). (If the optical power into the fiber is the only quantity known, as might occur with a pig-tailed optical source, this power would be plotted as the horizontal line and no coupling losses would be indicated.)

4. The receiver coupling loss (labeled "4") is measured vertically from the receiver power level and a new horizontal line is drawn, showing the required power in the fiber at the receiver.

5. This step is repeated with the aging allocation l_A (labeled "5"), producing the final value of the power in the fiber at the receiver.

A line is then drawn with a slope given by the fiber losses (see example in Figure 7.2). The line begins at the power level in the fiber and at the 0 distance position and drops downward to the right. The distance for the intersection of this line with the line representing the required receiver power is the maximum distance between the source and receiver that will achieve the BER of the specifications. Fibers with lower losses will achieve longer transmission distances; fibers with higher losses will have shorter distances. If the intersection is at a greater distance than required, the excess system margin can be used in various other ways—a higher-loss (i.e., cheaper) fiber might be used, a lower-sensitivity receiver might be substituted, or other changes can be made. If the extra system margin is retained, then the system BER will be smaller than the specified value. If the link distance is

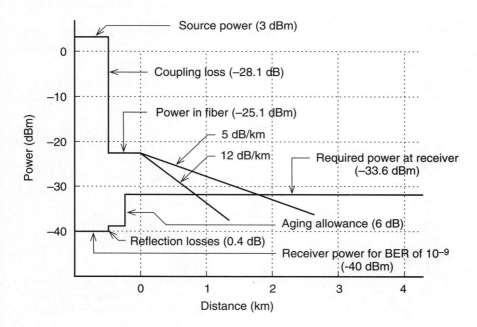

Figure 7.2 Design example.

not long enough for the specific application, then changes must be made. These could include lower-loss fiber, a more sensitive receiver, a more powerful source or some combination of changes. The graphical approach allows rapid iterative changes to the design and graphically presents the sensitivity of the design to the various changes that have been made.

Example: Let us design a system that operates at 830 nm at a data rate of 100 Mb · s^{-1}. We assume a BER of 10^{-9} so that we can use Figure 5.23. Assuming a pin-diode detector for simplicity, we find from the figure that we would require a receiver power level of approximately –40 dBm to achieve this error rate at 100 Mb · s^{-1} operating into a pin diode with a bipolar transistor amplifier. We draw the receiver power as the lowest horizontal line on Figure 7.2.

For the source we will select an LED that produces 2 mW (3 dBm) in a spot that is 225 µm in diameter. (This is representative of a commercial LED that incorporates an internal microlens to produce the spot.) A horizontal line is drawn at this value in Figure 7.2.

For the fiber, we will assume that we will use a parabolic graded-index fiber ($g = 2$) with a 50 µm core and a 125 µm outer diameter that has a numerical aperture of 0.25.

We note that the effective source radius ($r_s = 112.5$ µm) is larger than the fiber radius ($a = 25.0$ µm), so we use Equation 4.106 to calculate the coupling efficiency.

$$\eta = \left[NA(0)^2 \right] \left(\frac{a}{r_s} \right)^2 \left(\frac{g}{g+2} \right) \tag{7.4}$$

$$= (0.25)^2 \left(\frac{25}{112.5} \right)^2 \left(\frac{2}{2+2} \right) \tag{7.5}$$

$$= (0.0625)\,(0.0494)\,(0.50) \tag{7.6}$$

$$= 0.001543 \tag{7.7}$$

$$= 0.1543\% \,. \tag{7.8}$$

$$L_T = -10 \log \left(1.543 \times 10^{-3} \right) \tag{7.9}$$

$$= 28.1 \text{ dB} \,. \tag{7.10}$$

This loss is subtracted from the optical power on Figure 7.2 and a horizontal line is drawn at –25.1 dBm. (We note that the microlens has expanded the beam too much and, ideally, should be weaker in order to raise the coupling efficiency.)

To compute the losses at the receiver, we will assume that the detector area is bigger than the fiber core. The only losses, then, are the

Fresnel reflection losses at the fiber-air and air-detector interfaces (assuming that no index matching liquid is used). These losses are approximately 0.2 dB per interface for a total loss of 0.4 dB at the receiver. A horizontal line is drawn then at −39.6 dBm. We will include a representative 6 dB allowance to compensate for aging effects, resulting in a new horizontal line at −33.6 dB.

As an initial estimate, we will assume that there are no splices or connectors in the link, to get a feel for the representative length of the link with different fibers. Once we have computed these distances, we will iterate and include the splice effects. Representative losses of a series graded-index fibers might be 12 dB/km, 8 dB/km, and 5 dB/km. Using the first and third values as the extremes, we see in Figure 7.2 the resulting fiber-loss lines for the two losses.

From the intersections of the fiber-loss lines with the line representing the required power at the receiver, we find that the distances are approximately 1.7 km for the 5 dB/km fiber and 0.7 km for the 12 dB/km fiber for links with a system margin of 0 dB. We note that evaluation of Equation 7.2 would also allow us to algebraically solve for the distance L. (The reader is encouraged to do the calculations.) Note, however, that increased accuracy in the answer should not really be expected, since we are dealing with nominal values of each of the parameters.

We now allow for splices or connectors. For the distances calculated, three additional joints might be assumed as typical for the 1.7 km distance of the 5 dB/km fiber, one joint at both the transmitter and receiver and a joint splicing a 1-km piece and the remainder of the link. (The 700 meter distance with the 12 dB/km distance can be accommodated with only two joints, one at both the transmitter and receiver. We will discuss only the 5 dB/km case further.) If we assume splices with a loss of 0.1 dB per splice, then the 1.7 km distance is not changed very much, since a 0.3 dB decrement is not going to change the power required at the receiver by much. For one additional joint, there will be two connectors. For connectors with a loss of 2 dB, the diagram of Figure 7.3 applies. It is much like that of Figure 7.2, except that an additional 2 dB has been subtracted from the input power to account for the connector loss. From Figure 7.3, we note that the link distance for the 5 dB/km fiber is 1.3 km.

7.4 Dynamic Range

In Equation 7.2 we left a term for the system margin l_M. While adequate system margin should be built into the link, too much margin can cause problems with the dynamic range of the system. For example, too much optical power at the detector might saturate the receiver. A calculation of the dynamic range for system

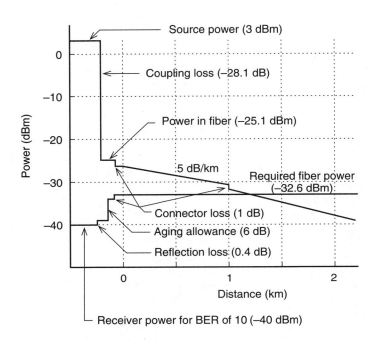

Figure 7.3 Design example.

operation may be required for some systems. From Equation 7.2, we can write the system margin l_M as

$$l_M = l_{TR} - l_{\text{system}} , \tag{7.11}$$

where l_{TR} is the ratio of the transmitter power to the required receiver power, expressed in dB (i.e., $10 \log (P_T/P_R)$), and l_{system} is the summation of the system losses, given by

$$l_{\text{system}} = \alpha_0 L + l_T + n l_s + l_R + l_A . \tag{7.12}$$

The *dynamic range* of the system is found by first computing the system margin with the maximum power ratio l_{TR} and the minimum system losses. Then, the system margin is calculated for the minimum power ratio and the maximum expected system losses. The two computations are summarized by

$$l_{M \max} = l_{TR \max} - l_{\text{system min}} \tag{7.13}$$

$$l_{M \min} = l_{TR \min} - l_{\text{system max}} . \tag{7.14}$$

The system dynamic range $DR(\mathrm{dB})$ is given by the difference in these values. Mathematically, we have

$$DR(\mathrm{dB}) = l_{\mathrm{M\,max}} - l_{\mathrm{M\,min}} \,. \tag{7.15}$$

The receiver must have an equivalent dynamic range for the system to work properly.

We are basically concerned with keeping the power at the receiver above the minimum detectable power of the detector $P_{\mathrm{R\,min}}$ and below the maximum-rated power of the detector $P_{\mathrm{R\,max}}$. We can find the power at the detector from Equation 7.2 if we calculate the power values in dBm (or dB μ). The power is

$$P_R(\mathrm{dBm}) = P_T(\mathrm{dBm}) - l_{\mathrm{system}} \,. \tag{7.16}$$

The transmitter power is a linear function of the drive current in the device. For a specified drive current there will be device-to-device variations in output power. There are similar variations in the fiber losses.

For example, the HFBR-1501 transmitter from Hewlett Packard is specified to produce a maximum power at a location 0.5 m into the fiber of –8.4 dBm with a drive current of 60 mA and to produce a minimum power of –14.8 dBm at the same current. Hewlett Packard's HFBR-3500 fiber-optic cable is specified to have a maximum loss of 0.63 dB/m at 665 nm (the wavelength of the HFBR-1501 source) with a minimum specified loss of 0.3 dB/m. The HFBR-2500 receiver used as a companion to the prior components requires a power level below –21.0 dBm to properly register a logical **0**, and requires a power level between a minimum of –21.0 dBm and a maximum of –12.5 dBm to avoid having a logical **1** confused with a logical **0**. (We will call these values $P_R(\mathbf{1})_{\mathrm{min}}$ and $P_R(\mathbf{1})_{\mathrm{max}}$, respectively.) Operating at or below the maximum power rating avoids overloading the detector. Figure 7.4 illustrates the locations of these power values.

If we assume that there are no connectors, no receiver losses, and no allowance for aging, then only the cable losses will contribute to the system losses. Since the optical power is specified at the output of a 0.5 m length of HFBR-3500 cable, we will have to represent the effective length of the link as $L - 0.5$ meters, where L is the total length of the link. The maximum/minimum conditions become

$$P_{\mathrm{Rmax}} = P_{\mathrm{T\,max}} - \alpha_{0\,\mathrm{min}}(L - 0.5) \le P_R(\mathbf{1})_{\mathrm{max}} \tag{7.17}$$

$$P_{\mathrm{R\,min}} = P_{\mathrm{T\,min}} - \alpha_{0\,\mathrm{max}}(L - 0.5) \ge P_R(\mathbf{1})_{\mathrm{min}}, \tag{7.18}$$

where the power levels are in dBm or dB μ. The first equation avoids overloading the receiver; the second equation ensures the minimum required power at the receiver.

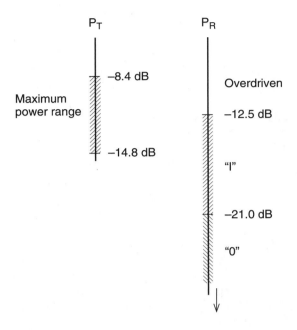

Figure 7.4 Location of power values for a Hewlett Packard data link.

Example: Let's consider a 2 m link using the Hewlett Packard components just described. We can use Equation 7.16 to calculate P_{max}.

We will assume that we transmit a logical **1**. We could be in danger of overdriving the receiver if we combined a source at maximum power output with a fiber that has minimum attenuation. We calculate the maximum transmitter power allowed in this case as

$$P_{T\,max} = P_R(\mathbf{1})_{max} + \alpha_{0\,min}(L - 0.5) \qquad (7.19)$$

$$= -12.5 + (0.3)\,(2.0 - 0.5) \qquad (7.20)$$

$$= -12.05 \text{ dBm} . \qquad (7.21)$$

Similarly, we could be in danger of underdriving the receiver when we combine the minimum transmitter power with a fiber having maximum attenuation. The expression for the minimum transmitter power allowed in this case is

$$P_{T\,min} = P_R(1)_{min} + \alpha_{0\,max}(L - 0.5) \qquad (7.22)$$

$$= -21.0 + (0.63)(2 - 0.5) \qquad (7.23)$$

$$= -20.1 \text{ dBm} . \qquad (7.24)$$

Hence the maximum transmitter power for a 2 m link is −12.05 dBm (62.3 μW) and the minimum power is −20.1 dBm (9.77 μW). From these values we now want to compute the drive current of the LED source required to produce each value of power.

For the source specified, we can produce a maximum output of −8.4 dBm (144 μW) at the maximum drive current of 60 mA. The output power of any LED is known to be linearly dependent on the drive current,

$$\frac{I}{I_{max}} = \frac{P}{P_{max}} . \qquad (7.25)$$

So, to produce the maximum allowed transmitter power of −12.05 dBm (62.3 μW), we can find the required drive current from the proportionality of

$$I = I_{max} \frac{P}{P_{max}} \qquad (7.26)$$

$$I_{f\,max} = (60 \times 10^{-3}) \left(\frac{62.3}{144}\right) \qquad (7.27)$$

$$= 26.0 \text{ mA} . \qquad (7.28)$$

The manufacturer recognizes that every device will not have the same output power at a given drive current, because of device-to-device variations. Hence the specifications indicate both the maximum power out at the maximum allowed drive current and the minimum power out at that current. The lowest rated output power at the 60 mA maximum-rated drive current is −14.8 dBm (33.1 μW). Hence, to produce the minimum output of 9.77 μW, we require a minimum drive current of

$$I = I_{max} \frac{P}{P_{max}} \qquad (7.29)$$

$$I_{min} = (60 \times 10^{-3}) \left(\frac{9.77}{33.1}\right) \qquad (7.30)$$

$$= 17.71 \text{ mA} . \qquad (7.31)$$

Thus we conclude that a drive current between the values of 17.71 and 26.0 mA will ensure proper operation of the 2 m link if the source and receiver meet specifications and if the other sources of system losses (other than fiber loss) are truly negligible.

If the same calculations are repeated for a 6m spacing, the results are a maximum optical power of -10.85 dBm and a minimum optical power of -17.35 dBm, a narrower range of values. As the distance increases, the values approach each other until the difference goes to zero. The link cannot operate at longer distances. The maximum length of the link is limited by the dynamic range of the receiver and the properties of the transmitter and fiber.

We can calculate this distance where we are dynamic-range limited by equating $P_{T\,max}$ with $P_{T\,min}$ in Equations 7.17 and 7.18. Combining these equations gives

$$P_R(\mathbf{1})_{max} + \alpha_{0\,min}(L_{max} - 0.5) = P_R(\mathbf{1})_{min} + \alpha_{0\,max}(L_{max} - 0.5) . \qquad (7.32)$$

Solving for $L_{max} - 0.5$,

$$(L_{max} - 0.5) = \frac{P_R(1)_{max} - P_R(1)_{min}}{\alpha_{max} - \alpha_{min}} \qquad (7.33)$$

$$= \frac{-12.5 + 21}{0.63 - 0.30} \qquad (7.34)$$

$$= 25.7 \text{ m} . \qquad (7.35)$$

Hence, we find that these components have a dynamic-range–limited transmission distance of 25.7 m. From Equation 7.33 we see that the way to increase this distance is to increase the receiver dynamic range (i.e., $P_R(\mathbf{1})_{max} - P_R(\mathbf{1})_{min}$) or to use a fiber with a tighter tolerance on the loss coefficient (i.e., $\alpha_{max} - \alpha_{min}$).

7.5 Timing Analysis

Once we have completed the power-budget analysis to ensure the proper maximum and minimum optical-power levels, the next step is to analyze the speed of the devices to ensure that the data-rate requirement can be met (Rodhe, 1985; Keiser, 1991). The usual technique is to analyze the rise time of the system in terms of the rise times of the components. The *rise time of the system* t_{sys} is given by the following relationship (Keiser, 1991),

$$\Delta t_{\text{sys}} = \sqrt{\sum_{i=1}^{N} \Delta t_i^2} \ , \tag{7.36}$$

where Δt_i is the rise time of each component in the system. Hence, we see that the system rise time is the square root of the sum of the squares of the system components. The four components of the system that can contribute to the system rise time are as follows:

■ The rise time of the transmitting source Δt_S. This quantity is usually found in the spec sheet of the emitter or is measured with a pulsed input and a fast detector circuit.

■ The rise time of the receiver Δt_R. This can be measured directly in the time domain with a pulsed signal or can be calculated from the bandwidth as measured in the frequency domain. If B_{3dB} is the 3-dB frequency bandwidth of the receiver, the rise time can be calculated as

$$\Delta t_R = \frac{0.35}{B_{\text{3dB}}} \ . \tag{7.37}$$

■ The material-dispersion time of the fiber Δt_{mat}. Equation 3.53 gives the dispersion relation as

$$\Delta t_{\text{mat}} = -\frac{L}{c} \frac{\Delta\lambda}{\lambda} \left(\lambda^2 \frac{d^2 n}{d\lambda^2} \right). \tag{7.38}$$

From this equation we see that the system rise time will be dependent on the length of the link (if this term is not negligible compared to the other terms).

■ The modal-dispersion time of the fiber link Δt_{modal}. This term is dependent on many variables: the excitation conditions, the fiber construction, the fiber length, and the effect of splices on the modal distribution. For a step-index fiber with length L, the modal-dispersion delay is given by

$$\Delta t_{\text{modal}} = \frac{L}{c}(n_1 - n_2) \ . \tag{7.39}$$

The modal delay time for a graded-index fiber is a more complicated expression. The delay time is a function of the index profile g and often g is optimized to reduce the delay. For a parabolic-index fiber ($g = 2$), the delay is estimated as

$$\Delta t_{modal} = \frac{L}{c} \frac{1}{8n_1^2} [NA(0)]^2 .$$ (7.40)

(In a single-mode fiber, of course, the modal delay is not present.)

In the above equations, we have assumed that there are no joints in the fiber, that the fiber was uniformly excited, and that modal equilibrium did not have a chance to occur. In an actual fiber link, any or all of these conditions might be violated. For these cases, empirical formulas have evolved which might be of use to the link designer.

For example, the bandwidth $B_M(L)$ associated with a length L of fiber can be extrapolated from the bandwidth of a 1 km length of that fiber $B_M(1$ km$)$ by the expression (Keiser, 1991)

$$B_M(L) = \frac{B_M(1 \text{ km})}{L^q} ,$$ (7.41)

where q is an empirically fit parameter. For lengths where a steady-state modal equilibrium has been reached, the value of q is 0.5. For short distances, where a steady-state modal equilibrium is *not* reached, the value of q is 1. For distances between these extremes, a reasonable estimate of q is 0.7. The relationship between the bandwidth of Equation 7.41 and the rise time of the fiber, as required by Equation 7.36, is given by

$$\Delta t_M = \frac{0.44}{B_M(L)} .$$ (7.42)

For the case where N sections of the same fiber are fastened together to form a long length, the total pulse broadening $\Delta t_M(N)$ can be estimated by (Keiser, 1991)

$$\Delta t_M(N) = \left[\sum_{n=1}^{N} (\Delta t_n)^{(1/q)} \right]^q ,$$ (7.43)

where Δt_n is the pulse spreading in the individual sections and q is the empirical value assigned as discussed in the previous paragraph.

Example: As an example of a rise time analysis, consider the 100 Mb $\cdot$ s^{-1} link that was described as the design example in the power-budget analysis previously. The LED postulated as a source might have a rise time of 8 ns and a spectral width of 40 nm. The pin diode might have a typical rise time of 10 ns.

For a silica fiber operating at a wavelength of 830 nm, the value of $\lambda^2(d^2n/d\lambda^2)$ is found to be approximately 0.024 (from Figure 3.5b). For a link distance of 2.5 km, the material-dispersion time (Equation 7.38) calculates as

$$\Delta t_{mat} = -\frac{L}{c}\frac{\Delta\lambda}{\lambda}\left(\lambda^2\frac{d^2n}{d\lambda^2}\right) \tag{7.44}$$

$$= -\left(\frac{2.5\times10^3}{3.0\times10^8}\right)\left(\frac{40}{830}\right)(0.024) \tag{7.45}$$

$$= -9.64\times10^{-9}\text{ s} \tag{7.46}$$

$$= -9.64\text{ ns .} \tag{7.47}$$

A typical intermodal dispersion for graded-index fibers is 3.5 ns/km. Hence for a 2.5 km link, we have $\Delta t_{modal} = 8.8$ ns. Calculating the system's rise time from Equation 7.36, we have

$$\Delta t_{sys}$$

$$= \sqrt{(\Delta t_S)^2 + (\Delta t_R)^2 + (\Delta t_{mat})^2 + (\Delta t_{modal})^2} \tag{7.48}$$

$$= \sqrt{8^2 + 10^2 + 9.64^2 + 8.8^2}\quad\text{(ns)} \tag{7.49}$$

$$= 18.3\text{ ns .} \tag{7.50}$$

We now must compare the system rise time with the required bit period T_B to achieve a bit rate B_R of 100 Mb $\cdot$ s^{-1}, where

$$T_B = \frac{1}{B_R}. \tag{7.51}$$

The 100 MB $\cdot$ s^{-1} bit rate of the previous example has a bit period of 10^{-8} s or 10 ns. The requirement on the system rise time depends on how we choose to encode the data. (Data encoding is described in the next section of this chapter.) Generally, the system rise time must be less than 70% of the bit period if NRZ (nonreturn-to-zero) coding is used, and it must be less than 35% of the bit period if RZ (return-to-zero) coding is used. These conditions are written as

$$\Delta t_{sys} \leq 0.7 T_B \quad\text{(NRZ coding)} \tag{7.52}$$
$$\Delta t_{sys} \leq 0.35 T_B, \quad\text{(RZ coding)}.$$

Example: The system rise time of the previous example was 18.3 ns. Using this value we can calculate the data rate that the system can support by inverting the previous equations:

For NRZ coding,

$$\Delta t_{sys} \le 0.7\, T_B \tag{7.53}$$

$$T_B \ge \frac{\Delta t_{sys}}{0.7} \tag{7.54}$$

$$B_R \le \frac{0.7}{\Delta t_{sys}} \tag{7.55}$$

$$\le \frac{0.7}{18.5 \times 10^{-9}} \tag{7.56}$$

$$\le 38.3 \text{ Mb} \cdot \text{s}^{-1}. \tag{7.57}$$

For RZ coding,

$$B_R = \frac{0.35}{\Delta t_{sys}} \tag{7.58}$$

$$= \frac{0.35}{18.3 \times 10^{-9}} \tag{7.59}$$

$$= 19.1 \text{ Mb} \cdot \text{s}^{-1}. \tag{7.60}$$

Neither coding will support the 100 Mb · s^{-1} data rate that is desired. To consider possible solutions, we look at Equation 7.48. Each term of the sum must *individually* be below the desired speed of response for the system. Obviously, the receiver speed and the material dispersion are too large. The modal-dispersion contribution is small because the distance is so short. The first step would be to use a faster detector. To reduce the material dispersion, inspection of Equation 7.38 reveals that one should reduce $\Delta\lambda$. Two methods of doing this would be

1. to use an LED with a longer wavelength (while keeping $\Delta\lambda$ constant), or
2. to use a laser source with its reduced value of $\Delta\lambda$. (The choice of a laser source would also allow increased distance in exchange for the increased cost and increased complexity.)

In either case, the power budget, as well as the timing analysis, should be recalculated to ensure proper operation of the system.

7.5.1 Dispersion-Limited Transmission Distance

The previous calculation illustrates an important result. In a fiber-optic system at long distances or high data rates, the system can be limited either by the losses (*attenuation limited*) or, assuming that the link is not limited by the source or detector speed, by the dispersion of the fiber (*dispersion limited*).

The calculation of the dispersion-limited transmission distances is easily accomplished. We ignore the rise times of the source and receiver, since we want to find the distances that are limited only by the fiber dispersion; then we isolate each dispersion factor and consider it as the sole source of dispersion.

Material-Dispersion–Limited Transmission

Consider material dispersion in a link using RZ coding. We require

$$\Delta t_{\text{mat}} \leq 0.35 T_B \, . \tag{7.61}$$

Hence,

$$0.35 T_B \geq \frac{L_{\text{max}}}{c} \frac{\Delta\lambda}{\lambda} \left(\lambda^2 \frac{d^2 n}{dn^2} \right) \tag{7.62}$$

$$L_{\text{max}} = (0.35 T_B) c \, \frac{\lambda}{\Delta\lambda} \frac{1}{\left(\lambda^2 \dfrac{d^2 n}{d\lambda^2} \right)} \, . \tag{7.63}$$

(The reader is encouraged to find the corresponding expression for NRZ coding.)

Modal Dispersion-Limited Transmission

For modal dispersion in a *step-index fiber*, we have

$$L_{\text{max}} = \frac{0.35 c T_B}{n_1 - n_2} \tag{7.64}$$

$$= \frac{0.35 c}{(n_1 - n_2) B_R} \tag{7.65}$$

$$= \frac{0.70 c n_1}{\text{NA}^2 B_R} \, . \tag{7.66}$$

For modal dispersion in a *graded-index fiber*, we have

$$L_{\max} = \frac{2.8 T_B c n_1^2}{[NA(0)]^2} \tag{7.67}$$

$$= \frac{2.8 c n_1^2}{[NA(0)]^2 B_R} . \tag{7.68}$$

(Again, the reader is encouraged to find the corresponding expression for NRZ coding.)

These latter three equations are useful for estimating the dispersion-limited transmission distances when waveguide dispersion is not significant (i.e., the results are not valid near 1300 nm).

Example: Calculate the modal-dispersion–limited transmission distance for a 50 Mb · s^{-1} data link using SI and GI fibers with $\Delta = 1\%$ and $n_1 = 1.45$. The coding is return-to-zero.

Solution:

$$n_1 - n_2 = \Delta n_1 \tag{7.69}$$

$$= 0.01(1.45) \tag{7.70}$$

$$= 0.0145 . \tag{7.71}$$

For the SI fiber,

$$L_{\max} = \frac{0.35 c}{(n_1 - n_2) B_R} \tag{7.72}$$

$$= \frac{(0.35)(3.0 \times 10^8)}{(0.0145)(50 \times 10^6)} \tag{7.73}$$

$$= 1.448 \times 10^2 \text{ m} . \tag{7.74}$$

For the GI fiber,

$$L_{\max} = \frac{2.8 c n_1^2}{\left[NA(0)\right]^2 B_R} \tag{7.75}$$

$$= \frac{2.8 c n_1^2}{2 n_1 (n_1 - n_2) B_R} \tag{7.76}$$

$$= \frac{1.4 c n_1}{(n_1 - n_2) B_R} \tag{7.77}$$

$$= \frac{(1.4)\,(3.0 \times 10^8)(1.45)}{(0.0145)\,(50 \times 10^6)} \tag{7.78}$$

$$= 840 \text{ m} . \tag{7.79}$$

The small transmission distances are caused by the use of a multimode fiber for a fairly high data-rate signal. Whether these transmission distances are acceptable depends on the application.

Attenuation-Limited Transmission

For comparison purposes, we frequently want to calculate the maximum link distance for a system limited only by the fiber attenuation. The formula for this is

$$L_{max} = \frac{P_T(\text{dBm}) - P_R(\text{dBm})}{\alpha_{fiber}} . \tag{7.80}$$

(One could also use dB μ for the units of power.) Here P_T is the power of the transmitter (usually independent of data rate, except for very high data rates), P_R is the power that the receiver requires to maintain the bit-error rate or the signal-to-noise ratio, and α is the fiber-attenuation value.

Example: Consider a graded-index fiber with $n_1 = 1.45$ and $\Delta = 1\%$ and a loss of 1 dB/km. It is used with an 850 nm source that produces an output power (in a fiber) of −10 dBm. The source linewidth is 60 nm. The receiver is a pin diode receiver that requires a power given by

$$P_R(\text{dBm}) = -65.0 + 20 \log\left[B_R(\text{Mb} \cdot \text{s}^{-1}) \right], \tag{7.81}$$

to maintain a BER of 1×10^{-9} where B_R is the data rate in Mb $\cdot$ s^{-1}. The coding is RZ. Set up the equations to find . . .

(a) . . . the material-dispersion–limited distance.

Solution: The equation for the material-dispersion–limited distance is

$$L_{max} = \frac{0.35 c \lambda}{B_R\, \Delta\lambda \left(\lambda^2 \dfrac{d^2 n}{d\lambda^2} \right)} . \tag{7.82}$$

The value of $\lambda^2(d^2n/d\lambda^2)$ is estimated to be 0.022 from Figure 3.5. Hence, we have

$$L_{max} = \frac{(0.35)\,(3.0 \times 10^8)\,(850 \times 10^{-9})}{(B_R)\,(60 \times 10^{-9})\,(0.022)}. \tag{7.83}$$

(b) ... the modal-dispersion–limited distance.

Solution: This distance is

$$L_{max} = \frac{2.8cn_1^2}{\left[NA(0)\right]^2 B_R} \tag{7.84}$$

$$= \frac{2.8cn_1^2}{2n_1(n_1 - n_2)\,B_R} \tag{7.85}$$

$$= \frac{1.4cn_1^2}{n_1^2\,\Delta\,B_R} \tag{7.86}$$

$$= \frac{1.4c}{\Delta\,B_R} \tag{7.87}$$

$$= \frac{1.4(3.0 \times 10^8)}{(0.01)\,(B_R)}. \tag{7.88}$$

(c) ... the attenuation-limited distance.

Solution: This distance is

$$L_{max} = \frac{P_T(dBm) - P_R(dBm)}{\alpha} \tag{7.89}$$

$$= \frac{-10 - (-65.0 + 20 \log{(B_R)})}{1.0}. \tag{7.90}$$

(d) Plot the results for a data-rate range from 1 kb $\cdot$ s^{-1} to 10 Mb $\cdot$ s^{-1}.

Solution: Figure 7.5 shows the plot of these curves. Note that for data rates below about 700 kb $\cdot$ s^{-1}, the link length is attenuation-limited. Above 700 kb $\cdot$ s^{-1}, the link is limited by the modal dispersion. The material-dispersion limit is slightly longer than the modal-dispersion limit.

7.6 Data Coding

In transmitting a digital signal, recovery of the data at the receiving end sometimes requires a sampling circuit that operates at the clock rate of the system. As seen in our discussion of the eye diagram in Chapter 5, timing information is required to sample the signal when the S/N ratio is maximized and to maintain proper pulse spacings. Low rate links use simple free-running clocks that require periodic resynchronization for long-term operation. The period between resynchronization can be extended by the use of crystal-controlled clocks. Many links choose to encode the clock into the data stream for recovery at the receiver. While such encoding requires increased date rates in the fiber, the advantages of being able to recover the clock can frequently outweigh this disadvantage. Another alternative is to transmit the clock signal over a separate channel. Several standard techniques have evolved for the data encoding that will be described next (Hewlett-Packard, 1978). (The various codes are represented in Figure 7.6 and Table 7.2.)

- The *nonreturn-to-zero (NRZ) code* is a code that is not required to return to the logical **0** level during the bit period. A string of logical **1**s will keep the output at the **1** level for the duration of the data string. The average (DC) output will vary with the signal content. While simple to generate and decode, this code has no clock-encoding capabilities or error detection or correction capability. This code requires the minimum bandwidth.

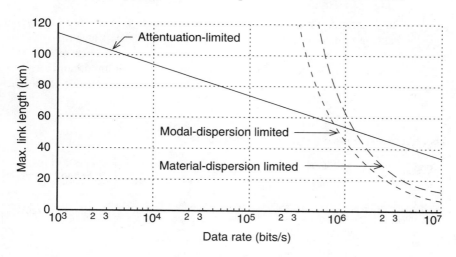

Figure 7.5 Transmission distance vs. data rate as affected by attenuation limits and dispersion limits.

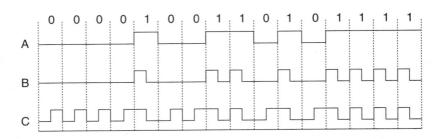

Figure 7.6 Examples of NRZ and RZ codes.

- The *return-to-zero (RZ) code* causes the output level to change from high to low (or low to high) within each bit period. (After encoding, we will call the period of the code that contains one bit of original data information the *bit period*. The rate of transmitting the original data will still be called the *data rate*; the rate of transmitting the encoded data will be the *baud rate*. The channel capacity required will be equal to the baud rate.) There are several variations of the RZ code. Curve B in Figure 7.6 uses the first half of the bit period to represent the level of the bit; the second half of the bit period is always the **0** level. We see that two bauds (or code intervals) are required to represent a bit. Again, the DC level depends on the message content, since a string of **1**s would have a different average value than a string of **0**s. No clocking information is transmitted.

- The *Manchester code* is an RZ code. As shown in Figure 7.6(c), the signal makes a transition in the center of the bit period. The transition is downward for a **0** and upward for a **1**. Since the average of each bit period is constant, the DC value of the output is constant. The important feature of the Manchester code is that it encodes the clock by the transition in the cen-

Table 7.2: Some simple digital communications codes.

Code	Description	Baud Rate (1 Mb · s^{-1} data)	Clock Encoded?	DC Level?
Nonreturn-to-zero (NRZ)	1 = High during entire bit period 0 = Low during entire bit period	1 Megabaud per Mb · s^{-1}	No	Yes
Return-to-zero (RZ)	1 = Momentarily high 0 = Low during entire bit period	2 Megabaud per Mb · s^{-1}	No	No
Manchester	1 = Positive transition 0 = Negative transition	2 Megabaud per Mb · s^{-1}	Yes	No

ter of the bit period. This transition is used at the receiver to recover the clock. (Integrated circuits are available that automatically encode and decode a data stream using the Manchester code.)

Many fiber-optic transmitter and receiver modules that are commercially available for the transmission of digital data require a specific coding of the signal (usually RZ). Inspection of the data sheets will reveal these requirements.

The increased bandwidth capability of a fiber-optic link allows increased freedom for the designer to incorporate coding for other purposes. Such coding usually inserts redundant bits into the data stream (thereby increasing the required channel capacity). This extra data can be used to encode the clock, check for errors in the data stream, or to correct errors at the receiver. These techniques frequently use *block codes*. In an *mBnB* code for example, a block (or word) of m bits is encoded into n bits ($m < n$). The encoded bits are transmitted by an NRZ or RZ code over the channel (with an increased channel-capacity requirement of n/m times the required capacity for the uncoded data). Integrated circuits are becoming available to generate various codes, primarily for error detection and correction.

Other types of data encoding are also possible with fiber links, including pulse position modulation (PPM), frequency-shift keying (FSK), or phase-shift keying (PSK). (These latter modulation formats are described in Chapter 9.) The simpler level-shifting methods are more frequently used.

7.7 Commercial Fiber-Optic Modules

For moderate data rates and moderate distance transmission, the link design reduces to the selection of commercially available transmitter and receiver modules. These devices are easily used. They provide the proper signal levels so that TTL or ECL input voltages will produce TTL or ECL output voltages. The user design is only in the selection of the proper modules and in the design of the electronic interface circuitry. The optical portion of the design has been optimized (the user hopes) by the manufacturer. Only a careful reading of the spec sheets and some experimentation (or a recalculation using our design procedure) will reveal if a particular module set can be successfully used.

7.8 Optical Amplifiers

Once the optical power in the link reaches the minimum detectable power, we need to find a way to increase the power (if we are not yet at the link destination). Two techniques have been used, repeaters and optical amplifiers.

A *repeater* consists of an optical detector, signal-recovery circuits, and an optical source. The optical detector transforms the optical signal into an electrical one. The signal-recovery circuits recover the clock from the data and detect the data stream in electrical form. The recovered data are retransmitted by the optical source. Many long-distance links use this technique, but it suffers the disadvantage of requiring a full electronic recovery system at each repeater.

An alternative approach is to amplify the optical signal without ever forming an electronic equivalent. This technique has potential economic advantages and is conceptually simple. Both semiconductor amplifiers and fiber amplifiers are being studied, with recent advances in fiber amplifiers showing great potential. The *ideal* optical amplifier would have the following properties:

- provide high gain (power gains on the order of 30 dB or more are desirable),

- have a wide spectral bandwidth (to allow several wavelengths to be transmitted simultaneously through the fiber and amplifier),

- provide uniform gain over amplifier spectral width (to maintain the relative strengths of the different spectral components),

- allow bidirectional operation (to allow bidirectional use of the optical fiber),

- add minimum noise from the amplifier,

- demonstrate lack of interference between multiple wavelengths within the spectral band (i.e., no crosstalk between spectral channels),

- provide a gain that is independent of the signal polarization (to maintain constant gain as the polarization of the signal varies due to environmental factors along the length of the single-mode fiber),

- have a low insertion loss when the amplifier is inserted into a fiber link (to avoid using significant amounts of the amplifier's gain just to overcome the insertion loss),

- have a gain that is applicable over a wide range of input power levels (i.e., a gain that does not saturate at high values of input power),

- use an amplifier pump source that is small and compact (on the order of the size of a diode laser), and

- have a good conversion efficiency (at converting the pump power input into amplifier gain).

The addition of amplifiers to a data link brings its own problems, however. The amplifiers add a different type of noise to the signal (due to their broadband

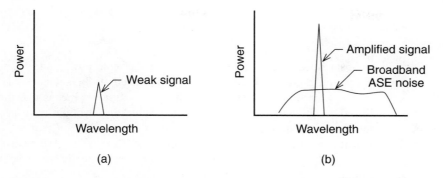

Figure 7.7 (a) Weak optical signal at amplifier input. (b) Amplified optical signal at amplifier output with amplified spontaneous emission (ASE) noise.

spontaneous emissions). The spontaneous emissions that are guided by the amplifier increase as they proceed, resulting in *amplified spontaneous emission* (ASE) noise. Figure 7.7 shows an example of the optical power entering and leaving an amplifier. At the output of the amplifier, the narrowband signal is joined by the broadband ASE noise. The added noise of an amplifier is represented by the noise figure F, defined as

$$F = \frac{\text{SNR}_{\text{in}}}{\text{SNR}_{\text{out}}}, \tag{7.91}$$

where SNR_{in} and SNR_{out} are the values of the signal-to-noise ratios at the input and output of the amplifier. (The SNRs are calculated by hypothesizing that the receiver is located at either position.)

Another problem is that the amplifier gain can be saturated by the introduction of too strong a signal. The gain coefficient g of a material (and, hence, the overall gain of a given length of the material) can be modeled (Giles and Desurvire, 1991a) by

$$g = \frac{g_0}{1 + \dfrac{P}{P_{\text{sat}}}}, \tag{7.92}$$

where g_0 is the unsaturated gain coefficient (whose value is determined by the material and the amplifier pump power), P is the strength of the optical wave being amplified, and P_{sat} is an amplifier parameter called the *amplifier saturation power*. The overall power gain of the amplifier is modeled as (Giles and Desurvire, 1991a)

$$G = G_0 \exp\left(\frac{(1-G)P_{\text{in}}}{P_{\text{sat}}}\right), \qquad (7.93)$$

where G_0 is the unsaturated gain of the amplifier, P_{in} is the input power, and P_{sat} is the amplifier saturation power. This nonlinear equation needs to be solved for G, once G_0, P_{in}, and P_{sat} are known or specified. This gain saturation means that the ASE noise will reduce the amplifier gain below the value that it might have had without the presence of the noise. The ASE noise robs the amplifier of part of its potential signal gain; the stronger the noise, the more gain it robs. The *gain saturation* is also detrimental because of nonlinear effects that can increase the crosstalk in a link carrying more than one optical signal.

Hence, the dynamic ranges of the signal and the noise in the link become important; too strong a signal will saturate the amplifier and too weak a signal will be lost in the noise. The presence of the noise leads to an interesting tradeoff. The noise becomes larger as the gain grows larger. The least noise is generated when the amplification is distributed over the entire length of the fiber (rather than occurring at discrete locations within the link). The optimum gain for this case is the gain required to just compensate for the fiber losses. In other words, the pump energy is used to make the fiber transparent (i.e., zero losses) to the signal. The problem is that we do not know how to make long lengths of weakly amplifying fiber efficiently, nor do we know how to maintain a uniform pump-power level over the entire length of the fiber (since we can introduce the pump only at the transmitter and receiver end of the fiber). The second best solution is to have a lot of small-gain amplifiers spaced closely together. This is uneconomic, since the amplifiers are currently quite expensive. To keep the total amplifier costs down, we have to space our amplifiers widely apart and accept the noise penalty.

Optical amplifiers and isolators also present problems in the use of optical time-domain reflectometers (see Chapter 13). Design efforts have begun (Nakagawa et al., 1992) to allow the amplifiers to be bypassed for measurement purposes.

Besides being used as in-line amplifiers to replace repeaters, optical amplifiers would also be useful as preamplifiers, immediately before the receiver, or to overcome losses when an optical signal passes through a 1×N splitter.

7.8.1 Erbium-Doped Fiber Amplifiers

An optical fiber doped with erbium ions has become a potential all-optical amplifier, called an *erbium-doped fiber amplifier* (Desurvire et al., 1987; Mears et al., 1987; Ainslie, 1991), operating in the 1520 to 1550 nm window of the fiber (near the fiber-optic attenuation minimum). Diode-laser pumps operating at 950 or 1480 nm are the pump source, with most amplifiers using a 1480 nm diode-laser source

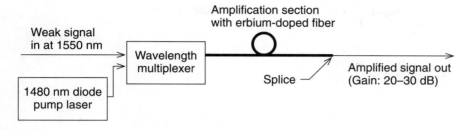

Figure 7.8 Erbium fiber laser geometry.

for a pump. Gains of 30 to 40 dB with output power levels of 1 mW (Kaminow, 1989) can be achieved in these fiber amplifiers with lengths of 10s of meters. The pump power required for such gains is a few 10s of mW, a level that is achievable with modern sources. Figure 7.8 illustrates the coupling of the pump power into the fiber amplifier. (When the pump enters the amplifier at the same end as the signal, the beams are *copropagating*. An alternative geometry is possible where the pump laser is introduced at the opposite end of the fiber amplifier and propagates in the opposite direction from that of the signal. The beams are then *counterpropagating*.) The insertion loss when the fiber amplifier is spliced into the link is low, typically less than 1 dB. Among the potential disadvantages of these amplifiers are that they are limited to operation near 1550 nm (although they are moderately broadband in the vicinity of that wavelength), that they require a non-trivial amount of pump power (40 to 50 mW) at 1480 nm, that they require lengths of fiber that exceed several meters (i.e., shorter lengths do not give as much gain), and that they add noise to the optical signal (Yariv, 1990; 1991).

Investigations to produce fiber amplifiers that operate at 1300 nm have centered on the praseodymium-doped fibers. One problem is that silica is not a good host for these ions; another host must be used. Efforts have been focused on fluoride glasses to serve as the host material. Gains of 20 to 30 dB have been demonstrated with these materials with pump-laser powers ranging from 100 to 300 mW.

The interaction between the signal beam, the amplifying medium, the pump beam, and the amplified spontaneous emission are described by models (Giles and Desurvire, 1991a; 1991b) that are beyond the scope of this text. We can, however, perform some simple analyses based on the results of those models.

7.8.2 Cascaded Amplifiers

A typical application of the amplifiers in a long-distance high–data-rate link is shown in Figure 7.9(a), where the fiber amplifiers act as in-line repeaters. The

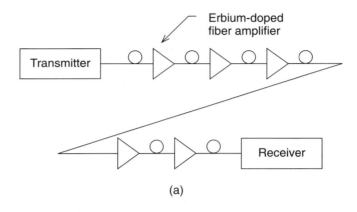

(a)

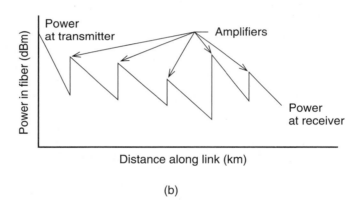

(b)

Figure 7.9 (a) Example of system using erbium-doped fiber
amplifiers. (b) Power levels in the fiber.

amplifiers serve to boost the signal strength after attenuation from traversing the
fiber between repeaters. The signal is successively amplified at each amplifier, as
shown in Figure 7.9(b). Unfortunately, the process cannot be carried out indefinitely,
as the amplifiers all add noise to the signal.

The primary noise that is added to the signal is due to light from spontaneous emissions occurring within the fiber amplifier. A portion of the spontaneous
emitted light is in the same direction as the signal and is amplified the remaining
length of the fiber. (Generally, the copropagating pump geometry will introduce less
noise than the contrapropagating pump beam.) This added light, called *amplified
spontaneous emission* or ASE, is characterized as having a wide spectrum compared

to the signal as seen in Figure 7.7(b). The spectral width of this light can be reduced by including a narrow optical filter centered on the signal wavelength. (This filter will increase the insertion loss of the amplifier, however.)

To perform our analysis we will follow the work of Giles and Desurvire (1991a). We want to consider a string of N amplifiers. To make the analysis tractable, we will assume that the amplifiers are equally spaced with an equal fiber loss of L between amplifiers. Each amplifier can have a different gain, in general, of G_i. (The size of the gain is controlled by the amount of pump power and the length of the amplifying fiber. We note, in passing, that, for a given fiber and pump strength, there is an optimum fiber length to maximize the gain.) We will assume that each amplifier will amplify the signal and the noise from the previous amplifiers and will also add its own ASE power to the output. We assume that the output of each amplifier has an optical filter of bandwidth B_0, centered on the signal wavelength, to reduce the broadband ASE noise. At the output of the i-th amplifier, the output signal power $P_{s,i,out}$ is

$$P_{s,i,out} = G_i P_{s,i,in} \tag{7.94}$$

$$= G_i L P_{s,i-1,out}, \tag{7.95}$$

where $P_{s,i,in}$ is the optical signal power entering the i-th amplifier, and $P_{s,i-1,out}$ is the signal power out of the previous amplifier. The total ASE power out of the i-th amplifier $P_{ASE,i,out}$ is given by (Giles and Desurvire, 1991a)

$$P_{ASE,i,out} = G_i P_{ASE,i,in} + b P_{ASE,i}(G_i, n_{sp}) \tag{7.96}$$

$$= G_i L P_{ASE,i-1,out} + b P_{ASE,i}(G_i, n_{sp}) \tag{7.97}$$

where $P_{ASE,i,in}$ is the total ASE power from previous amplifiers at the input to the i-th amplifier, $b = B_o/\Delta v$ (with B_o being the the bandwidth [in Hz] of the optical bandpass filter placed after each amplifier and Δv being the spectral width of the amplifying medium), and $P_{ASE,i}$ is the portion of the forward-propagating ASE power that is captured and guided down the fiber. This latter power is a function of the gain of the amplifier and n_{sp}, the ratio of the ASE power at the output of the amplifier $P_{ASE}(l)$ and $2hv \, \Delta v \, (G-1)$, i.e.,

$$n_{sp} = \frac{P_{ASE,out}}{2hv \, \Delta v \, (G-1)}. \tag{7.98}$$

This denominator is used because the expression for the ASE optical power P_{ASE} in a single mode is (Yariv, 1991)

$$P_{ASE} = \mu h\nu \, \Delta\nu \, (G - 1) , \qquad (7.99)$$

where μ is a measure of the population inversion efficiency within the amplifying medium (with a maximum value of 1). Hence, we see that n_{sp} is a normalized ASE power. In Equation 7.97, the initial values are $P_{s,0,out} = P_s$, where P_s is the signal power in the fiber at the transmitter and $P_{ASE,0,out} = 0$ (i.e., there is no ASE power present at the transmitter end of the link).

Adding these power contributions, we find that the total power out of the the i-th amplifier, $P_{total,i,out}$, is

$$P_{total,i,out} = LG_i P_{total,i-1,out} + 2n_{sp}(G_i - 1)h\nu B_o . \qquad (7.100)$$

Example: As an example of the analysis of a cascaded chain of amplifiers, we consider the case where the total output power of each amplifier is the same (i.e., $P_{total,i,out} = P_{total,in} = P_{s,0}$). We will use the following values: $P_{sat} = 8$ mW, $P_{s,0} = 9$ mW, $G_0 = 35$ dB, $LG_0 = 3$, and $n_{sp} = 1.3$. We will also assume that the optical filter has a bandwidth of $B_o = 126 \times 10^9$ Hz (or 1 nm), that the optical amplifier has a spectral width of 25 nm ($\Delta\nu = 3.10 \times 10^{12}$), and that the signal wavelength is 1545 nm.

(a) Find the value of the gain G_i required for each amplifier.

Solution: The value of gain is found from Equation 7.100 as

$$G_i = \frac{P_{s,0} + 2n_{sp}h\nu B_o}{2n_{sp}h\nu B_o + LP_{s,0}} . \qquad (7.101)$$

For the case where $P_{s,0} \gg 2n_{sp}h\nu B_0$, the equation reduces to

$$G \approx \frac{1}{L} . \qquad (7.102)$$

This gain is what is required to just balance the attenuation incurred in the fiber between stages, as we would intuitively expect. We are given that $G_0 = 35$ dB or $G_0 = 3.16 \times 10^3$, so

$$L = \frac{LG_0}{G_0} \qquad (7.103)$$

$$= \frac{3}{3.16 \times 10^3} \qquad (7.104)$$

$$= 9.49 \times 10^{-4} . \qquad (7.105)$$

The gain of each stage, then, is

$$G_i \approx \frac{1}{L} \tag{7.106}$$

$$\approx \frac{1}{9.49 \times 10^{-4}} \tag{7.107}$$

$$\approx 1.049 \times 10^3 \tag{7.108}$$

$$\approx 30.2 \text{ dB} . \tag{7.109}$$

(b) Find the value of P_{sat} required to achieve this gain.

Solution: To find the saturation power at each stage, we solve Equation 7.93 for P_{sat} to find

$$P_{sat,i} = \frac{(1 - L)P_{s,in}}{\ln(LG_0)} \tag{7.110}$$

$$= \frac{(1 - 9.49 \times 10^{-4})\,(9 \times 10^{-3})}{\ln(3)} \tag{7.111}$$

$$= 8.19 \times 10^{-3} \text{ W} \tag{7.112}$$

$$= 8.19 \text{ mW} . \tag{7.113}$$

(c) Calculate and plot $P_{s,i,out}$, $P_{total,i,out}$, and $P_{ASE,i,out}$ for each of 100 stages. These powers are, respectively, the signal power out of each stage, the total power out of each stage, and the ASE power out of each stage.

Solution: The equation for $P_{total,i,out}$ is easy; it is

$$P_{total} = P_{s,0,in} \tag{7.114}$$

$$= 9 \times 10^{-3} \text{ W} \tag{7.115}$$

$$= 9 \text{ mW} . \tag{7.116}$$

The ASE power is found from Equation 7.98 as

$$P_{ASE,i,out} = i\,(2n_{sp})\,(G_i - 1)h\nu B_0 . \tag{7.117}$$

Using an iterative approach, we can find and plot $P_{ASE,i,out}$. (One way to perform the iterative calculations is to use a spreadsheet program.) Figure 7.10 shows a plot of the resulting ASE power.

The signal power $P_{s,i,out}$ is obtained by subtracting the ASE power $P_{ASE,i,out}$ from the total power $P_{total,i,out}$. Figure 7.10 also shows a plot of the resulting power.

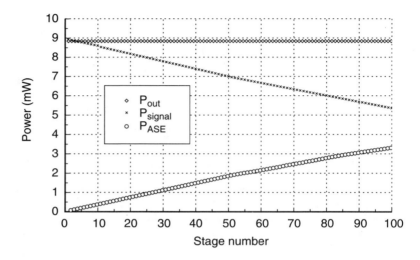

Figure 7.10 Power levels in a sample 100-amplifier fiber link (see example for link parameter values). Powers shown are the total power out of each amplifier, the signal power out of each amplifier, and the power in amplified stimulated emission at the output of each amplifier. The link is designed to keep the signal power out of each stage the same.

Looking at the figure, we note the following:

1. As required by our desired goal of keeping the total power at the amplifier outputs constant, the plot of $P_{total,i,out}$ is flat.
2. After the first amplifier, the ASE power begins to grow due to the subsequent amplification of the attenuated ASE power from the prior amplifier(s) and the addition of a fixed amount of ASE power by each amplifier.
3. The signal power steadily falls as it traverses the chain of amplifiers. The increasing ASE power is using an increasing portion of the gain, and the signal uses the declining remainder of the gain. (The signal is decreasing much slower than it would without the amplifiers, however.)
4. Eventually, the ASE power would equal or exceed the signal power, if more stages were added.

Receiver Signal-to-Noise Ratio

We have seen from our example that, as a signal passes through the amplifier chain, the signal power decreases and the ASE power grows. Eventually, we suspect, the signal will be lost in the noise; where this occurs depends on the detection of light at a receiver. The hypothetical receiver can be placed at the output of any amplifier (or anywhere else in the link).

When we detect light that is the combination of the signal and the ASE, several noise sources are possible (Yariv, 1991):

- shot noise due to the signal,
- thermal noise associated with the detector load resistance,
- noise due the mixing (or "beating") of the signal and the ASE (which produces frequencies within the electronic bandwidth of the receiver), and
- noise due to the mixing of different ASE components with each other (which can also produce frequencies within the receiver bandwidth).

The latter two sources are new noises introduced by the ASE light.

The mean-square noise currents associated with these latter noise sources are (Giles and Desurvire, 1991a)

$$<i_{sig-ASE}^2> = 2I_{ASE}I_{sig}\frac{B_e}{B_o} \tag{7.118}$$

$$<i_{ASE-ASE}^2> = I_{ASE}^2\frac{B_e}{B_o}, \tag{7.119}$$

where B_e is the electrical bandwidth of the receiver. (The electrical bandwidth can be estimated as one-half the data rate.) The currents in these expressions are found from

$$I_{ASE} = \mathcal{R}P_{ASE} \tag{7.120}$$

$$I_{sig} = \mathcal{R}P_{sig}, \tag{7.121}$$

where $\mathcal{R}$ is the responsivity of the detector at the ASE wavelengths (assumed constant over the band of ASE wavelengths) and signal wavelength, P_{ASE} is the ASE power in the optical bandwidth B_o at the receiver, and P_{sig} is the signal power at the receiver.

If we define the ratio of the optical filter bandwidth to the electrical receiver bandwidth as

$$R_B = \frac{B_o}{B_e} \qquad (7.122)$$

and the ratio of the ASE power to the signal power as

$$R_{\text{ASE}} = \frac{P_{\text{ASE}}}{P_{\text{sig}}}, \qquad (7.123)$$

then the value of the Q-parameter (defined and discussed on page 212) required to meet the link's BER requirement is (Giles and Desurvire, 1991a)

$$Q = \frac{I_{\text{sig}}}{\sqrt{I_{\text{N1}}^2} + \sqrt{I_{\text{N0}}^2}}, \qquad (7.124)$$

where I_{N1}^2 (I_{N0}^2) is the noise in the receiver when a logical **1** (logical **0**) is received. Assuming Gaussian noise and that only the ASE-signal and ASE-ASE beat noises are important, then (Giles and Desurvire, 1991a)

$$Q = \frac{2\sqrt{R_B}}{R_{\text{ASE}} + \sqrt{4R_{\text{ASE}} + R_{\text{ASE}}^2}}. \qquad (7.125)$$

We now want to illustrate the application of this formula.

Example: Consider a link that uses a cascaded series of optical amplifiers with the properties of the previous example on page 319. The data rate is to be 2.5 Gb $\cdot$ s^{-1} and the BER is to be 10^{-9}. We want to see if we will have enough signal power (compared to the ASE noise) to achieve the desired BER after traversing 100 amplifiers and the fiber between them.

Solution: The electrical bandwidth is estimated as

$$B_e \approx \frac{B_R}{2} \qquad (7.126)$$

$$\approx \frac{2.5 \times 10^9}{2} \qquad (7.127)$$

$$\approx 1.25 \times 10^9, \qquad (7.128)$$

and the ratio R_B is

$$R_B = \frac{B_o}{B_e} \tag{7.129}$$

$$= \frac{126 \times 10^9}{1.25 \times 10^9} \tag{7.130}$$

$$= 100.8 . \tag{7.131}$$

From Table 5.3, we find that a BER of 10^{-9} requires a Q of 6.0, so

$$Q = \frac{2\sqrt{R_B}}{R_{ASE} + \sqrt{4R_{ASE} + R_{ASE}^2}} \tag{7.132}$$

$$6.0 = \frac{2\sqrt{100.8}}{R_{ASE} + \sqrt{4R_{ASE} + R_{ASE}^2}} \tag{7.133}$$

$$R_{ASE} = 1.046 . \tag{7.134}$$

Figure 7.11 shows a plot of R_{ASE} for the cascaded amplifiers of the previous problem. We find that the value of R_{ASE} stays below 70% after 100 stages. We could add appreciably more stages before we would achieve the ratio of 104% where the BER would reach a value of 10^{-9}.

From these examples (and problems at the end of the chapter), we find that each cascaded amplifier allows the signal power to be increased. The penalty imposed for the amplification is an addition of ASE noise. Eventually, the ASE-to-signal power ratio is appreciable and exceeds the upper limit required by the BER and data rate of the link.

7.8.3 Cascaded Amplifiers: Other Effects

An additional effect of N cascaded amplifiers will be an accompanying decrease in the bandwidth of the channel. Any element with a bandpass response will contribute to this effect; hence, the amplifier spectral response and the optical filter response contribute to the bandwidth reduction. If the responses are mathematically modeled, the net effect is that the value for B_o that is used in the calculations must be replaced by an effective bandwidth for the cascaded units. This effective bandwidth will reduce the noise contributions from the ASE noise of previous stages; only amplifiers later in the chain contribute the most noise (Giles and Desurvire, 1991a). In addition, nonuniform spectral response complicates the spectrum of the ASE noise; some parts are attenuated more heavily than other parts.

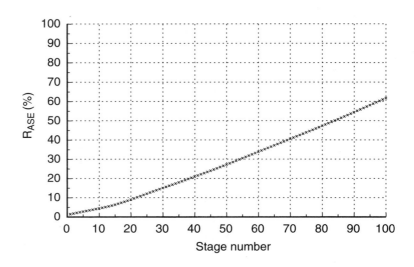

Figure 7.11 Ratio of P_{ASE} to $P_{out,i,signal}$ for cascaded amplifiers in a fiber-optic link (see example for link specifications). This ratio is R_{ASE}.

Complicated models based on measured spectral response can be used (Giles and Desurvire, 1991b).

Also, large amplifier gains call for the use of optical isolators after the amplifiers to keep reflections from being amplified and upsetting the data link. These optical isolators do not allow bidirectional links and do not allow optical time domain reflectometers (discussed in Chapter 13) to operate without a special apparatus to subvert the isolators, when desired.

Several demonstration projects have been done showing long-distance transmission (Giles et al., 1989; Urqhart and Whitley, 1990; Saito et al., 1991; Nakagawa et al., 1992) and in the transmission of solitons. (*Solitons* are pulses of moderate power that induce nonlinear interactions in the fiber to counter the dispersion in the fiber, allowing ultra-low dispersion transmission. Multi–gigabit-per-second signals have been sent through millions of kilometers of [simulated] fiber, using solitons.) Work is proceeding to incorporate fiber amplifiers in undersea telecommunication cable systems.

7.8.4 Diode Amplifiers

Semiconductor diode amplifiers can also be used to amplify the light (O'Mahony, 1988). One of the problems, however, is that amplitude light modulation in one channel (if present) can modulate the gain characteristics for other channels, causing crosstalk. Nonlinearities in the gain medium can also affect the amplification process (Kaminow, 1989). Additionally, the insertion losses of these diode amplifiers tend to be 6 dB or more. Due to the increased popularity of the erbium-doped fiber amplifiers, few applications of diode amplifiers are available.

7.9 Summary

In this chapter, we have considered the methodology used in the design of a fiber-optic data link. Since a link can be either attenuation-limited or dispersion-limited, calculations of the power budget and the system rise time are required to assure that each meets the performance requirements of the link (as usually expressed by the desired data rate, BER, and approximate link distance). Additionally, we have seen that the channel capacity required to meet the data rate performance depends on the type of data encoding used—for either clock encoding or error correction.

Optical amplifiers, especially erbium-doped fiber amplifiers, allow us to amplify the signal without detecting it and electronically regenerating it, as is done in repeaters. Although the signal is amplified, we found that noise is also generated within the amplifier and subsequently amplified by the following amplifiers. The signal-to-noise ratio gradually decreases as the signal proceeds through a chain of concatenated amplifiers until the bit–error-rate requirement can no longer be met. These amplifier techniques are proving increasingly popular in long-haul high–data-rate systems.

7.10 Problems

1. Consider a 50/125 step-index fiber with a fractional index difference of 1.5% with $n_1 = 1.45$. Calculate the approximate bandwidth-distance product at 820 nm of this fiber if the chromatic pulse spreading is $t_M = 1.5$ ps/km (and is dominant).

2. A short-range system is to operate at 700 Mb $\cdot$ s^{-1} at a 1500 nm wavelength. Choose a source and detector for the system and explain your choices.

3. A system operates at 1550 nm with the following components:
   □ LED: –13 dBm coupled into a fiber pigtail (of negligible length).
   □ Pin diode: –45 dBm incident power is required in the detector's connectorized pigtail to achieve link requirements.
   □ Fiber: 1.3 dB/km losses. The fiber is available in any length up to 1 km.
   □ Connectors: 2 dB loss.

Calculate the maximum length of the link as determined by attenuation, allowing 6 dB margin for aging, effects, etc.

4. Consider a system operating at a 1300 nm wavelength at a 100 Mb $\cdot$ s^{-1} data rate with the following components:

□ Transmitter: InGaAs LED producing 50 µW into a fiber pigtail. The spectral width is 40 nm.

□ Connectors: 1 dB losses.

□ Fiber:

● Losses = 1.5 dB/km.

● Intermodal bandwidth-distance product as measured in a 1 km sample is 800 MHz- $\cdot$ km.

● Mode mixing—moderate mode mixing observed (q = 0.7).

● Lengths available—any length is available.

□ Receiver:

● InGaAs pin photodiode with a connectorized pigtail.

● Sensitivity required for a BER = 10^{-9} is given by

$$P_R(\text{dBm}) = (11.5 \log B_R) - 60.5,$$

where P_R is in dBm and B_R is the data rate *in Mb $\cdot$ s*$^{-1}$.

● Allowance for aging, etc.: 6 dB.

● Coding: NRZ.

(a) Calculate the maximum link distance L assuming that the link is attenuation-limited.

(b) Calculate the maximum link distance L assuming that the link is dispersion-limited.

5. A receiver has the following performance:

□ $P_R < -30$ dBm is a logical **0**.

□ -30 dBm $< P_R < -15$ dBm is a logical **1**.

□ $P_R > -15$ dBm is not allowed due to detector saturation.

This receiver is to be used with a fiber that has an attenuation specified as being between a minimum of 3 dB/km and a maximum of 4 dB/km. If the source has a power into the fiber of –3 dBm, calculate the dynamic range (in dB) of a 2 km link (neglecting all other losses except fiber losses).

6. Consider a link that uses a source with a rise time of 1 ns and a receiver with a rise time of 2 ns operating at a data rate of 50 Mb · s^{-1} with RZ coding.

   ☐ If the link has moderate mode conversion ($q = 0.7$), find the maximum distance between the source and receiver for a modal-dispersion–limited fiber (i.e., material dispersion is negligible). Assume that the measured bandwidth of a 1 km sample of this fiber is 400 MHz.

   ☐ If a single-mode fiber is used and the source operates at a wavelength of 900 nm with a spectral linewidth of 3 nm, calculate the dispersion-limited distance between this source and this receiver.

7. In the example of a cascaded chain of equally spaced fiber amplifiers, we assumed that the *total* output power of each amplifier was constant. Consider the case where the *signal* power out of the amplifiers is kept constant. The output signal power is

   $$P_{s,i,out} = P_{s,0,in} \, ,$$

   and, hence, $LG_i = 1$ and

   $$P_{total,i,out} = P_{s,0,in} + 2in_{sp}G_i\frac{hc}{\lambda}B_o \, .$$

   The following parameters apply: $LG_0 = 3$, $G_0 = 35$ dB, $P_{s,0,in} = 5$ mW, $n_{sp} = 1.3$, $\lambda = 1545$ nm, and $B_o = 1$ nm ($= 126$ GHz).

   (a) Show that the saturation power required at the i-th stage is (Giles and Desurvire, 1991a)

   $$P_{sat,i} = \frac{1-L}{\ln(LG_0)}\left[P_{s,0,in} + 2in_{sp}\left(\frac{1}{L} - 1\right)\frac{hc}{\lambda}B_o\right].$$

   (b) Using a computer, plot $P_{s,i,out}$, $P_{sat,i}$ and $P_{ASE,i,out}$ as a function of the stage number for 100 stages (similar to Figure 7.10).

8. Consider the prior problem, where the output signal power of each amplifier is kept constant.

   (a) Using Table 5.3, find the value of R_{ASE} required to achieve a BER of 10^{-14} for a data rate of 2.5 Gb · s^{-1}.

   (b) Using your results of the previous problem, find the number of amplifiers that can be used in the link.

(c) If the fiber loss is 0.5 dB · km^{-1}, calculate the distance between amplifiers (using the values of the previous problem) and the end-to-end distance of the link.

9. In the example of a cascaded chain of equally spaced fiber amplifiers, we assumed that the *total* output power of each amplifier was constant. Consider the case where there is no attempt made to regulate the output power of the amplifiers, but the saturation power $P_{sat,i}$ of each amplifier is equal. The following parameters apply: $LG_0 = 3$, $G_0 = 35$ dB, $P_{sat} = 8$ mW, $P_{s,0,in} = 1$ mW, $n_{sp} = 1.3$, $\lambda = 1545$ nm, and $B_o = 1$ nm ($= 126$ GHz).

(a) Using a computer, find and plot the value of the saturated gain G_i at each stage of a 100-stage amplifier chain.

(b) Using a computer, plot $P_{s,i,out}$, $P_{total,i,out}$, and $P_{ASE,i,out}$ as a function of the stage number for 100 stages (similar to Figure 7.9).

10. An optical filter used after a fiber amplifier has a spectral response $f(\lambda)$ that is modeled by (Giles and Desurvire, 1991a)

$$f(\lambda) = \frac{1}{1 + \left(\dfrac{\lambda - \lambda_c}{B_1}\right)^6},$$

where λ_c is the center wavelength of the passband and B_1 is the 3-dB spectral width of the filter.

(a) The frequency response of a cascade of N filters is equal to $[f(\lambda)]^N$. Using a computer, plot the spectral response of a cascade with N equal to 1, 2, 5, 10, 20, 50, and 100 if $B_1 = 1$ nm. (The horizontal axis should be $\lambda - \lambda_c$ and should extend from -1 nm to $+1$ nm. The vertical axis should be in dB (relative to the response at the center wavelength) and should extend from 0 to -20 dB.)

(b) The 3-dB spectral bandwidth after N stages is reduced by a factor of $(\ln(2)/N)^{1/6}$ (Giles and Desurvire, 1991a). Estimate (from your plot) and calculate the spectral bandwidth after 50 and 100 filters.

References

Ainslie, B. J., "A review of the fabrication and properties of erbium-doped fibers for optical amplifiers," *J. Lightwave Technology*, vol. 9, no. 2, pp. 220–227, 1991.

Desurvire, E., J. Simpson, and P. Becker, "High-gain erbium-doped traveling-wave fiber amplifier," *Optics Letters*, vol. 12, no. 11, p. 888, 1987.

Giles, C., E. Desurvire; J. Talman; J. Simpson; and P. Becker, "2 Gbit/s signal amplification at $\lambda = 1.53$ µm in an erbium-doped single-mode fiber amplifier," *J. Lightwave Technology*, vol. 7, no. 4, p. 651, 1989.

Giles, C. and E. Desurvire, "Propagation of signal and noise in concatenated erbium-doped fiber optical amplifiers," *J. Lightwave Technology*, vol. 9, no. 2, pp. 147–154, 1991a.

Giles, C. R. and E. Desurvire, "Modeling erbium-doped fiber amplifiers," *J. Lightwave Technology*, vol. 9, no. 2, pp. 271–283, 1991b.

Hewlett Packard. *Digital data communications with the HP fiber optic system*. Hewlett-Packard, Palo Alto CA, 1978. Technical report.

Kaminow, I. P., "Non-coherent photonic frequency-multiplexed access networks," *IEEE Network*, pp. 4–12, 1989.

Keiser, G., *Optical Fiber Communications,* 2nd Edition. New York: McGraw-Hill, 1991.

Mears, R.; J. Jauncey; and D. Payne, "Low-noise erbium-doped fiber amplifier operating at 1.54 µm," *Electronics Letters*, vol. 23, no. 19, p. 1206, 1987.

Nakagawa, K.; K. Aida; K. Aoyama; and K. Hohkawa, "Optical amplification in trunk transmission networks," *IEEE Lightwave Telecommunications Systems (LTS)*, vol. 3, no. 1, pp. 19–26, 1992.

O'Mahony, M., "Semiconductor laser optical amplifiers for use in future fiber systems," *J. Lightwave Technology*, vol. 6, no. 4, pp. 531–544, 1988.

Rodhe, P. M., "The bandwidth of a multimode fiber chain," *J. Lightwave Technology*, vol. LT-3, no. 1, pp. 145–154, 1985.

Saito, S.; T. Imai; and T. Ito, "An over 2200-km coherent transmission experiment at 2.5 Gb/s using erbium-doped-fiber in-line amplifiers," *J. Lightwave Technology*, vol. 9, no. 2, pp. 161–169, 1991.

Urqhart, P. and T. Whitley, "Long-span fiber amplifiers," *Applied Optics*, vol. 29, no. 24, p. 2503, 1990.

Yariv, A., "Signal-to-noise considerations in fiber links with periodic of distributed optical amplification," *Optics Letters*, vol. 15, no. 19, p. 1064, 1990.

Yariv, A., *Optical Electronics,* 4th Edition. New York: Holt Rinehart and Winston, 1991.

Chapter 8

Fiber-Optic Networks

8.1 Introduction

In recent years several emerging trends have been challenging the telecommunication industry:

- Telecommunication volume and requirements demand an ever increasing bandwidth. Table 8.1 shows some representative bandwidth services offered (or planned) by telecommunications companies to meet these expanding requirements.
  In the computer world, the data rate of local area networks has also been constantly increasing. Table 8.2 indicates the increasing data rates of some present and proposed standard computer interconnections.

- Telecommunications systems now carry a mixture of analog signals (e.g., voice, TV channels) and digital data (e.g., computer interchanges). While the communications trunk system is thoroughly digitized, the user connections are typically analog (i.e., a twisted pair of wires connected to the microphone and speaker in the handset). Worldwide efforts are being made to provide total digital service throughout the network as the proportion of digital traffic increases steadily.

Table 8.1: Bandwidth requirements for various data service needs.

Data Transmission Rates	Application
300 b/s	modems
1200/2400/9600 b/s	modems
1.544 Mb/s	T1 data transmission or
	ISDN (23 64-kb/s channels and 1 16- [or 14]-kb/s channel)
44.783 Mb/s	T3 data transmission

- The telecommunications industry is increasingly global in scope. Worldwide standards need to be agreed upon so that national telecommunications systems can be interconnected.

- Because of the advent of personal computers and the increasing popularity of workstations, there is currently a trend away from centralized computers to a confederation of "peer" computers (or servers with clients) and their peripherals. Along with this trend comes the need for intercommunication of data between the elements of the confederation.

- New trends toward video conferencing and picture transmission are arising that will tax the current bandwidth capabilities of communications networks. Digital picture transmission in real time is a voracious consumer of bandwidth.

Table 8.2: Data rates of representative computer networks.

Data Rate	Network
9600 b/s	RS232 printer interface
4 or 16 Mb/s	Token Ring Network
10 Mb/s	Ethernet
50 Mb/s	coax Hyperchannel
100 Mb/s	fiber Hyperchannel
100 Mb/s	FDDI
600 Mb/s	GaAs FDDI (proposed)
800 Mb/s	HSC channel (used in Cray supercomputers)

■ Also, increased bandwidth is required for token passing (or random access strategies), greater complexity in packet headers, use of cyclic error-correcting (CRC) codes, greater complexity of network control overhead as network use increases, higher security requirements, and requirements for graceful network degradation.

Standards need to be set to allow interconnection of units on a network. Users are unwilling to commit major capital resources to a network unless assured that the units will intercommunicate and will communicate with networks of other users.

Networks can be divided into the following roughly defined categories:

1. *Local Area Networks (LANs)* interconnect users within a department, a building, or a campus.
2. *Metropolitan Area Networks (MANs)* interconnect users within a city or the metropolitan area surrounding a city.
3. *Wide Area Networks (WANs)* can interconnect users within a wide geographic area (e.g., within a country).

The wide bandwidth offered by optical fibers is helping to lead the way into innovative data networks (Cochrane and Brain, 1988) that will be capable of carrying the mixture of traffic demanded by the users, including video signals, voice, and high-speed data. Multiplexing techniques (Su et al., 1986) can include time-division multiplexing (TDM), frequency-division multiplexing (FDM), code-division multiple-access multiplexing (CDMA) (Lee and Un, 1987; Foschini and Vannucci, 1988; Salehi, 1989; Salehi and Brackett, 1989), and wavelength–division multiplexing (WDM). (The latter topic is considered in more detail in Chapter 10.) The protocols that are used to control access to the network can include (Personick, 1985) central control (frequently with controlled switches (Midwinter, 1988)), token passing (Gerla et al., 1985), and contention arbitration schemes similar to the carrier-sense multiple access/collision detection (CSMA/CD) scheme used in Ethernet (Rawson, 1985; Reedy and Jones, 1985).

In this chapter we want to introduce some of the topologies that are possible for fiber-optic networks and to consider the tradeoffs that network designers must analyze to meet their goals. This is again done through the power budget analysis. This analysis shows that the star configuration has some advantages over the linear data bus for applications with many users. We then introduce two new standards that are evolving for fiber-optic networks, the Fiber Distributed Data Interface (FDDI) for local area networks and the Synchronous Optical NETwork (SONET) for interoperability of equipment on wide area networks.

8.2 Network Topologies

A few topologies of networks have become popular. These include the star network, the linear bus network, and the ring network (see Figure 8.1).

In the *star network* (Figure 8.1[a]), the transmission lines are brought together at a common point and distributed among the other lines, and the receiver lines branch out from that point. The star has the disadvantage of requiring a central node, which can disrupt the entire network if damaged. It has the advantage of making it relatively simple to add terminals to the network, especially if spare lines have been designed into the node at its inception.

In the *linear bus network* (Figure 8.1[b]), the network takes the form of a backbone, with individual stations receiving or adding data as required. In electrical form it is very easy to add stations by tapping into the line; in fiber form it is more difficult to splice in a station, as the network must be shut down.

In the *ring network* (Figure 8.1[c]), the stations are arranged into a continuous circuit. Each station receives a message from its upstream neighbor and, if it is not the addressee, repeats the message to its downstream neighbor. The topology is characterized by its closure at the ends. It suffers from the disadvantage that damage to any station on the ring will disrupt the entire network unless bypass switches or network redundancy are built in.

Other topologies exist, as do combination topologies. Figure 8.2 shows a "leaf" topology that combines the linear bus and star networks. Such a configuration would be useful for interconnecting isolated groups of users with fairly large distances between them.

8.3 Network Design Tradeoffs

We now want to consider some of the tradeoffs that are needed to analyze networks, so we will compare the star network and the linear bus network.

8.3.1 Data Bus Power Budget

The power budget for a data network (Barnoski, 1981; Keiser, 1991) is important because we have to ensure that we have enough power in the source to reach the farthest station without saturating the nearest station. (This is the dynamic range of the source.) We will also find that, everything else being equal, the star network has an inherent lower loss advantage over the linear bus network as the number of stations grows large.

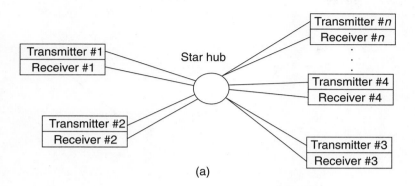

(a)

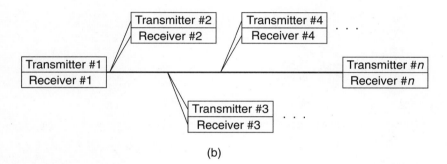

(b)

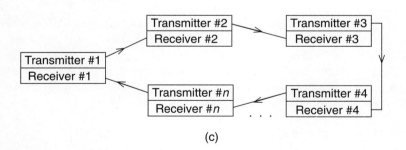

(c)

Figure 8.1 Typical network configurations. (a) A star network.
(b) A linear bus network. (c) A ring network.

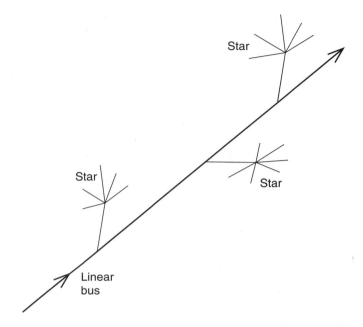

Figure 8.2 A combination of a linear bus network and star networks into a "leaf" topology.

Star Network Power Budget

We begin by calculating the power budget for a star network. Consider the power entering a fiber at the input to a coupler as P_F and the power required by the receiver is P_R. We will assume an insertion loss of L_{insert}, a power splitting loss of $L_{\text{pwr split}}$, a connector loss of L_C, a system margin of L_M, and a fiber loss of α (dB/km). If L is the distance from the star coupler to each station, the path of the power from the transmitter to the receiver would be as follows:

- P_F in the fiber at the transmitter,
- a fiber loss of αL in going from the transmitter to the star,
- a loss of L_C as the light passes through the fiber/coupler connector pair to enter the star coupler,
- a loss of $L_{\text{pwr split}}$ as the power is divided among the output fibers of the star,
- a loss of L_C as the power passes through coupler/fiber connector pair leaving the star coupler,

- an additional loss of L_{insert} due to the insertion loss of the star coupler, and
- a fiber loss of αL in going to the receiver. (We assume ideal coupling into the receiver.)

Adding up all of these dB losses, the link power budget would be

$$10 \log \left(P_F/P_R \right) = L_{\text{pwr split}} + \alpha(2L) + 2L_C + L_{\text{insert}} + L_M \qquad (8.1)$$

$$= 10 \log N + \alpha(2L) + 2L_C + L_{\text{insert}} + L_M \qquad (8.2)$$

where L_M is the link margin. From the right-hand side of Equation 8.2, we note that, as additional stations are added, the loss increases as $10 \log N$, where N is the total number of stations.

Linear Bus Power Budget

We now consider a linear data bus made up of tee couplers. Each station has two couplers. One coupler (the "tap out" coupler [shown in Figure 8.3] couples a fraction C_T of the incident light into the "local out" arm. The other "tap in" coupler adds the "local in" signal to the data bus. The losses include the following:

- The transmission for light in the fiber bus passing either tap will be $1 - C_T$; expressed as a loss, this is $L_{\text{thru}} = -10 \log (1 - C_T)$. Since each coupler has two taps, the total transmission loss for light passing through the coupler on the data bus will be $2L_{\text{thru}}$.
- Light that is coupled from the data fiber to the receiver arm will have a transmission factor of C_T; the loss is given by $L_{\text{tap}} = -10 \log (C_T)$.

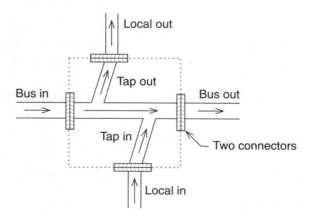

Figure 8.3 Tee coupler parameters.

- Light that comes from the "tap in" arm is assumed to add its power to the bus with C_T coupling efficiency. The loss it encounters is also $L_{tap} = -10 \log (C_T)$.

- There is a connector loss of L_C for each pair of connectors associated with coupler.

- Finally, there is an insertion loss, L_{insert}, associated with the device to account for the excess losses.

We want to consider a total of N stations in the linear bus, each separated a distance L by fibers with losses given by α (in dB/km). The smallest losses are encountered when transmitting from one station to its nearest neighbor station that is located a distance L away. The largest losses will occur when a station transmits to the N-th station, located a distance of $(N-1)L$ away. We want to write a power budget for each of these cases.

We begin by calculating the power transmitted from adjacent stations, say from station #1 to station #2. We let P_F be the power available into the fiber from the source and neglect any coupling loss from the source fiber (typically a fiber pigtail) into the coupler fiber. The power path in this case is the following:

- We begin with the power P_{F1} being coupled into the transmitter arm of coupler #1.

- There is a connector pair when entering coupler #1 (loss of L_C).

- The power couples to the data bus with C_T efficiency (loss of L_{tap}).

- There is a connector pair when leaving coupler #1 (loss of L_C).

- There is an insertion loss associated with the "local in" to "bus out" path (loss of L_{insert}).

- There is a fiber loss between adjacent stations (loss of αL).

- There is a connector pair upon entering coupler #2 (loss of L_C).

- The power couples into the receiver arm of coupler #2 (loss of L_{tap}).

- There is a connector pair on leaving the receiver arm of the coupler on the way to the detector (loss of L_C).

- There is an insertion loss associated with the "bus in" to "local out" path (loss of L_{insert}).

Thus, the power budget for this path is written as,

$$10 \log (P_{F1}/P_{R2,1}) = \alpha L + 2L_{tap} + 4L_C + 2L_{insert} , \tag{8.3}$$

where $P_{R2,1}$ is the power received at station #2 from station #1.

The power budget for the case of going from transmitter #1 to the receiver on station #N is determined from the following path analysis:

- We begin with power P_{F1} in the fiber at the transmitter.
- There is a connector pair upon entering coupler #1 (loss of L_C).
- The power couples to the data bus with C_T efficiency (loss of L_{tap}).
- There is a connector pair when leaving coupler #1 (loss of L_C).
- There is an insertion loss associated with the "local in" to "bus out" path (loss of L_{insert}).
- There are $(N-1)$ pieces of fiber, each of length L (loss of $[N-1]\alpha L$).
- There are $(N-2)$ couplers between the first and last station, and *each* coupler will introduce a transmission loss (losses of $[L_{insert} + 2L_C + 2L_{thru}]$).
- There is a connector pair upon entering coupler #N (loss of L_C).
- The power couples into the receiver arm of coupler #N (loss is L_{tap}).
- There is a connector pair on leaving the receiver arm of the coupler on the way to the detector (loss of L_C).
- There is an insertion loss associated with the "bus in" to "local out" path (loss of L_{insert}).

The power budget for this link is

$$
\begin{aligned}
10 \log \left[(P_{F1}/P_{RN,1}) \right] = (N-1)\alpha L & \quad \text{(the fiber losses)} \\
+ (N-2)(2L_C) + 4L_C & \quad \text{(the connector losses)} \\
+ (N-2)(2L_{thru}) & \quad \text{(the coupler losses)} \\
+ (N-2)L_{insert} + 2L_{insert} & \quad \text{(the insertion losses)} \\
+ 2L_{tap} & \quad \text{(data bus and rcvr arm coupling losses)} \\
= N \left(\alpha L + 2L_c + L_{insert} + 2L_{thru} \right) & \\
- \alpha L + 2L_{tap} - 4L_{thru} \, . & \qquad\qquad (8.4)
\end{aligned}
$$

From this last equation (with the exception of the constant terms), we see that the losses of the linear bus increase *linearly* with an increasing number of stations, N.

The linear increase for the linear bus compares with a log N increase for the star bus, implying that systems with a large number of stations should use the star configuration. The actual number of stations for which the star configuration becomes advantageous depends on the relative value of the different losses of the busses.

Example:

(a) Consider a star network with a connector loss L_C of 1.5 dB per connector pair and an insertion loss (for each channel) of 0.75 dB. Calculate the system losses for $N = 3$ and $N = 50$ stations on the fiber. Ignore the fiber loss and the system margin.

Solution: The system loss for a star network is

$$L_{total} = 10 \log (N) + \alpha(2L) + L_{insert} + 2L_C + L_M \qquad (8.5)$$

$$= 10 \log (N) + 0 + 0.75 + 2(1.5) + 0 \qquad (8.6)$$

$$= 10 \log (N) + 3.75 \qquad (8.7)$$

$$= 8.52 \text{ dB} \quad \text{(for } N = 3\text{)} , \qquad (8.8)$$

$$= 20.7 \text{ dB} \quad \text{(for } N = 50\text{)} . \qquad (8.9)$$

(b) Consider a linear data bus that taps 10% of the light into the arms of the tee couplers that are used. The insertion loss per tee coupler is 0.5 dB. Calculate the system losses for $N = 3$ and $N = 50$ stations on the fiber. Ignore the fiber loss and the system margin.

Solution: For the linear network, we first find

$$L_{tap} = -10 \log (C_T) \qquad (8.10)$$

$$= -10 \log (0.10) \qquad (8.11)$$

$$= 10 \text{ dB} \quad \text{and} \qquad (8.12)$$

$$L_{thru} = -10 \log (1 - C_T) \qquad (8.13)$$

$$= -10 \log (0.90) \qquad (8.14)$$

$$= 0.458 \text{ dB} . \qquad (8.15)$$

The system loss for a linear network is

$$L_{total} = N\left(\alpha L + 2L_c + L_{insert} + 2L_{thru}\right)$$
$$- \alpha L + 2L_{tap} - 4L_{thru} \qquad (8.16)$$

$$= N(0 + 2(1.5) + (0.5) + 2(0.458))$$
$$- 0 + 2(10)) - 4(0.458) \qquad (8.17)$$

$$= 31.4 \text{ dB} \quad \text{(for } N = 3\text{)} , \qquad (8.18)$$

$$= 239 \text{ dB} \quad \text{(for } N = 50\text{)} . \qquad (8.19)$$

(c) Plot the losses of both networks as a function of the number of stations on the network.

Solution: Figure 8.4 compares the losses of the star configuration and linear configuration for all values of N between 3 and 50. We note that the losses of the linear network continually increase as stations are added to the network, but the losses of a star configuration level off and increase at a much slower rate. This tendency is generally true of linear and star networks and provides an advantage to star configurations.

Dynamic Range of Receivers in Networks

Local area networks or data buses are susceptible to dynamic-range limitations because of the wide range of possible distances between stations. Since $P_{R2,1}$ (the power received at station #2 from its next neighbor, station #1) is the maximum power received at a receiver, and $P_{RN,1}$ (the power received at station #N from station #1) is the minimum power between stations, we can compute the required *dynamic range DR* of the receiver from the ratio of these quantities,

$$DR = 10 \log \left(\frac{P_{R2,1}}{P_{RN,1}} \right) \tag{8.20}$$

$$= 10 \log P_{R2,1} - 10 \log P_{RN,1} . \tag{8.21}$$

We now can find the dynamic range required by the receivers of a linear tee network. Subtracting Equation 8.5 from Equation 8.3, we find

$$DR = -2\alpha L + 4L_{insert} + 4L_{thru} + 4L_C - N(\alpha L + 2L_C + L_{insert} + 2L_{thru}) . \tag{8.22}$$

Assuming that all receivers on the network are the same, each must have the dynamic range of Equation 8.22 to avoid being saturated by a nearby transmitter or underdriven by a far away transmitter.

8.4 Proposed Standard Fiber Networks

Prior to the breakup of AT&T in 1984, the Bell system was such a dominant force in US telecommunications that it set *de facto* standards. Since then, setting of standards for telecommunications networks in the United States has followed a different path, involving many operating companies, long-distance service suppliers,

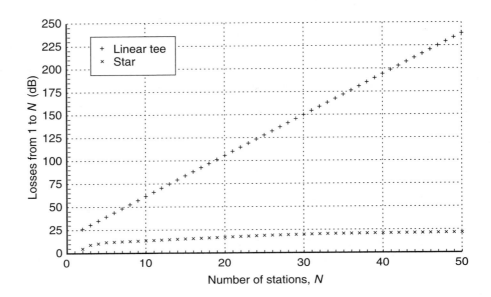

Figure 8.4 Comparison of losses of representative star and linear data bus
configurations. (Parameter values are from the example problem.)

and equipment manufacturers. (In most other countries the telecommunications
industry is administered by the post, telegraph, and telephone government-organiza-
tions [the "PTTs"].) International standards are set by the International Telegraph
and Telephone Consultative Committee (CCITT). The primary US standards are
proposed and discussed in Committee T1 (and its technical subcommittees) of the
American National Standards Institute (ANSI). Administrative support comes from a
trade association, the Exchange Carrier Standards Association (ECSA). Committee
T1 has been charged by ANSI with developing the North American Telecommunications standards and with developing reports and American recommendations for
CCITT. The committee has developed the Integrated Services Data Network (ISDN)
standard, the SS7 switch services, and the Synchronous Optical Network (SONET)
standard (discussed later in this chapter). A standard can take from one to three years
to evolve due to multiple layers of discussion and compromise among the many par-
ties to the committee. Once approved by the committee, the proposed standard is
sent to the governing body of ANSI for approval and dissemination.

In the international arena, CCITT deals only with governmental bodies, so
the US Department of State has a US National Committee for CCITT (US CCITT)
to present the T1 Committee proposals for international standards. A T1 Committee
proposal is sent to one of US CCITT's four study groups. Once approved, the State

Department submits them to CCITT headquarters in Geneva as an official US proposal. CCITT deliberates on the proposal, discusses pros and cons, and, if approved, issues its official Recommendations. These Recommendations can have a legal status in member countries that varies from mandatory observance to voluntary observance. (In the US, observance of ANSI standards and CCITT standards is voluntary, but failure to follow a standard is at the company's risk.)

Computer standards in most countries are set by the private industry without interaction with the government. Data networks, traditionally the realm of the computer industry, are starting to merge with the telecommunications network, leading to new problems. In the United States, computer-network standards are developed and set by Committee X3 of the Computer Business Manufacturers Association. Local area network standards, however, are set by the Institute of Electrical and Electronic Engineers (IEEE) Committee 802. It has set the standard for the Ethernet LAN and the Token Ring Network. Committee 802, the IEEE, and ANSI work with the International Standards Organization (ISO) to develop international computer standards.

Potential conflicts between telecommunications standards and computer-network standards call for close cooperation between the standards-setting groups, usually obtained by the overlapping membership of the groups.

Two fiber-optic data networks have been proposed as international standards. The Fiber Distributed Data Interface (FDDI) network is a wide-bandwidth local area network (LAN) operating at 100 Mb/s, and the Synchronous Optical NETwork (SONET) is a proposed wide area network (WAN) operating at a base rate of 155 Mb/s (with expansion capability to achieve data rates of several Gb/s). We now want to consider these proposed standard networks in some detail.

8.5 FDDI Networks

In the 1980s it became apparent that the 10 Mb/s data rate associated with the Ethernet standard was limiting some wideband applications. A faster standard bus was required for two primary applications:

- High-performance interconnections between mainframes, peripherals, and other mainframes, and
- High-speed backbones for use in interconnecting lower speed LANs and devices.

Efforts to specify a standard were first initiated in 1982 to exceed links that were then being standardized (i.e., Ethernet and Token Ring Network) and to incorporate the data-handling requirements of digital PBXs (private branch exchanges)

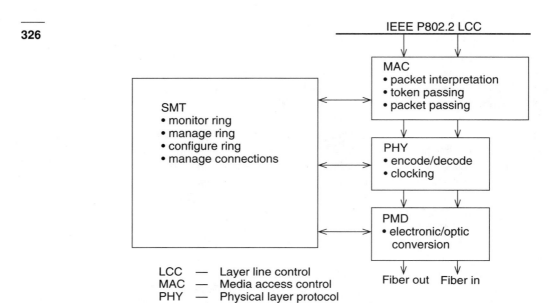

Figure 8.5 Block diagram of the FDDI layers.

that were widely introduced into the business environment to handle voice, fax, video, and sensor data streams. At the same time, the Open Systems Interconnection (OSI) standard was formalized. The primary advantage of the OSI standard was that it defined the layers of the interconnection so they could be *separately* designed and standardized.

The standard for FDDI is specified by the X3T.9 working group of ANSI (American National Standards Institute). In terms of the OSI layer nomenclature (Walrand, 1991), it specifies the MAC (Media Access Control) layer, the PHY (Physical) layer protocol, the PMD (Physical Medium Dependent) layer, and the SMT (Station Management) Document (see Figure 8.5). The function of each of these is as follows:

■ SMT (Station Management) provides the overall control of a ring by monitoring, managing, and configuring the ring. It determines the logical connection between the stations and automatically reconfigures the ring in case of a failure.

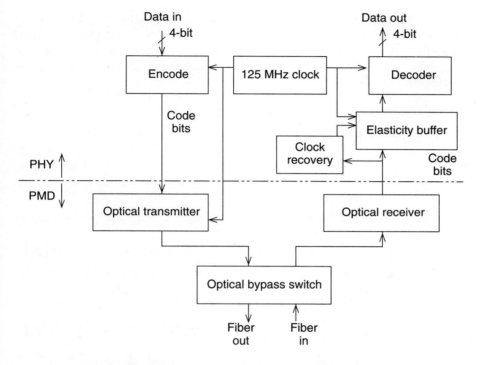

Figure 8.6 Detail of functions performed by the PHY and
PMD layers of the FDDI network.

- MAC (Media Access Control) performs packet interpretation and controls token passing and packet framing. It interfaces with the local data through a standard Logical-Link Control (LLC).

- PHY (Physical) layer protocol performs clock recovery as well as encoding and decoding input signals from the PMD layer.

- PMD (Physical Medium Dependent) defines and characterizes the optical sources and detectors, cables, connectors, optical bypass provisions, and physical hardware characteristics. A detailed view of the separate functions implemented by the PHY and PMD layers is shown in Figure 8.6.

- The FDDI unit is placed on a data bus conforming to the IEEE P802.2 Logical-Link Control (LLC) data-bus standard.

The outcome of the standards work was the Fiber Distributed Data Interface (FDDI). The topology of the network is primarily a dual counter-rotating ring. Concentrators will allow star arrangements of single-unit stations off of the ring. Properties (discussed in more detail in later subsections) of the proposed standard include the following:

- Data transmission rate: 100 Mb/s.

- Optical source: 1300 nm LEDs.

- Data encoding: The data are encoded using a 4-out-5 code into a *code group*. (This implies an actual transmission rate of 125 Mb/s to achieve the desired *data* rate of 100 Mb/s.) Each code group is called a *symbol*. There are thirty-two symbols possible:

  □ Sixteen of the symbols represent data sets (i.e., 4 bits of ordered data representing values from 0,1,2,...,15).

  □ Three symbols (called J, K, and L) are used for starting delimiter (JK).

  □ One symbol (called T) is used for the ending delimiter (TT).

  □ Two symbols (called R [reset] and S [set]) are used for control indicators.

  □ Three symbols (called QUIET, IDLE, and HALT) are used for handshaking between stations.

  □ The remaining seven symbol combinations are invalid.

- Signal Format: This data coding does not allow an unbalanced code (i.e., a long string of **1**s or **0**s with an average value that increases with the length of the string). (One of the advantages of Manchester encoding is that it is a balanced code with a constant DC value.) The code chosen for FDDI use has to lie within a ±10% range to minimize this DC effect on the receiver electronics. Empirical measurement indicates that the use of these codes adds only about a 1 dB power penalty to the link under worst-case conditions (Burr, 1986).

- Number of stations: Up to 1000 connections are allowed. (This implies that up to 500 dual-connectored stations are allowed.)

- Total fiber path: Up to 200 km of total fiber length are allowed. (Incorporating dual counter-rotating rings, up to 100 km of fiber would be allowed between stations.) The maximum distance between stations is 2 km in the early proposal; efforts have begun to offer a long-distance FDDI link.

- Frame size: The frame size is ≤ 4500 bytes. Also, the frame time cannot exceed the minimum time that it takes a token to make a trip around the ring (i.e., with no data traffic on the ring).

It is expected that the standardized FDDI network will be popular for both commercial (Strohl, 1991) and military (Halloran et al., 1991; Kochanski and Paige, 1991) applications.

8.5.1 Frame and Token Formats

The data and overhead bits are transmitted as a frame. A *frame* contains the data and ancillary information; a *token* is used to signify the right to transmit on the link. Only the possessor of the token is allowed access to the link.

FDDI Data Frame

The contents of a *data frame* (see Figure 8.7) are

- *Preamble (PA) field:* consists of IDLE line-state symbols, which occur at the maximum frequency and are used to establish and maintain clock synchronization.
- *Preamble Starting Delimiter (SD) field:* a sequence of two delimiter sequences (JK). This field is recognized as a unique boundary of a frame.
- *Preamble Frame Control (FC) field:* defines the type of frame and its characteristics, such as
  - □ synchronous or asynchronous frames,
  - □ the length of the address field (i.e., 16 or 48 bits),
  - □ the kind of frame (i.e., a Logical Link Control [LLC] or a Station Management [SMT] frame), or
  - □ whether the frame is a token.
- *Preamble Destination Address (DA) field:* a 16-bit or 48-bit representation of the destination. (A group address that is recognized by more than one station is possible.)
- *Source Address (SA) field:* a 16-bit or 48-bit representation of the source, originating the message.
- *Information field:* contains the data symbols.
- *Preamble Frame Check Sequence (FCS) field:* a 32-bit field that contains the information from a cyclic redundancy check. This information is used at the destination to check the message for errors.
- *Preamble Ending Delimiter (ED) field:* one delimiter symbol (T).

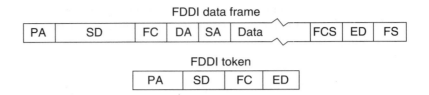

Figure 8.7 Structure of a FDDI data frame and token.

- *Preamble Frame Status (FS) field:* a minimum of three control symbols. These symbols are modified by a receiving station and indicate
  - □ whether the addressed station has recognized its address,
  - □ whether the frame has been copied, and
  - □ whether any station has detected an error in the frame.

FDDI Token

The contents of a *token* (see Figure 8.7) are

- *Preamble (PA) field* (as discussed in the frame description),
- *Preamble Starting Delimiter (SD) field* consisting of two delimiters,
- *Preamble Frame Control (FC) field* with a token description,
- *Preamble Ending Delimiter (ED) symbol* consisting of two end-of-token symbols (TT).

Two types of tokens are possible, depending on the contents of the FC field. They are the *restricted token* and the *unrestricted token*.

8.5.2 Network Operation

Each station repeats the message to its downstream neighbor. The receiving station has to decide which station has control of the medium and when to place information on the network. These ring control functions are controlled by the MAC.

Upon arrival of a message from its upstream neighbor, a station must first decide if the message is a token or a data frame, based on the frame contents.

- If the message is a token, the station then decides if it has data to transmit.
  - □ If the station has no data to transmit, it rebroadcasts the token to its downstream neighbor.
  - □ If the station does have data to transmit, it must decide if is authorized to capture the token. This authorization scheme is a mechanism to allo-

cate priority assignments to stations. Because their data are critical, some stations will have a higher priority than others to capture the token; a means of setting and recognizing this priority must be incorporated in the network. The mechanism for setting the priority is in the rules for capturing the token. (More information on priority implementation is found in Section 8.5.4.)

- If the station is not authorized to capture a token it receives, it must repeat the token to its downstream neighbor (while keeping track of the number of times repeated). If no other station wants to transmit after the number of repetitions set by the priority scheme, the station is free to capture the token and transmit its message.

- The station captures the token by removing it from the network (i.e., failing to rebroadcast the token to the downstream neighbor). The MAC generates the station's data frame, and the station transmits the frame to its downstream neighbor. When receipt of the message has been successfully acknowledged by the receiving station (as described below), the transmitting station regenerates the token and transmits it to the downstream neighbor. Following transmission of a token, the transmitter sends a string of IDLE symbols until the starting delimiter (SD) of a new frame is received. At this point the MAC analyzes the frame and proceeds as described above.

   In an alternative scheme, the MAC appends the token at the end of the frame for the use of the receiving station. For this *early release token,* the transmitting station does not wait for the return of the frame and network efficiency is increased (if there are not significant errors).

- If the arriving message is a data frame (i.e., it is not a token), the station checks the destination address (DA) portion of the frame to see if the message is addressed to it. (Note that a station can have more than one address.)

   □ If the message is not addressed to the station, it transmits the unchanged frame to its downstream neighbor.

   □ If the destination address (DA) of the message matches one of the receiving station's addresses, the frame is copied into the station's buffer, error checking is performed, and the MAC notifies the LLC (or SMT) that a message has arrived for processing. The MAC changes the frame status (FS) symbols of the data frame to show that the address has been detected and that the frame has been copied, or that the received frame was in error. The modified frame then proceeds around the ring until it arrives at the transmitting station.

☐   After receiving a message whose source address (SA) fields match its own address (i.e., after receiving a message that it has originated), the transmitting station examines the contents of the FS symbols.

- If an error message is indicated, then the original message is regenerated and retransmitted.

- If the message is acknowledged (i.e., received without error), then the transmitting station removes all of its frames from the ring. This process is called *frame stripping*. During the stripping, the source transmits IDLE symbols to the network. It should be noted that, by the time that a transmitter decides that the frame has been received and acknowledged, the transmitter has already been repeating the beginning of the frame. This results in a remnant of a stripped frame appearing on the network. This remnant will not contain valid ending delimiters, so it will not be recognized as a valid message by any of the network stations and will be removed from the network. Following the frame stripping, the station regenerates the token and transmits it to its downstream neighbor.

8.5.3 Station Types

Three types of stations (see Figure 8.8) can be used on the FDDI network:

- A *dual-attachment station* (also called a *Class A node*) has two physical layer entities and two MAC entities to accommodate a dual counter-rotating ring arrangement. (These are called "P" and "M" in Figure 8.8.) These stations can allow a wrap-around configuration in case of damage to the ring as discussed in Section 8.5.7. Figure 8.9 shows a representative ring configuration after the network has reconfigured itself to avoid a broken fiber cable.

  This wrap-around is configured by the SMT layer, which also determines the presence of the problem on the ring. Failure of the station allows an optical bypass mode to be initiated. In this mode an optical switch is activated, which routes the optical signal around the receiver and transmitter of the station. (The optical bypass also allows these stations to be powered down for maintenance without disrupting ring operation. This maintenance function is optional in the proposed standard.)

- A *single-attachment station* (also called a *Class B node*) has only one P entity and one M entity and is connected to the concentrator, which is connected to the ring. In this fashion, failure of a Class B node will not affect the network operation; only that node is affected.

- A *concentrator* allows single-attachment stations to be joined together in a star configuration and then the combination is attached to the ring. It also serves as a bridge to slower networks, such as an Ethernet network, and allows single-attachment stations to be added to the ring without compromising the speed of the main ring.

8.5.4 Prioritization Schemes

One feature of token ring networks is the need for a *prioritization scheme* to ensure that stations with important traffic have a chance to obtain control of the token, rather than having the token controlled by a station or stations transmitting data of lesser priority. The control mechanism is implemented as part of the MAC. This is done by determining *a priori* the priority that a station will have. The implementation of the priority is controlled by the use of a *Timed Token Rotation (TTR) protocol*.

Generally two types of traffic are envisioned on the network:

- *Synchronous traffic* is controlled by a master synchronization scheme (e.g., the telephone voice system). The data must be transmitted during certain prescribed time periods because it is time-sensitive (e.g., a telephone conversation requires that the channel be revisited every 20 ms).
- *Asynchronous traffic* is "bursty" by nature; the communications occur sporadically. When an element has data to transmit, it can wait to assume network control to transmit its message. Small, reasonable delays (which depend on the message's importance and the station's priority) are allowable. Most computer data transmissions can be asynchronous.

In a network mixing synchronous and asynchronous traffic, the synchronous traffic is usually guaranteed a certain minimum amount of the network bandwidth (or, equivalently, a minimum data delay). Asynchronous traffic is allowed only in the bandwidth that exceeds this minimum that is allocated to the synchronous traffic. This idea is implemented by introducing the idea of *token timers* to keep the *token rotation time* within specific bounds. The token rotation time is simply the time that it takes the token to make one round trip on the network. The *minimum token rotation time* is the time that it takes the token to circulate one round trip when each node has no data to send. It is a function of the length of the ring, the token processing time at each node, and the number of nodes on the network. The *maxi-*

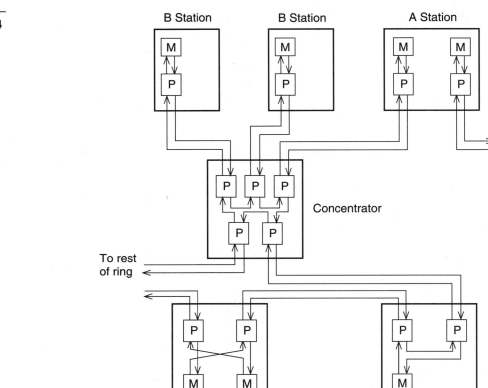

Figure 8.8 Representative interconnection of dual-attachment stations (A stations), a concentrator, and Class B stations (B stations) attached to the ring through the concentrator.

mum token rotation time is the time that we want to control; we want to avoid very long rotation times.

The timing of the token possession needs to be carefully regulated. Each station is guaranteed a minimum amount of time and uses this time to transmit synchronous traffic, if it has any. Each station has two timers associated with it. One is the *token rotation timer* (TRT), which is used to measure the time that it takes the token to circulate around the ring, and the other is the *token holding timer* (THT), which measures the time that the station has controlled the token. The THT is used with the TRT to control the transmission of asynchronous traffic.

During network initialization, a *Target Token Rotation Time* (TTRT) is assigned to the ring, based on the expected traffic. This TTRT is set to the minimum time that will be allocated to each station. (The setting of the TTRT is a nontrivial exercise, since it is critical to network throughput and delay times.) The choice of TTRT depends on the length of the ring and the number of stations.

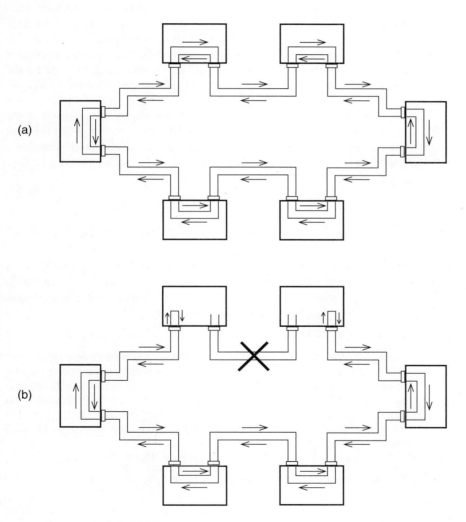

Figure 8.9 Example of link reconfiguration. (a) Original link.
(b) Reconfigured link with failure point isolated.

Using its TRT, each station measures the amount of time since the last token was received. The two classes of services are implemented as follows:

- The *synchronous class* allows the MAC to capture the token whenever the MAC has a string of synchronous frames ready to transmit. The station can transmit up to the maximum transmission time allocated to it.

- The *asynchronous class* allows the capture of a token *only if* the time since the last token was received (TRT) does not exceed the TTRT. If the TRT value exceeds the TTRT, transmission of asynchronous traffic is postponed. (Multiple priority levels can be implemented by adding more restrictive TRT requirements for token acceptance.) In the worst case, the time for the reception of two consecutive tokens will never exceed twice the TTRT.

Jain (1991) describes some of the tradeoffs that are necessary in setting the TTRT. Suppose that we have a ring with N stations attached to it and that the total length of fiber in the ring is L. The *latency D* of the ring is the amount of time that it will take the token to pass through the ring when there is no traffic. This is given by

$$D = \frac{L}{(c/n)} + NT_s \qquad (8.23)$$

$$= (5.085 \times 10^{-6})L + N(1 \times 10^{-6}) \text{ s} , \qquad (8.24)$$

where L is in kilometers and T_s is the token processing time (in microseconds) at a station. A represeative value that has been assumed is $T_s = 1$ μs. For longer links, we note that the propagation delay (the first term of Equation 8.23) is dominant (unless there are a very large number of attached stations).

Example: We will consider three network configurations described by Jain (1991). The first is a "typical" network with 20 stations on the network and a total of 4 km of fiber in the ring. The second is a "big" network with 100 stations attached to a ring with 200 km of fiber (the most fiber allowed in an FDDI ring). The third configuration is the "largest" network, which contains the maximum of 500 dual-attachment stations (i.e., the equivalent of 1000 stations) attached to 200 km of fiber.

We want to calculate the ring latency D for these cases.

Solution: For the "typical" network, we have

$$D = \frac{L}{(c/n)} + NT_s \qquad (8.25)$$

$$= (5.085 \times 10^{-6})L + N(1 \times 10^{-6}) \text{ s} \qquad (8.26)$$

$$= (5.085 \times 10^{-6})\,(4) + (20)\,(1 \times 10^{-6}) \tag{8.27}$$
$$= 40.3 \times 10^{-6}\ \text{s} \tag{8.28}$$
$$= 40.3\ \mu\text{s}\ . \tag{8.29}$$

The latency for the "big" network is

$$D = \frac{L}{(c/n)} + NT_s \tag{8.30}$$
$$= (5.085 \times 10^{-6})L + N(1 \times 10^{-6})\ \text{s} \tag{8.31}$$
$$= (5.085 \times 10^{-6})\,(200) + (100)\,(1 \times 10^{-6}) \tag{8.32}$$
$$= 1.117 \times 10^{-3}\ \text{s} \tag{8.33}$$
$$= 1.117\ \text{ms}\ . \tag{8.34}$$

Finally, the latency of the "largest" network is

$$D = \frac{L}{(c/n)} + NT_s \tag{8.35}$$
$$= (5.085 \times 10^{-6})L + N(1 \times 10^{-6})\ \text{s} \tag{8.36}$$
$$= (5.085 \times 10^{-6})\,(200) + (1000)\,(1 \times 10^{-6}) \tag{8.37}$$
$$= 2.02 \times 10^{-3}\ \text{s} \tag{8.38}$$
$$= 2.02\ \text{ms}\ . \tag{8.39}$$

Two parameters are used to characterize the ring performance. The first is the *efficiency* of the ring, given by Jain (1991) as

$$\text{Efficiency} = \frac{N(T - D)}{NT + D}\ , \tag{8.40}$$

where T is the TTRT assigned to the ring. The second is the *maximum access time delay* (T_{max}), which is given by Jain (1991) as

$$T_{max} = (N - 1)T + 2D\ . \tag{8.41}$$

We want to maximize the ring efficiency and to minimize the access time delay (or at least not have too large a value). Both of the above expressions assume that the link is carrying asynchronous traffic only.

Example: We continue the previous example, which began with our latency calculations. We want to calculate and plot (a) the efficiency as a function of assigned TTRT for the three network configurations, and then (b) the maximum access delay time.

Solution:

(a) For the "typical" network, we have

$$\text{Efficiency} = \frac{N(T-D)}{NT+D} \qquad (8.42)$$

$$= \frac{(20)\,(T-40.3\times10^{-6})}{20T+40.3\times10^{-6}} \; . \qquad (8.43)$$

For the "big" network, we have

$$\text{Efficiency} = \frac{N(T-D)}{NT+D} \qquad (8.44)$$

$$= \frac{(100)\,(T-1.117\times10^{-3})}{100T+1.117\times10^{-3}} \; . \qquad (8.45)$$

For the "largest" network, we have

$$\text{Efficiency} = \frac{n(T-D)}{nT+D} \qquad (8.46)$$

$$= \frac{(200)\,(T-2.02\times10^{-3})}{200T+2.02\times10^{-3}} \; . \qquad (8.47)$$

Figure 8.10 shows the desired plot of network efficiency.

(b) Performing the similar calculations for the maximum access delay, we have, for the "typical" network,

$$T_{max} = (N-1)T+2D \qquad (8.48)$$

$$= (20-1)T+2(40.3\times10^{-6}) \; . \qquad (8.49)$$

For the "big" network, we have

$$T_{max} = (N-1)T+2D \qquad (8.50)$$

$$= (100-1)T+2(1.117\times10^{-3}) \; . \qquad (8.51)$$

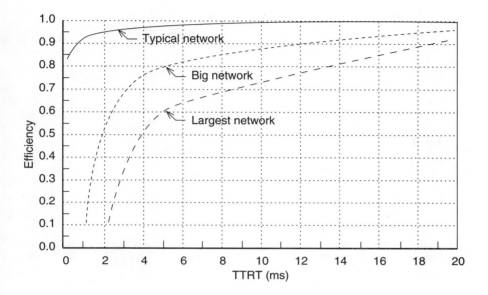

Figure 8.10 Plot of FDDI network efficiency vs. assigned
TTRT for three network configurations.

For the "largest" network, we find

$$T_{max} = (N - 1)T + 2D \tag{8.52}$$
$$= (1000 - 1)T + 2(2.02 \times 10^{-3}) . \tag{8.53}$$

Figure 8.11 shows the desired plot of the maximum access delay time for
the three network configurations.

Looking at these plots, we see the penalties of setting TTRT too
small or too large. From the plot of the network efficiency, we see that setting
TTRT too small lowers the efficiency, especially as the load on the network
grows. In the second figure, we note that setting TTRT too big will cause the
access delay to grow too long, especially if the network is heavily loaded.

We have seen from the example that setting the TTRT too big or too small
will affect network performance. The FDDI specification establishes a minimum
TTRT of 4 ms and a maximum of 165 ms to avoid extreme cases. The network man-
ager, however, gets to select the actual value between these extremes. Jain (1991)

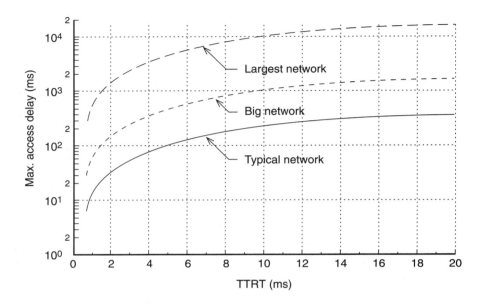

Figure 8.11 Plot of FDDI maximum access delay time vs. assigned TTRT for three network configurations.

recommends a starting value of 8 ms as the default value of TTRT. If synchronous traffic is to be carried, then the TTRT needs to be at least one-half the revisit time requirement. For example, voice traffic (that requires a revisit every 20 ms) will require an allocation time T_{synch} of 10 ms. The ring latency D, the token time T_t (19 bytes for a total of 0.88 μs), and the maximum frame time $T_{f\,\text{max}}$ (4500 bytes for a total of 0.360 ms) are added to this to ensure

$$TTRT \geq T_{\text{synch}} + D + T_t + T_{f\,\text{max}}. \tag{8.54}$$

With a maximum latency of $D = 1.73$ ms, the TTRT should equal or exceed $2.3 + T_{\text{synch}}$ ms.

   An additional scheme to establish priorities in FDDI is the use of *restricted and nonrestricted tokens*. Restricted tokens allow only certain stations, designed to recognize the special token, to receive and add data. Cooperating stations can, after prior agreement, enter into a restrictive token mode of operation. Use of this mode with multiple restricted tokens allows the stations to vie for available channel bandwidth adaptively, rather than on a pre-allocated basis.

Four kinds of traffic have been identified for use in this system of prioritization; all can be used on an FDDI network.

- The highest priority is given to circuit-switched data in an *isochronous channel*. *Isochronous traffic* is data that *must* be delivered in fixed units at fixed time intervals. The destination and source must be established by an *a priori* agreement. This traffic can be handled only by FDDI-II stations on the network and only after an isochronous channel has been assigned. (See the brief discussion in Section 8.5.8 for more information on FDDI-II.)

- The second highest priority is for *synchronous data traffic* (i.e., data that are delivered in fixed units at regular time intervals). In the FDDI network, data must be delivered at intervals that will not exceed twice the TTRT value, to avoid exceeding the TTRT limit. The address and sender are encoded in the frame, and either a restrictive or a nonrestrictive token can be used.

- The third highest priority is for asynchronous data after reception of a restrictive token. As described earlier, *asynchronous traffic* is data that are delivered with no time regularity. (The receiving station is expected to buffer the data until the complete message is received.) The bandwidth of the FDDI link is allocated first to isochronous and synchronous demands. The remaining bandwidth is allocated to asynchronous data. These data are transmitted in packets after capture of a token.

- The lowest priority is for asynchronous traffic transmitted after reception of a nonrestrictive token. The station vies for bandwidth on a packet basis with all of the other asynchronous traffic on the network.

8.5.5 Station Management

The SMT controls station functions including initialization, performance monitoring, activation and deactivation, error control, and maintenance. It also is in communication with the other SMTs on the network, controlling the network administration in such areas as address assignment, bandwidth allocation, and network configuration and reconfiguration in case of a failure. A special function of the SMT is management of the Connection Management (CMT) function. The CMT controls the logical interconnection of the PHY and MAC functions within a station and establishes the logical connection with the adjacent stations upstream and downstream.

To establish a connection with a neighbor after a request from the SMT, the CMT checks for the existence of an operating link to its neighbor. To do this, it has the PHY transmit a sequence of continuous QUIET, HALT, and IDLE signals (Ross, 1986).

- A return stream of QUIET symbols (or the lack of any response at all) indicates that no link is available, due to either a network partition that has occurred or a failure in the link or station.
- A return string of HALT symbols or a string of alternating HALT and QUIET symbols indicates a willingness of a neighboring concentrator to make a connection.
- A stream of IDLE symbols indicates the acceptance of the requested connection.

Once the physical connection is made and acknowledged through these handshake signals, the CMT needs to logically configure the paths between the MAC and the PHY. There are several possibilities (depending on the complexity of the station); this flexibility allows the station manufacturer to adapt the station to various applications. The A stations in Figure 8.8 show three alternative internal connections between the MAC and PHY available to the CMT.

8.5.6 FDDI Optical Components

The PMD specifies the dimensions and tolerances of the components to ensure interoperability.

Fiber Media

In the original FDDI specification, only optical fiber was allowed, although the size of the fiber can vary. (Work has begun to specify a shielded twisted-wire electrical version of FDDI for short-distance transmission.) The recommended fiber size is 62.5/125 μm, although it is expected that 50/125, 85/125, and 100/140 μm fibers will also be used. The best compromise between fiber dispersion and power coupling is in 62.5/125 and 85/125 fibers. Typically, the fibers will be required to have bandwidth-distance products of 400 MHz · km or more and attenuations of less than 2.5 dB/km (Burr, 1986). Fibers of other sizes will change the maximum distance between stations.

Sources

The operating wavelength is specified to be 1300 nm. The data rate is specified at 100 Mb/s with an expected maximum transmission distance of 2000 m. This combination can be met most economically with 1300 nm LEDs for the distance specified. (Long-range FDDI systems under development could use lasers.) While short-wavelength sources and receivers would be more economical for short-distance use, the standards designers decided to avoid the confusion that a second wavelength would bring, so the standard allows 1300 nm sources only (Burr, 1986). The minimum allowable power for the transmitter (into the fiber) is −16 dBm.

Receivers

The standard calls for pin diodes to be used in the link. While avalanche photodiodes would be more sensitive and would result in more link margin, the designers felt that pin diodes would be a more mature technology and would result in a lower cost receiver. (This decision to use LEDs combined with pin-diode receivers reduces the acceptable losses of the system to a fairly low value of 11 dB, as described in the next section.) To achieve the required bit-error rate with a pin diode requires a receiver power level of at least −27 dBm.

Losses

The maximum optical loss, including cable loss, connector loss, and optical bypass devices, over the 2 km maximum distance is 11 dB. This loss-budget goal can be distributed among the components in any fashion desired. Short links within buildings will typically have mostly connector losses; long-distance links will have appreciable cable loss. (The cable must also be designed to meet the 100 Mb/s data rate for the 2 km distance specified.) A worst-case cable loss of 2.5 dB/km will give 5 dB of cable loss for the 2 km maximum transmission distance, allowing a 6 dB loss allocation for the rest of the link, including splices, connectors, and bypass switches.

FDDI Connectors

The cable receptacle defines the boundary between the cable and the station. The cable connector is a duplex plug (see Figure 8.12). The design of the single-station connector is different than that of the dual-attachment station, since they should never be connected together; the single-attachment station should be connected only to the concentrator. The connector is a dry, lensless connector with the fiber typically epoxied into a ferrule. The grind-and-polish method is used to smooth the ends. Connectors will exhibit losses in the 0.2 to 1 dB range (Burr, 1986). While connectors at the station are specified in the standard, other connectors in the cable system are unspecified and left for the user to choose, as long as the 11 dB total loss budget is met.

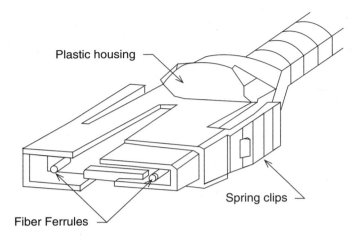

Figure 8.12 FDDI connector with two fibers.

An additional specification is made on the optical bypass switch included in the dual-port stations. The losses of this switch in the closed (or bypass-activated) position cannot exceed 3 dB. (Recall that this switch was required to allow us to bypass a faulty station.)

Jitter

The phenomenon of *jitter* is movement of the data transitions slightly from their theoretically predicted positions in time. Jitter can be caused by random components, such as thermal noise in the receiver. Sources of random jitter can be combined to form an rms jitter J_{rms} as

$$J_{rms} = \sqrt{\sum_i \Delta \tau_i^2}, \tag{8.55}$$

where $\Delta \tau_i$ is the individual jitter component from any given source. There is also a component of jitter that is data-dependent. An example of this is the recognition of the fact that a logical **1** voltage only asymptotically approaches its value, it never quite gets there. The voltage level where a downward transition begins, therefore, depends on how many **1**s precede it. The more **1**s there have been, the higher is the initial starting voltage. This small difference in starting voltage will cause a small difference in the time when the transiting voltage achieves the voltage threshold, causing some jitter. This data-dependent jitter cannot be added together in an rms fashion, since it is dependent on the previous data.

The FDDI standard specifies the maximum allowed jitter at the optical output. (The designer decides the tradeoffs that must be made within the station to achieve less than the required amount of jitter.) The amount of jitter is difficult to measure and to design. Systems designers will need more work and experience to gain familiarity with how to meet a jitter specification. New instrumentation is also becoming available to measure and characterize the various jitter components.

Other Specifications

The bit-error rate (BER) of the network is 4×10^{-11}, allowing for high data-transmission integrity (Chlamtac and Franta, 1990). The maximum number of nodes is 500.

8.5.7 FDDI Reliability Provisions

Reliability of a data network is of paramount importance. Token ring networks are, by their nature, susceptible to reliability problems, since any defect that prevents transmission of the token can cause the net to fail. (Accidental loss of the token due to, for example, data transmission errors will cause the network to eventually reinitialize itself through the SMT functions.) The architecture of the FDDI net and its built-in reconfigurability strengthen the network's reliability. Features that increase the reliability include the following:

- Station-bypass switch: This switch allows an optical bypass of a station. Broken and powered-down stations can be taken off the network for maintenance or replacement without affecting the network operation.

- Counter-rotating ring connection: All stations directly attached to the link must be dual-connectored. The second counter-rotating link can be used as a standby link or for concurrent transmission. If a station fails or the link fails, the two rings will automatically be reconfigured into one ring (with twice the original length) to maintain link continuity (see Figure 8.9). In this reconfiguration, the stations on either side have noted the break in the ring and have used their bypass switches to reroute traffic over the secondary ring, maintaining continuity on the network. It should be emphasized that this reconfiguration is done automatically by the network (using the control path) and that the link will automatically return to the original configuration once the repair has been made. Several breaks in the link at different locations will cause the ring to segment into several independent sub-rings.

- Concentrators: These are used to bring single-connected stations into the link. Failure of a single-connected station plugged into a concentrator will not affect the network. Failure of a concentrator will drop all attached stations, but the network will either bypass the failed unit (with the optical bypass switch) or reconfigure around it.

8.5.8 FDDI-II

An additional variant of the FDDI standard has been proposed for switched traffic such as PBX (private branch exchange) operations within a business or other telephone applications. Special FDDI-II equipment would be developed especially for this application.

8.6 SONET/SDH

The proliferation of optical trunk-networks has given rise to the requirement for a standard format for these signals so that they may be shared between networks (e.g., so that a signal can passed from an MCI trunk to an AT&T trunk) without any need for an interface. (This is the so-called "mid-fiber meet.") The Synchronous Optical NETwork (*SONET*) standard has been proposed by the T1 committee of the American National Standards Institute (ANSI) and is also being considered by the CCITT as an international standard (Ballart and Ching, 1989). (The international version is called the *Synchronous Digital Hierarchy* or [*SDH*].) (A *synchronous network* has a master clock that controls the timing of events all through the network.) The extension of SONET to become an international standard (SDH) holds the promise of breaking down an incompatibility between North American data rates, European data rates, and Japanese data rates (see Figure 8.13).

The features of the standard include (Ballart and Ching, 1989)

- families of standardized digital interfaces to accommodate future data rate increases (see Figure 8.14);
- a base rate of 51.84 Mb/s to carry all signals in the North American hierarchy up to DS3 (44.7363 Mb/s);
- expandability to higher data rates;

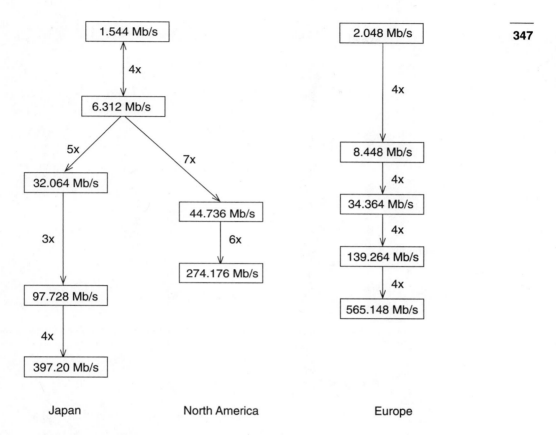

Figure 8.13 Current telecommunications multiplexing schemes. (Products are not exact because of the addition of overhead bits to the data stream.)

- synchronous multiplexing and demultiplexing for simple combining of channels, for simplified access to data payloads on the network, and for increasing synchronization of the network; and
- sufficient overhead channels to accommodate maintenance of the network.

The standard has several parts (Ballart and Ching, 1989): the optical interface that specifies the operating wavelengths, power levels, etc.; operations specifications; and the rate and format specifications that give the data format, the frame size, etc.

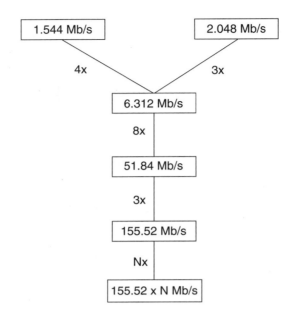

Figure 8.14 Proposed international standardized telecommunications multiplexing scheme. (Products are not exact due to the addition of overhead bits to the data stream.)

8.6.1 SONET Optical Specifications

Some of the optical parameter specifications (Boehm et al., 1986) are

- nominal wavelength: 1310 nm (within a range of 1280 to 1340 nm),
- maximum spectral width: 9 nm,
- signal optical extinction ratio: 10:1,
- coding: NRZ,
- pulse shape: to be determined.

8.6.2 SONET Rate and Format Specifications

The key challenge that arose in setting the proposed standard was to allow SONET to work in a pleisiochronous environment and *still* to maintain the synchronous nature of the network. (Signals are considered to be *pleisiochronous* if their "significant instants" occur at nominally the same rate; any allowed variations in the rate are constrained to be within predetermined limits.) The SONET solution was to

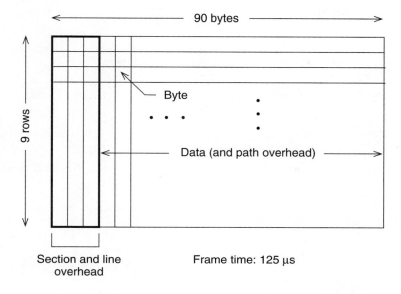

Figure 8.15 The STS-1 frame structure of 9 rows by 90 columns. Each block is an 8-bit byte. The first three columns contain overhead data for section and line management. The remaining 87 columns contain data and path layer management data.

incorporate a feature that gives the network a high degree of flexibility, the use of payload "pointers" to indicate the phase of the data payload within the SONET frame.

The basic building block for a US telecommunications system is the Synchronous Transport Signal–Level 1 (STS-1) transmitted at 51.84 Mb/s. The frame structure can be modeled as a 90-column by 9-row structure of 8-bit bytes (see Figure 8.15). One frame is transmitted every 125 μs (set by the 64 kb/s sampling rate for voice signals, which produces one byte every 125 μs). The first three columns are section overhead and line overhead bytes. The remaining 87 columns × 9 rows contain the *Synchronous Payload Envelope* (SPE) which carries the data plus nine bytes of path overhead. The format of the frame is set up so that the frame can carry a DS3 signal (at 44.736 Mb/s) or a mixture of lower-rate signals such as DS1 and DS2.

8.6.3 SONET Overhead Channels

The *overhead channels* are divided into section overhead, line overhead, and path overhead and are used to manage the communications network. Each overhead channel is designed to be used by different pieces of equipment in the network.

Section Overhead

The *section overhead* contains information used by all of the SONET equipment and breaks down as follows (see Figure 8.16):

- Two framing bytes (A1 and A2) that locate the start of a frame,
- An STS-1 identification byte (C1),
- An 8-bit Bit-Interleaved-Parity (BIP-8) byte (B1) used to check for errors in the section-overhead information,
- An orderwire byte (E1) used by the network system to control the channel,
- A byte reserved for user applications (F1), and
- Three bytes for section-overhead data (D1, D2, and D3), such as maintenance data or supply support data.

Line Overhead

The *line-overhead* bytes are processed by all SONET equipment except the signal regenerators (i.e., the repeaters). The bytes are allocated as follows (see Figure 8.16):

- An STS-1 pointer (H1, H2, and H3) that provides information about the location of the data relative to the STS-1 frame (explained in more detail later in this section),
- An 8-bit Bit-Interleaved-Parity (BIP-8) byte (B2) used to check for errors in the line-overhead information,
- A 2-byte autoprotection switch (APS) message protection channel (K1 and K2) that will switch to backup components or otherwise implement the user's protection scheme,
- Nine bytes of line-overhead data (D4–D12),
- Two bytes reserved for future uses (Z1 and Z2), and
- An orderwire channel (E2) for control of the channel by the SONET equipment.

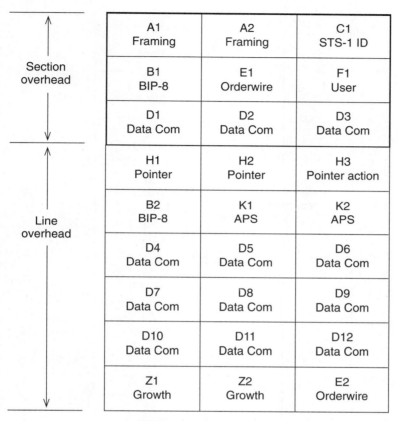

First three columns of each STS-1 frame

Figure 8.16 Section and line overhead data
bytes in first three columns of each STS-1 frame.

Path Overhead

The *path overhead* is processed by the SONET payload terminating equipment at the destination and is associated with the data package. Its byte assignments are as follows (see Figure 8.17):

- A trace byte (J1) used to trace the location of the signal through the various elements of the network,

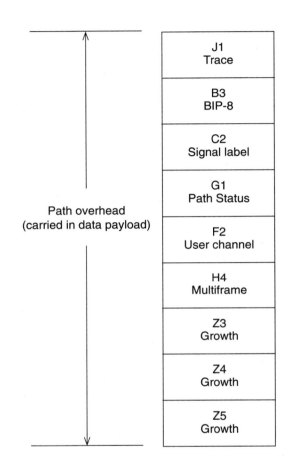

Figure 8.17 Path-overhead data bytes in first column of
a Synchronous Payload Envelope (SPE) frame.

- An 8-bit Bit-Interleaved-Parity (BIP-8) byte (B3) used to check for errors in the path overhead information,
- A signal-label byte (C2) used to identify the type of payload being carried,
- A path-status byte (G1) used to carry path-maintenance signals,
- A byte reserved for user applications (F2),
- A multiframe byte (H4) used to indicate the DS0 signaling-bit phase when carrying DS0 signals, and
- Three bytes (Z3, Z4, and Z5) reserved for future growth of the standard.

8.6.4 SONET Payload Pointer

The *payload pointer* carried in the line overhead is a new feature introduced in SONET. Its purpose is to provide multiplexing synchronization of the pleisiochronous frames and to align the frames in an STS-N signal.

Low–data rate signals are combined (multiplexed) into higher data-rate signals for trunk transmission. The usual multiplexing strategy of the telecommunications industry is shown in Figure 8.18 (Jacobs, 1986). The incoming data streams have already been multiplexed and might represent a DS3 signal (at 44.736 Mb/s) in the US, a CEPT-4 signal (at 139.264 Mb/s) in Europe, or a fifth-level signal (at 397.20 Mb/s) in Japan. The *multiplexer* provides synchronization and the addition of bits (*bit stuffing*) for performance monitoring and maintenance functions. The *multiplexing function* is a simple interleave of the tributary channels. The *high-speed signal conditioner* balances the statistics into the data stream (e.g., to avoid long strings of **0** values or **1** values or provide for the insertion of block codes). The conditioned signal is then passed on to the optical source for high-speed transmission.

Two types of conventional multiplexing have evolved, bit stuffing and fixed-location mapping.

Bit Stuffing

In the *bit-stuffing method*, extra bits are stuffed into the higher-speed channel to fill the data stream to capacity, as the signals are multiplexed together (e.g., four 1.544 Mb/s DS1 *tributary channels* being multiplexed into a 6.312 Mb/s DS2 stream). Bit-stuffing indicators are included in the frame at fixed positions and tell whether the stuffing bits contain some useful system information or are just dummy bits. The method has the advantage of being able to accommodate signals with widely varying clock speeds, but has the disadvantage of requiring de-stuffing of the bits at the demultiplexer end.

Fixed-Location Mapping

In the second multiplexing method, the tributaries are mapped into fixed, preassigned locations in the main data stream. *Fixed-location mapping* requires that synchronized data arrive at the multiplexer at just the right instant for inclusion in the channel. In networks that are highly synchronized (e.g., long-distance carriers), this required synchronization of the tributaries is no problem, as it is already included in the network. Even in these networks, however, there can be phase misalignments due to propagation delays and other effects, so 125 μs buffers (called

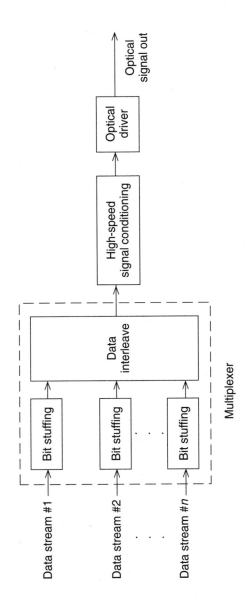

Figure 8.18 Typical multiplexing technique.

"slip buffers") are included in the multiplexers to align the signals with their respective slots. In systems that are to multiplex signals from sources with separate synchronizing clocks, this is a major problem. The floating payload in the SONET frame is a potential solution.

There are two ways to align the signal within the STS-1 frame—to "float" the payload at any arbitrary location within the frame (and to use the payload pointer to record the payload starting position) or to rigidly fix the location of the payload and require that the system timing be stringent enough to meet the subsequent requirements.

Floating Payload

The floating payload is an innovation of SONET and has an advantage when dealing with signals that are not set up by a master clock. In SONET the DS3 signal is mapped into the STS-1 frame directly. No data buffers or master synchronization of the signals are required. The payload pointer is contained in the H1 and H2 bytes of the STS-1 line overhead and indicates the starting-byte location of the payload. A late-starting payload is allowed to overlap into the following STS-1 frame (see Figure 8.19 [Ballart and Ching, 1989]). The payload is allowed to float both vertically and horizontally within the STS-1 frame (Ballart and Ching, 1989) (also portrayed in Figure 8.19). While offering high flexibility to data stream handling, the technique does have the disadvantage of requiring pointer data generation, reading, and interpretation. This new manipulation hardware and software must be designed and integrated into SONET equipment.

The payload pointer also has the function of accommodating slippage in the data due to slight differences in the data clock and the STS-1 clock. Figure 8.20 shows the case where the data clock is slightly faster than the frame clock. The payloads in frames #1 and #2 are at a given location. In frame #3, the faster clock of the data places the payload at a location that is one byte earlier than the previous frames. A stuff byte is moved into the H3 location of the payload pointer in this frame and the payload pointer location (in bytes H1 and H2) is decreased by one to point to the new locations of the payload. In frame #4 of Figure 8.20, the payload has adjusted to its new position. This method of handling slippage in the data without lengthy data delays is one of the primary advantages of the floating payload structure.

Figure 8.21 (Ballart and Ching, 1989) shows a similar adjustment when the data clock is slightly slower than the frame clock. In frame #3 the payload has moved to one byte later in the frame. A stuff byte has been added immediately after the H3 location and the payload pointer is increased by 1.

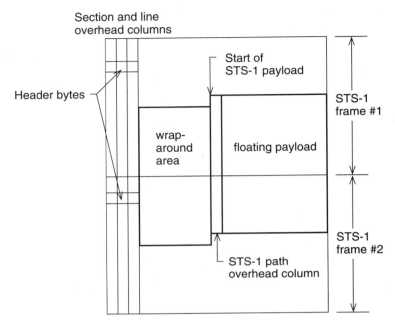

Figure 8.19 Representative alignment of a floating payload within two contiguous STS-1 frames.

Fixed Payload

An alternative use of the frame is to assign the payload to a rigidly defined, fixed location in the STS-1 frame. The payload is not allowed to float, but must fit into exactly the same location in each and every frame. This is accomplished by rigid synchronization of the data clock with the frame clock, as sometimes found in single-provider telecommunications networks. The data are synchronized to be at the input at the proper time for inclusion at the proper location. As long as the synchronization is correct, the data will be assigned their proper location.

8.6.5 Broadband Signal Handling

One of the advantages of SONET is its capability in handling higher data-rate signals than today's standard DS3 signals. (In fact, the European standard transmission of 139.264 Mb/s means that the European network cannot use the STS-1 frame structure alone for its signals; the STS-1 frame does not have a high enough

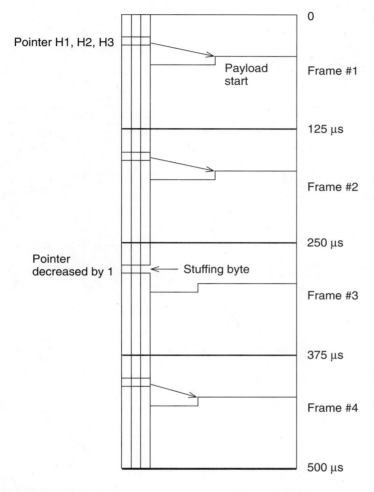

Figure 8.20 Payload pointer adjustment for fast data clock. (After Ballart, R. and Ching, Y., "SONET: Now it's the standard optical network," *IEEE Communications Magazine,* pp. 8–15, March 1989. © 1989 IEEE.)

data rate. This upward compatibility allowed SONET to be proposed as the SDH international standard.) Higher data-rate signals are obtained by byte interleaving N STS-1 frames into a so-called STS-N frame (Ballart and Ching, 1989) (see Figure 8.22). The frame is now $N \times 90$ columns (bytes) wide by 9 rows high. The time duration of the frames is $155N$ μs. For example, three STS-1 frames can be multiplexed to reach the 139 Mb/s rate preferred by the European networks.

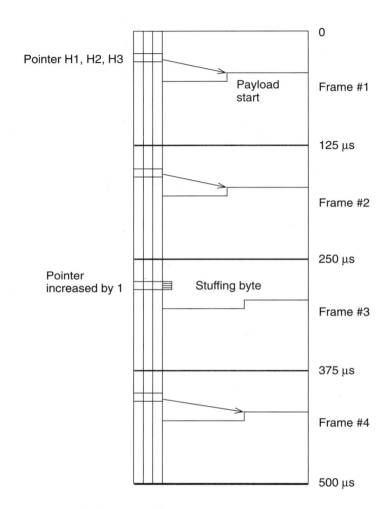

Pointer H1, H2, H3

Payload
start

Frame #1

0

125 μs

Pointer
increased by 1

Stuffing byte

Frame #2

250 μs

Frame #3

375 μs

Frame #4

500 μs

Figure 8.21 Payload pointer adjustment for slower data clock.
(After Ballart, R. and Ching, Y., "SONET: Now it's the standard optical network,"
IEEE Communications, pp. 8–15, March 1989, © 1989 IEEE.)

The payload pointers are used to multiplex the *N* STS-1 frames to STS-N
levels. When multiplexed, the section and line overhead bytes of the first STS-1
frame are used for the overhead information; the overhead bytes in the rest of the
multiplexed frames are not used. The STS-N frame is then scrambled (to avoid the

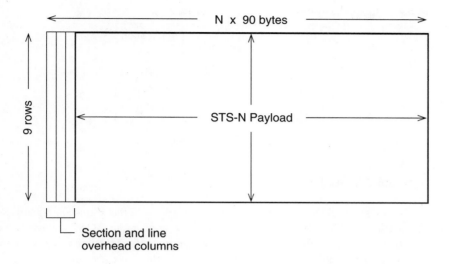

Figure 8.22 An STS-N frame used for carrying wider
bandwidth payloads than the STS-1 frame.

possibility of long strings of **0**s or **1**s) and converted to an Optical Carrier–Level N
(OC-N) signal. The line rate of the OC-N signal will be exactly N times the rate of
the OC-1 signal rate. The OC-N levels allowed by the standard are shown in Table
8.3. These higher data-rate signals provide upward compatibility of future systems.

Table 8.3: SONET signal rates.

Level	Line Rate (Mb/s)
OC-1	51.84
OC-2	155.52
OC-9	466.56
OC-12	622.08
OC-18	933.12
OC-24	1,244.16
OC-36	1,866.24
OC-48	2,488.32

Table 8.4: Virtual tributary descriptions and carrying-capacity.

Name	No. of Columns	No. of Bytes	Application
VT1.5	3	27	DS1
VT2	4	36	CEPT-1
VT3	6	54	DS1C
VT6	12	108	DS2

8.6.6 Virtual Tributaries

Just as the STS-1 frames can be combined for higher data-rate signals, so they can also be subdivided for signals that are low in data rate. SONET allows for a variety of fixed-size smaller payload units (called *virtual tributaries* or *VTs*). There are four allowed sizes of VTs as shown in Table 8.4; this table also shows the carrying-capacity of each VT in terms of the present communications standards. A *VT group* is a 12-column × 9-row payload structure that can carry

- four VT1.5s *or*
- three VT2s *or*
- two VT3s *or*
- one VT6.

Seven VT groups (i.e., 84 columns), one path overhead column, and two unused columns are byte interleaved to form one STS-1 payload. This payload is then placed into the STS-1 frame. Figure 8.23 shows 28 VT1.5s put together to form a payload. When added to the three columns of section and line overhead, they form the complete STS-1 frame. VT groups can also be mixed when they are combined to form a frame; they do not all need to be of the same type.

The VT payload can either float within the VT or be assigned a fixed position (called a "locked" position).

- The *floating VT payload* is allowed to float in a fashion similar to the STS-1 payload; the VT pointer is used to show the starting byte location within the VT payload. Again, the primary advantage is that no data buffers (with subsequent data delay) are required to handle asynchronous data streams.
- The *locked VT payload* is assigned a fixed location within the STS-1 payload (which can, itself, be floating or fixed within the STS-1 frame). Again, the locked mode provides a simple scheme for handling the multiplexing of

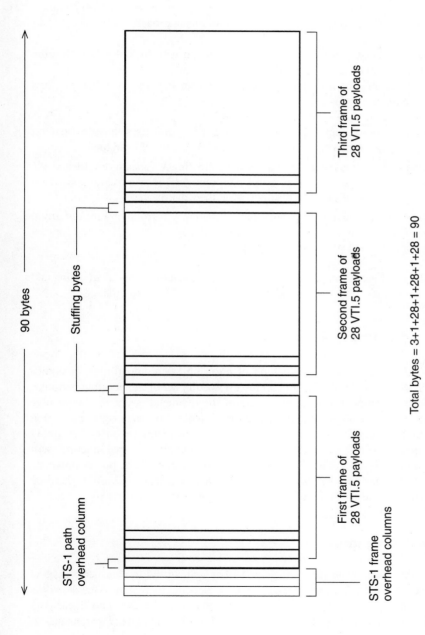

Figure 8.23 Combination of 28 VT1.5 packages to form an STS-1 payload. (After Ballart, R. and Ching, Y., "SONET: Now it's the standard optical network," *IEEE Communications Magazine*, pp. 8–15, March 1989. © 1989 IEEE.)

channels *if* a master synchronization is maintained. This scheme is primarily useful in the telecommunications industry where several DS0 channels (in the United States) can be handled with this approach.

It should be noted that floating VTs may not be mixed with locked VTs within the STS-1 payload. All have to be either floating or fixed.

The bundling of information channels within the VTs is called *payload mapping*. Five types are possible.

- Asynchronous mapping into a floating VT group uses conventional bit stuffing to multiplex asynchronous channels into the VT payload.
- Byte-synchronous mapping into either a floating VT group or a locked VT group (counts as two separate mapping methods) is proposed for synchronous mapping of the DS0 channels (and their overhead).
- Bit-synchronous mapping into either a floating VT group or a locked group (also counted as two separate mapping methods) has been proposed for carrying unframed synchronous signals.

The most useful mappings will probably be the asynchronous mapping and the byte synchronous mode, both used with the floating VT group mode.

8.6.7 SONET Compatibility

Although proposed as a method of ensuring intercompatibility of equipment, the SONET standard does not completely specify *all* of the possible permutations and combinations. For example, some of the overhead is left for future user applications. Users must ensure that the equipment will follow their own application but disregard the application data from "foreign" systems. How can this be ensured? Protective codes are included in the SONET overhead, but each manufacturer will handle a protection-provoking event differently (e.g., one might power-down its source while another may activate an optical switch). Much work still needs to be done to ensure and demonstrate intercompatibility of SONET equipment.

8.7 Summary

In this chapter we have considered fiber optics used in data networks. As the long-haul applications for fiber optics have become saturated, data networks offer opportunities for fiber-optic systems. We have seen that star and linear data busses have proved to be popular configurations. For fiber-optic networks, an analysis of the power budget showed that the star configuration has advantages over the linear bus for applications with many stations on the network. The FDDI and SONET standards represent efforts by the local area network community and the telecommunications community to establish standards that incorporate fiber optics as the media of choice. It is important to realize, however, that optical transmission is only part of the breakthroughs represented by these standards. The novel architectures proposed and being adopted offer equally revolutionary concepts.

8.8 Problems

1. Consider a 20 station linear data bus with connector losses of 1 dB per connector, coupling fractions of 5% at each arm of the coupler, insertion losses of 2 dB per coupler, fiber losses of 3 dB/km, and station spacings of 1 km.

   (a) Calculate the transmissivity of each connector.

   (b) Calculate the transmission loss (in dB) of each arm of the coupler (L_{arm}).

   (c) Calculate the ratio of $P_{1,2}/P_f$ in dB.

   (d) Calculate the ratio of $P_{1,20}/P_f$ in dB.

   (e) Calculate the dynamic range required of the receiver used in this network.

2. Consider a star network that operates with sources that produce + 3 dBm of output power in a fiber. Assume that the fiber loss is 0.6 dB/km and that the station-to-star distance is 2 km. The required receiver sensitivity is −30 dBm. Calculate the number of stations that can be on this network if the connector loss is 1 dB per connector, the excess loss of the star is 3 dB (from any input to any output), and the link margin is 0 dB.

3. Consider the same transmitter and receiver as the previous problem used in a linear network. Assume that 5% of the light is coupled into the arm of the tee coupler and that insertion loss is 1 dB per tee coupler. Calculate the number of stations that can be on this network if the connector loss is 1 dB per connector and the link margin is 0 dB.

4. Consider the network of Figure 8.9(a). Suppose that the node on the left side of the ring fails. Draw a sketch of the ring after reconfiguration has occurred.

5. Consider an FDDI link with a transmitter power of −16 dBm and a receiver sensitivity of −27 dBm. The loss in the fiber has a worst-case value of 2.5 km and the loss of the connectors is 1 dB per connector. The losses of the bypass switch in the station is 2 dB when bypassed. The link length is 2 km.

(a) Ignoring all other losses, how many stations can be shut down in the network at the same time?

(b) Assuming that no stations are shut down and that there are two connectors per station, calculate the maximum number of stations that can be connected to the network.

(c) Repeat the previous calculation if the link length is 500 m.

(d) If the connector losses are reduced to 0.2 dB per connector and the network length is 500 m, how many stations can be on the network?

(e) Repeat the previous calculation if the link length is negligible.

References

Ballart, R. and Y. Ching, "SONET: Now it's the standard optical network," *IEEE Communications Magazine*, vol 26, no. 3, pp. 8–15, March 1989.

Barnoski, M., "Design considerations for multiterminal networks," in *Fundamentals of Optical Fiber Communications*, (M. F. Barnoski, ed.), pp. 329–351, New York: Academic Press, 1981.

Boehm, R. J., Y. Ching, C. G. Griffith, and F. A. Saal, "Standardized fiber optic transmission system—a synchronous optical network view," *IEEE J. on Selected Areas in Communications*, vol. SAC-4, no. 9, pp. 1424–1431, 1986.

Burr, W., "The FDDI optical data link," *IEEE Communications Magazine*, vol. 24, no. 5, pp. 8–23, 1986.

Chlamtac, I. and W. R. Franta, "Rationale, directions, and issues surrounding high speed networks," *Proc. IEEE*, vol. 78, no. 1, pp. 94–120, 1990.

Cochrane, P. and M. Brain, "Future optical fiber transmission technology and networks," *IEEE Communications Magazine*, vol. 23, no. 11, pp. 45–60, November 1988.

Foschini, G. J. and G. Vannucci, "Using spread-spectrum in a high-capacity fiber-optic local network," *J. Lightwave Technology*, vol. 6, no. 3, pp. 370–379, 1988.

Gerla, M., P. Rodrigues, and C. Yeh, "Token-based protocols for high-speed optical-fiber networks," *J. Lightwave Technology*, vol. LT-3, no. 3, pp. 449–466, 1985.

Halloran, F., L. A. Bergman, E. G. Edgar, R. Hartmayer, and J. Jeng, "An FDDI network for tactical applications," *IEEE Lightwave Communications Systems (LCS)*, pp. 29–35, February, 1991.

Jacobs, I., "Design considerations for long-haul lightwave systems," *IEEE J. on Selected Areas in Communications*, vol. SAC-4, no. 9, pp. 1389–1395, 1986.

Jain, R., "Performance analysis of FDDI token ring networks: Effect of parameters and guidelines for setting TTRT," *IEEE Lightwave Telecommunications Systems (LTS)*, vol. 2, no. 2, pp. 16–22, 1991.

Keiser, G., *Optical Fiber Communications,* 2nd Edition. New York: McGraw-Hill, 1991.

Kochanski, R. J. and J. L. Paige, "SAFENET: The standard and its application," *IEEE Lightwave Communications Systems (LCS)*, vol. 2, no. 2, pp. 46–51, February, 1991.

Lee, J. and C. Un, "A code-division multiple-access local area network," *IEEE Trans. on Communications*, vol. COM-35, no. 6, pp. 667–671, 1987.

Midwinter, J., "Photonic switching technology: component characteristics versus network requirements," *J. Lightwave Technology*, vol. 6, no. 10, pp. 1512–1519, 1988.

Personick, S., "Protocols for fiber-optic local area networks," *J. Lightwave Technology*, vol. LT-3, no. 3, pp. 426–431, 1985.

Rawson, E. G., "The Fibernet II Ethernet-compatible fiber-optic LAN," *J. Lightwave Technology*, vol. LT-3, no. 3, pp. 496–501, 1985.

Reedy, J. W. and J. R. Jones, "Methods of collision detection in fiber optic CSMA/CD networks," *IEEE J. on Selected Areas in Communications*, vol. SAC-3, no. 6, pp. 890–896, 1985.

Ross, F. E., "FDDI—A tutorial," *IEEE Communications Magazine*, vol. 24, no. 5, pp. 10–17, 1986.

Salehi, J. A., "Code division multiple-access techniques in optical fiber networks—Part I: Fundamental principles," *IEEE Trans. on Communications*, vol. 37, no. 8, pp. 824–833, 1989.

Salehi, J. A. and C. A. Brackett, "Code division multiple-access techniques in optical fiber networks—Part II: Systems performance analysis," *IEEE Trans. on Communications*, vol. 37, no. 8, pp. 834–842, 1989.

366

Strohl, M. J., "High performance distributed computing in FDDI networks," *IEEE Lightwave Telecommunications Systems (LTS)*, vol. 2, no. 2, pp. 11–15, 1991.

Su, S., L. Jau, and J. Lenart, "A review on classification of optical switching systems," *IEEE Communications Magazine*, vol. 24, no. 5, pp. 50–55, 1986.

Walrand, J. *Communications Networks: A First Course.* Homewood IL: Aksen Associates and Richard D. Irwin, Inc., 1991.

Part Four

Portents

Chapter 9

Coherent Techniques

9.1 Introduction

So far, we have been concerned with intensity-modulated signals followed by direct detection. The source power has been modulated by the information to be transmitted and the received power has been sensed by the detector. No information has been present in the phase or frequency of the optical signal. Modern radio communications, on the other hand, can modulate the phase or frequency of the carrier to great advantage and use coherent detection techniques. Much optical-fiber communication compares to the spark-gap transmitter of early radio, limited to transmitting only Morse code. While incoherent techniques have the advantage of simplicity and its resulting low cost, coherent modulation and detection techniques (Okoshi, 1986; Kimura, 1987; Okoshi, 1987; Smith, 1987; Stanley et al., 1987; Wagner et al., 1987; Nosu and Iwashita, 1988; Nosu, 1989; Okoshi and Kikuchi, 1988) represent a possible advancement in state-of-the-art optical-fiber communications.

In this chapter we want to discuss the advantages that coherent optical communications present in the areas of increased sensitivity and increased frequency selectivity. Both heterodyne detection and homodyne detection are described, as well as various angle-modulation techniques including frequency-shift–keying modulation and phase-shift–keying modulation. The components required to achieve useful coherent communication systems, including narrow-linewidth sources, stabilized tunable sources, modulators, and polarization manipulation devices, are also presented.

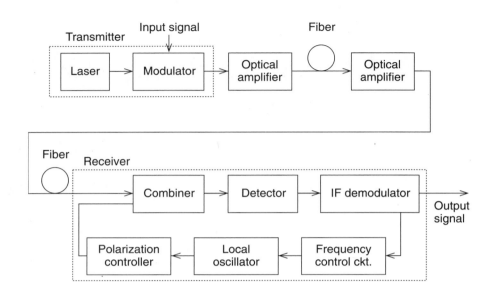

Figure 9.1 Representative coherent system.

9.2 Coherent Systems

Coherent fiber systems hold the possibility of two primary advantages, increased frequency selectivity (allowing more channel capacity) and increased receiver sensitivity. A typical fiber-optic medium has a potential bandwidth of 20,000 GHz. Systems that directly detect only the optical power use a small fraction of this bandwidth. Heterodyne detection uses a local oscillator to coherently detect the amplitude and phase of the optical signal, to produce an intermediate frequency (IF) at rf or microwave frequencies, and to allow use of filtering and modulation schemes available to radio and microwave communications for decades.

A possible coherent system is illustrated in Figure 9.1. In this system, light from a tunable ultra-stable laser source is phase-modulated externally. (The laser might also be current-modulated, as discussed later.) The optical signal can be optically amplified at the source to strengthen it and to make up for losses encountered in the modulator. In the transmission path, the optical signal can be amplified by optical amplifiers, as appropriate, to overcome the fiber losses. At the receiver, a local oscillator optical beam is combined with the signal at the beam combiner. The local oscillator operates at a frequency that is offset from the source by a predetermined frequency (called the *IF frequency*). The polarization controller located before the combiner ensures that the local oscillator beam and the signal are added with the same polarization to maximize interference between the beams. (Other detection schemes can relax this polarization requirement.) The detector produces an intermediate frequency (IF) rf signal which is demodulated by the IF demodulator. The IF demodulator can use *synchronous demodulation* (i.e., the IF carrier is detected) or *asynchronous demodulation* (i.e., only the envelope of the IF carrier is detected). The frequency control circuit ensures that the local oscillator frequency is properly offset from the signal frequency, allowing temperature and aging stabilization as well as a receiver channel-tuning capability.

Figure 9.2 illustrates two detector configurations that can be used. The 2×2 combiner shown in the figure is a fiber-coupler device (see Chapter 6) that divides the power on either input fiber equally into the output fibers (i.e., ideally 50% of the signal power and 50% of the local oscillator power will appear on each output fiber). In addition, there is a 180-degree phase difference between the signals on the output fibers. In the unbalanced detector configuration of Figure 9.2(a), half of the signal power and local oscillator power is wasted on the unused output. By adding a symmetric detector, as in Figure 9.2(b), to make a *balanced detector*, this power can be recovered and the efficiency of the receiver is increased. (We perform a difference operation because the signals from the detectors are 180 degrees out of phase.) An additional benefit is that common noise mechanisms in the detectors will be canceled (or reduced) by this balanced detector configuration. It should be noted that, while the rest of the detector configurations in the figures of this chapter show an unbalanced detector for simplicity, they could be drawn with the balanced detector configuration, as well.

Coherent detection of optical waves allows us to recover modulated information, not just in the amplitude or power of the signal wave, but also in the frequency or phase of the wave. Analog modulation techniques, such as frequency modulation (FM) or phase modulation (PM), can be also used for analog signals. For digital data we can consider using the following modulation formats:

- Frequency-shift–keying (FSK) modulation, where the optical signal frequency is one value for a logical **1** and a different value for a logical **0**,

- Binary-phase-shift–keying (BPSK) modulation, where the optical signal phase is one value (e.g., 0 degrees) for a logical **1** and a different value (e.g., 180 degrees) for a logical **0**,

- Multilevel FSK or PSK, used to encode bytes of data (e.g., any particular two-bit word could be encoded as one of four frequencies or phases), and

- Other exotic modulation formats, such as differential-phase-shift–keyed (DPSK) where the phase is changed upon a *change* in the data signal.

Each modulation format has its own sensitivity advantages.

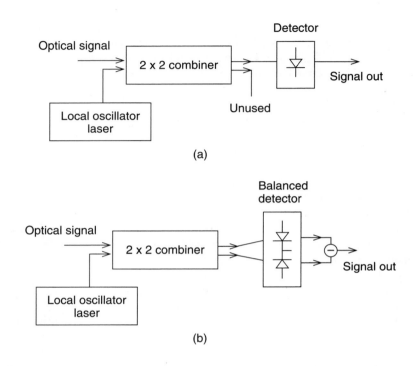

Figure 9.2 (a) Unbalanced detector configuration. (b) Balanced detector configuration.

Obviously, some key technology devices (such as the single-frequency tunable laser, the external modulator, etc.) will need to be developed commercially to make this architecture possible. We will discuss this technology later, but first we want to isolate and focus on the coherent detection capability and its potential advantages.

9.3 Coherent Detection

Figure 9.3 shows block diagrams that define the three primary detection methods.

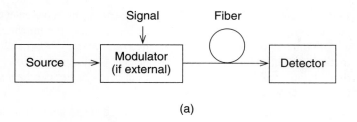

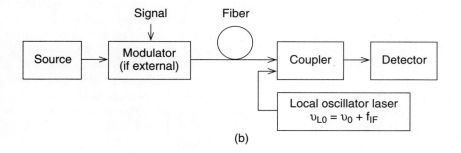

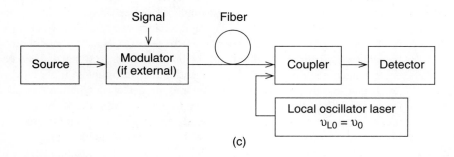

Figure 9.3 Detection methods. (a) Direct detection. (b) Heterodyne detection with offset-frequency local oscillator. (c) Homodyne detection with phase-locked local oscillator.

- In *direct detection*, the modulated optical carrier falls on the detector, which converts the carrier into a voltage signal that contains the information (usually modulated on an rf subcarrier). The voltage is amplified and the information is demodulated from the detected subcarrier. This electrical demodulation of the information from the rf subcarrier can be either asynchronous or synchronous.

- In *heterodyne detection*, the signal beam is added to a beam from a coherent local oscillator with considerably more strength. The local oscillator must be coherent with the signal, implying a high degree of monochromaticity and the same polarization in the beams. (This requirement to frequency-lock the local oscillator to the transmitter beam is the primary disadvantage of heterodyne detection.) The combined beam is allowed to fall on the detector (typically, a pin photodiode). The two waves form an interference pattern, or *fringe pattern*, at the detector. If the local oscillator is at a different frequency than the signal beam, then the fringes are moving with a velocity that is proportional to the frequency difference between the waves. (If the waves are at the same frequency, the fringe pattern is stationary, except for any frequency or phase modulation. This detection scheme is called homodyne detection and will be discussed in the next paragraph.) The detector is sized smaller than a fringe in the interference pattern and behaves (ideally) like a point detector. As the interference fringes run across the detector, the output current is modulated at the difference frequency, the intermediate frequency (IF). The IF frequency is found from $f_{IF} = v - v_{LO}$, where we will use v to indicate optical frequencies and f to indicate sub-optical frequencies. The IF frequency can be at an rf frequency in the range extending from several MHz to several GHz. Any modulation present on the optical signal will also be present on the IF signal and can be demodulated electronically with conventional rf techniques. It should be noted that one disadvantage of the heterodyne detection technique is that the bandwidth of the receiver will have to equal at least the IF frequency *plus* the signal bandwidth (which depends on the signal rate and the modulation format used), rather than just the signal bandwidth alone.

- In *homodyne detection*, the local oscillator frequency is controlled so that it is always equal to the frequency of the incoming optical signal. The phase of the local oscillator should remain 90 degrees from the phase of the incoming signal for maximum sensitivity. Usually, an optical phase-locked loop is required to maintain this frequency and phase alignment, making homodyne detection more difficult to implement.

Of the three detection techniques, homodyne detection is the most sensitive and the most difficult to implement. Heterodyne detection is a bit less sensitive than homodyne detection but can be fairly readily implemented. Heterodyne detection is considerably more sensitive (at least 10 dB) than direct detection (which is the easiest detection method to implement, since no local oscillator is required).

9.4 Improvement in System Performance

9.4.1 Sensitivity Improvement

With coherent detection, the transmitted information can be encoded in a changing amplitude, frequency, or phase of the optical signal. (It should be noted that coherent detection is not limited to the detection of phase- and frequency-modulated signals only; a conventional intensity-modulated signal can also be detected with heterodyne or homodyne detection with increased sensitivity.) Table 9.1 shows the relative sensitivity of the various modulation formats (Okoshi and Kikuchi, 1988). From this table the following conclusions can be drawn:

- The largest theoretical gain in sensitivity (10 to 25 dB) is made by changing from direct detection to coherent detection. (The performance of direct detection can be improved by including an optical amplifier just before the detector.)
- Using FSK modulation brings 3 dB more sensitivity when compared to intensity modulation (i.e., on-off keying).

Table 9.1: Receiver sensitivities (relative to the intensity modulation/direct detection case) for various combinations of modulation and detection. (After Okoshi, T. and Kikuchi, K., *Coherent Optical Fiber Communications*, Tokyo; KTK Scientific Publishers, 1988.)

Modulation/Detection	Relative Sensitivity
Intensity Modulation/Direct	0 dB
Intensity Modulation/Heterodyne	−10 to −25 dB
FSK/Heterodyne	−13 to −28 dB
4-Level FSK/Heterodyne	−16 to −31 dB
8-Level FSK/Heterodyne	−18 to −33 dB
PSK/Heterodyne	−16 to −31 dB
PSK/Homodyne	−19 to −34 dB

- Multilevel FSK modulation provides 3 to 5 dB more sensitivity than binary FSK, with the amount of increase dependent on the number of levels implemented.
- PSK provides the most sensitivity with homodyne detection having a 3 dB edge over heterodyne detection (but homodyne detection is currently considerably harder to implement).

The increase in sensitivity of coherent detection lies partly in the improved receiver sensitivities of heterodyne and homodyne detection and partly in the improved bit-error rate performance of coherent modulation formats that are allowed by the use of coherent detection.

The improved sensitivity of heterodyne detection is because the noise of the detection process can always be limited by the shot noise associated with the local oscillator, by turning up the power in the local oscillator (subject, of course, to the power dissipation limits of the detector). Local oscillator power levels of 0 dBm (easily achievable with modern diode laser sources) are required to reach this range of operation. In the case where we cannot provide enough local oscillator power to make the circuit shot-noise limited (e.g., when there is a large amount of dark current in the detector), we want to use an APD rather than a photodiode to help us achieve the best sensitivity. (For the case where we have enough local oscillator power to achieve shot-noise limited operation, there is no advantage of an APD over a photodiode; in fact, if we set too high a gain on the APD, the excess noise might actually raise the minimum amount of power required.)

The ideal signal-to-noise ratio for a shot-noise limited detector is (Okoshi and Kikuchi, 1988)

$$\frac{S}{N} = \frac{P_s}{2h\nu B},\tag{9.1}$$

where P_s is the average power in the optical signal, ν is the optical signal carrier frequency, and B is the bandwidth of the receiver.

For a direct-detection receiver, we have seen in our chapter on receivers that the noise is determined by the shot noise associated with the dark current of the APD used (especially for the noisier long-wavelength detectors) and by the noise from the preamplifier. For long-wavelength operation, only coherent-detection techniques have the potential of allowing us to operate near the shot-noise-limited signal-to-noise ratio of Equation 9.1.

The increase in the sensitivity for heterodyne detection over direct detection is also a function of the modulation format used (e.g., frequency-shift keying [FSK], phase modulation [PM], binary-phase-shift keying [BPSK], quaternary-phase-shift keying [QPSK]) and the data rate. Analysis, such as that found in Barry and Lee (1990), needs to be performed to calculate the signal-to-noise ratio for the

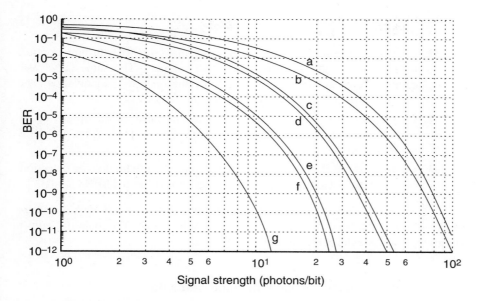

Figure 9.4 Typical bit-error rate curves vs. required signal optical power at
the detector. (The detection method for each curve is described in the text.)
(After Barry, J. R. and Lee, E. A., *Proc. IEEE*, vol. 78, no. 8, pp. 1369–1394, 1990.)

different formats. Usually, the resulting sensitivities are graphed in "waterfall" plots,
similar to Figure 9.4, with the bit-error rate (BER) plotted against the required signal
power (or number of photons) at the detector. The overall sensitivity of the receiver
is found to be dependent on the optical modulation format, the optical detection
method, and the subcarrier detection method. The curves in Figure 9.4, ranging from
the least sensitive to the most sensitive, are (Barry and Lee, 1990):

- Curve a: Amplitude-shift–keying (ASK) modulation, heterodyne optical
  detection, and asynchronous subcarrier (or IF signal) detection. (80 pho-
  tons/bit for BER $= 1 \times 10^{-9}$.)

- Curve b: Amplitude-shift–keying (ASK) modulation, heterodyne optical
  detection, and synchronous subcarrier detection. (72 photons/bit for BER
  $= 1 \times 10^{-9}$.)

- Curve c: Frequency-shift–keying (FSK) modulation, heterodyne optical
  detection, and asynchronous subcarrier detection. (40 photons/bit for BER
  $= 1 \times 10^{-9}$.)

- Curve d: Frequency-shift–keying (FSK) modulation, heterodyne optical
  detection, and synchronous subcarrier detection. The same curve also

results from amplitude-shift–keying (ASK) modulation with homodyne detection. (36 photons/bit for BER = 1×10^{-9}.)

- Curve e: Quantum limit for intensity-modulation (IM) with direct optical detection (i.e., no device thermal noise sources). The same curve also results from differential-phase-shift–keying (DPSK) modulation and homodyne optical detection. (20 photons/bit for BER = 1×10^{-9}.)

- Curve f: Phase-shift–keying (PSK) modulation, heterodyne optical detection, and synchronous subcarrier detection. (18 photons/bit for BER = 1×10^{-9}.)

- Curve g: Phase-shift–keying (PSK) modulation with homodyne detection, and asynchronous subcarrier detection. (9 photons/bit for BER = 1×10^{-9}.)

Achieved laboratory experimental sensitivities are already within a factor of two of the theoretical values (Okoshi and Kikuchi, 1988). The main difficulties are in implementing the coherent receivers.

9.4.2 Frequency Selectivity Improvements

The second advantage of coherent techniques is the ability to carry many closely spaced carriers to utilize the fiber bandwidth more efficiently. The improved frequency sensitivity of the coherent detection scheme results from our ability to design an IF amplifier and filter with a frequency bandpass response that is superior to any response of an optical filter. In particular, the steepness of the IF filter cutoffs is much higher, allowing the channels to be spaced closely together. This allows the use of *frequency-division multiplexing* or *FDM* with narrow channel spacing. Optical filters are much broader in their width in the frequency domain and can only be used to separate fairly widely spaced wavelengths, as is done in *wavelength-division multiplexing* (discussed in detail in the next chapter). In frequency-division multiplexing, a frequency-agile local oscillator laser at the receiver allows tuning of any one of the channels through the IF amplifier (in a fashion similar to your radio), or a bank of such receivers could be implemented, each tuned to a different IF signal (as shown in Figure 9.5). The adjacent channels are rejected by the IF amplifier/filter. Such close packing of the IF channels should enable high–data-rate, high-capacity networks. (One problem that remains in the FDM network of Figure 9.5 is to overcome the $1/N$ splitter loss. An optical amplifier, inserted in front of the splitter, might help [if N is not too large].)

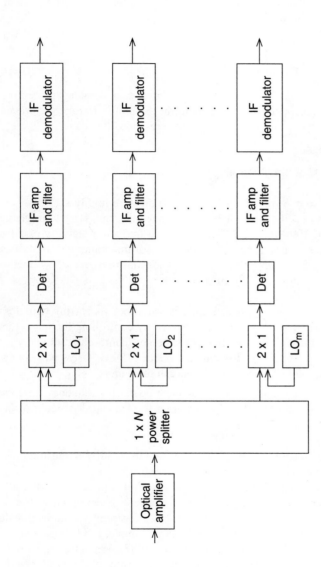

Figure 9.5 Power division and reception for frequency-division multiplexed (FDM) networks.

9.5 Component Requirements

We now discuss some of the performance requirements for the system components to be used in coherent operations.

9.5.1 Sources

The sources used in coherent systems have extra demands placed on them in the stability of their operating wavelength, in the ability to tune the wavelength controllably, and in the decreased allowable source linewidth.

Source Frequency Stability

The laser source frequency (Lee, 1991) must be highly stable for use in a transmitter or a local oscillator. To achieve the tunability necessary to implement an IF channel width of 200 MHz to 2 GHz requires a frequency stability of 1 part per 100,000 to 1 part per million. The usual technique for maintaining stability is with a feedback loop that cancels any frequency variations in the source. Two methods of controlling the operating frequency are available.

- *Temperature control*—Slow frequency variations (occurring with durations on the order of seconds) can be canceled by controlling the temperature of the source through the thermoelectric cooler incorporated in many modern laser diodes. Generally, distributed feedback (DFB) lasers will change frequency smoothly as the device temperature changes.
- *Current control*—Faster variations in the operating frequency (up to microseconds) can be controlled by changing the biasing drive current into the laser.

Figure 9.6 illustrates some techniques used to stabilize a transmitter (Figure 9.6[a]) and a local oscillator at a receiver (Figure 9.6[b]).

In the transmitter scheme, the optical discriminator can be either a Fabry-Perot interferometer (i.e., a fiber Fabry-Perot interferometer) tuned to the desired wavelength or an absorption cell containing a gas that strongly absorbs at a desired wavelength (Têtu et al., 1989; Chung, 1990). For example, H_2O absorbs at 824 nm, $Cs-D_2$ at 852 nm, and NH_3 at 1.5196 μm (Kimura, 1987). When the signal out of the detector varies, the bias current or the cooler drive current of the laser, or both, are changed to restore the detector output to its former value. Obviously, the key feature in determining the achieved stability is the passband of the optical discriminator.

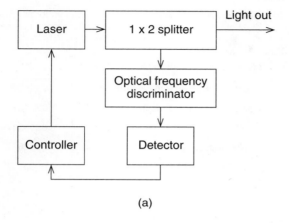

(a)

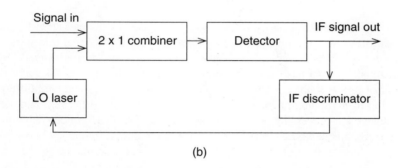

(b)

Figure 9.6 Techniques for stabilizing operating frequencies. (a) For a coherent system transmitter. (b) For a coherent system receiver.

At the receiver we want to offset the local oscillator laser frequency from the signal by a preset frequency f_{IF}. The setup of Figure 9.6(b) shows the usual arrangement. The combiner is a 2×1 coupler that adds the local oscillator wave to the signal. The combination falls on the detector (or a balanced detector configuration), which produces an electrical output consisting of some DC terms (which are discarded) and the IF frequency. The IF signal is split. One part (labeled "IF signal out" in Figure 9.6[b]) goes to the signal processing circuitry. The other portion goes to an IF discriminator which measures the IF frequency. If the IF frequency is not the desired value, an error signal is generated, which is applied to the local oscillator laser (or its cooler) to adjust the offset frequency until it equals the desired IF frequency.

Table 9.2: Predicted laser frequency linewidth requirements (in terms of data rate DR) for coherent detection of different modulation formats with postdetection processing as indicated. (After Kazovsky, L. G., *Optical Engineering*, vol. 25, no. 4, pp. 575–579, 1986.)

Modulation/Detection/Processing	$\Delta\nu$
Intensity/Heterodyne/Asynchronous	0.09 DR
FSK/Heterodyne/Asynchronous	0.09 DR
DPSK/Heterodyne/Asynchronous	0.00165 DR
PSK/Heterodyne/Synchronous	0.00226 DR
PSK/Homodyne/ —	0.00031 DR

Source Spectral Purity

The spectral purity of the diode laser sources (as measured by the source linewidth) must be improved to avoid overlapping the channel transmitters and to avoid frequency and phase changes that corrupt the desired signal. Kazovsky (Kazovsky, 1986) has studied the predicted linewidth required for some combinations of different detection methods (i.e., heterodyne and homodyne), different modulation formats (i.e., intensity modulation, FSK, PSK, and DPSK), and the type of postdetection processing applied to the IF signal (i.e., synchronous demodulation or asynchronous demodulation). In each case, the required frequency linewidth can be expressed as a fraction of the data rate, DR, of the information in the fiber, as in Table 9.2. Other similar predictions are found in Kimura (1987), and Okai et al., (1990). From these predictions, we find that we may require linewidths as narrow as 1 MHz or less to achieve optimum operation,. The linewidths lie below those presently achieved in commercial diode lasers (5 to 50 MHz). Research is being intensively pursued to narrow these linewidths (Okoshi and Kikuchi, 1988; Lee, 1991).

It should be noted that the linewidth requirements depend on the choice of modulation formats and detection methods. (As nature would have it, the combination leading to the most receiver sensitivity also requires the smallest linewidths.) Other modulation formats, such as heterodyne detection of amplitude-shift–keying (ASK) signals (e.g., the conventional on-off–keying [OOK] digital waveform), FSK signals, and polarization-shift–keying signals, are less demanding on the linewidth required than the PSK modulation format.

In a semiconductor laser, the linewidth is typically inversely proportional to the power of the laser output; that is,

$$\Delta\nu\, P = \text{constant} . \tag{9.2}$$

So, for any given laser, we want to operate at close to the maximum power output to minimize the linewidth. (This high-power operation will also have the deleterious effect of reducing the laser lifetime.) The construction of the laser affects the linewidth-power product, with typical values of $\Delta v\,P$ being in the range of 10 to 200 MHz · mW.

Several techniques to reduce the laser linewidth have been explored. They include addition of an external cavity, the use of injection locking, and the use of DFB, DFR, and quantum-well lasers.

- *External cavity*—The reasons for the relatively large linewidths of semiconductor lasers are because the optical resonator is too short or the mirror reflectivities are too low (i.e., the resonator Q is too low). Further improvements in reducing the linewidth come from increasing the cavity Q. This can be done by increasing the cavity length L or increasing the reflectivity of the mirrors. Since the latter method has only a weak effect on decreasing the linewidth and reduces the output power of the laser, the primary method has been to increase the effective length of the cavity (while keeping the size of the active semiconductor the same). The drawbacks of this method are an increased propensity of the laser to mode hop (i.e., to suddenly change frequencies), an increased vulnerability to poor mechanical stability (e.g., frequency changes due to vibrating components), and a loss of the compactness offered by a diode laser.

  An *external cavity laser* (Olsson and Van der Ziel, 1987; Koch and Koren, 1990; Lee, 1991) is shown in Figure 9.7. Here the diode laser facets are low-reflectivity surfaces, usually anti-reflection coated (as are the lens elements). A grating is used as the totally reflecting mirror. The lens elements are chosen to collimate the light at the reflecting elements and to focus the light into the mode structure of the laser. (Both graded-index [GRIN] rod lenses and fibers have also been used to implement the external-cavity laser.) Linewidth values as low as a few kHz have been demonstrated for cavity lengths of several 10s of cm, but with a penalty of difficult alignment of the components and an impractical increase in the laser-cavity size.

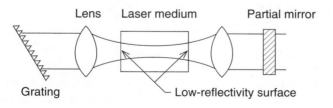

Figure 9.7 External-cavity laser.

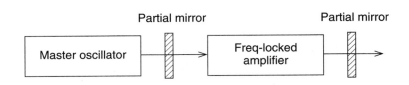

Figure 9.8 Injection-locked laser.

- *Injection locking*—An *injection-locked laser* is shown in Figure 9.8. The amplifier's reflecting surfaces are now highly reflecting with a partially reflecting mirrors added around it. The linewidth is determined by the resonator frequencies and the phase of the light reflected back into the laser resonator. Problems that can occur are a tendency to mode hop whenever there is a temperature change or bias current change, and mechanical stability problems due to the presence of the external element that is not part of the semiconductor device.

- *DBR and DFB lasers*—Distributed Bragg reflector (DBR) lasers, distributed feedback (DFB) lasers, and quantum-well lasers have demonstrated narrow linewidths and frequency stability under pulsed operation (Koch and Koren, 1990; Lee, 1991). In the DFB laser, the addition of a phase change to the grating (i.e., a one-time change in the grating structure) of 90 degrees leads to stable single-frequency operation. The DBR laser with multiple electrodes (typically three electrodes) has also proved to be a useful structure. With the proper design, the current in one electrode is used to control the output power, the current in a second electrode is used to tune the operating frequency (the current density changes the index of refraction of a portion of the resonator medium), and the current in the third electrode adjusts the phase of the light. Linewidths as low as 580 kHz have been achieved with DBR lasers (Koch and Koren, 1990). For DFB lasers, linewidths as low as 470 kHz have been measured (Koch and Koren, 1990). External-cavity versions of these lasers are also studied. Quantum-well lasers have also exhibited low linewidths with linewidth-power products in the range of 1 to 50 MHz · mW (Lee, 1991).

9.5.2 Local Oscillator Requirements

The need for a coherent local oscillator to use with FDM channelization sets new requirements for frequency setting and stability. A multi-channel coherent system will use precisely established frequency allocations (similar to radio and TV broadcasting channel assignments) and the signal oscillator will be required to be highly stable. In other applications, we will want to change the wavelength of the laser local oscillator in order to change channels. Once the wavelength has been changed, however, we will want it to remain stable during reception of the channel.

In addition, the transmitted signal frequency can shift due to temperature changes, indicating that we might need a means to control changes in the frequency of the local oscillator so we can track environmental changes in the signal frequency. The technique for doing this tracking has been addressed previously in the discussion of Figure 9.6(b).

9.5.3 Modulators

Simple modulation techniques must be introduced that do not add noise to the high purity sources that are required. These modulators tend to be external to the laser, since they usually have little effect on the source laser if care is taken to isolate the laser from reflections from the modulator.

Direct current modulation can be used to frequency-modulate the laser; typically, external modulators are used to phase-shift or amplitude-modulate the source light. Some of the modulation techniques that can be used are shown in Figure 9.9.

- *Amplitude modulation*—We can amplitude-modulate an optical carrier to follow an analog signal or, for digital data, we can shift the amplitude between two levels (called *on-off–keying [OOK] modulation* or *amplitude-shift–keying [ASK] modulation*). This can be done either by modulating the current through the device, as described in Chapter 4, or by using an external modulator for the light after it has left the source. (This latter technique has the advantage of not introducing any frequency chirp to the signal.) External modulators based on the electro-optic effect have bandwidths up to 20 GHz.

- *Phase modulation*—For phase-modulation techniques, such as PSK, an external electro-optic modulator can be used, as shown in Figure 9.9(a). This device is an electro-optic crystal that can change the phase of an optical wave by an amount that can be electronically controlled by the voltage applied to the device. External phase, frequency, and amplitude modulators have been successfully implemented using lithium niobate (LiNbO$_3$) devices. These devices use the electro-optic effect and can have bandwidths up to 30 GHz.

- *Frequency modulation*—Frequency modulation of the laser signal can be done by varying the current applied to the source, as shown in Figure 9.9(b). The increase in charge carriers in the active region changes the

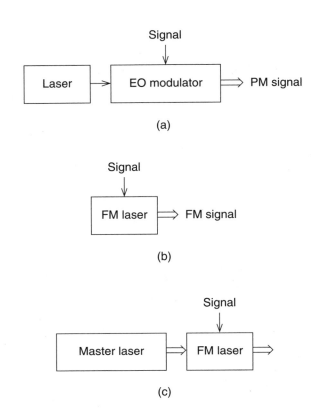

Figure 9.9 Modulation techniques useful in coherent systems.
(a) External phase modulation. (b) Internal current-driven frequency modulation.
(c) Phase modulation with injection locking.

index of refraction in the resonator, thereby changing the effective optical length of the resonator. This change in the effective resonator length causes a comparable change in the resonator mode frequency, causing the laser output to change frequency. (This same frequency chirp leads to undesirable frequency spreading of amplitude modulation schemes. For a direct-detection link, the frequency change is ignored; only the power change is detected.) Frequency changes of up to 1 GHz/mA can be induced with this technique (Okoshi and Kikuchi, 1988) with total frequency shifts of up to 10 MHz to 1 GHz being possible (Kimura, 1987). In coherent detection, the power variation is (ideally) ignored (as long as the signal-to-noise ratio is adequate) and the frequency modulation is detected.

9.5.4 Phase-Locking Techniques

Homodyne detection requires that the local oscillator be locked to both the frequency *and* the phase of the signal. Two techniques are available to do this, the optical phase-lock loop and injection locking.

■ *Optical phase-lock loop*—In the optical phase-lock loop (Kazovsky and Atlas, 1990), shown in Figure 9.10(a), the optical signal into the receiver adds to the local oscillator (in the 2×1 combiner) and the two beams fall on the detector. The detector converts the power to electric form consisting of DC terms and a Δf term (assuming that frequency locking has not yet occurred). The DC is blocked and the Δf term is fed to the loop filter. An error signal from the loop filter is fed to the slave laser that corrects the frequency (and phase) of the local oscillator until they are matched to the incoming signal. While the analysis of this technique is beyond the scope of this book, Okoshi and Kikuchi (1988) show that, if the loop gain is large enough and the original Δf is within the locking range of the phase-lock loop, the slave laser will eventually (dependent on the loop response time and the slave-laser response time) lock to the frequency of the master laser and, then, to the phase of the master laser. For the optical phase-lock loop to

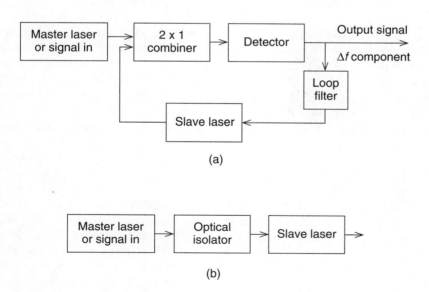

(a)

(b)

Figure 9.10 Techniques for phase-locking a laser source.
(a) An optical phase-lock loop. (b) Injection locking.

be successful, the feedback loop requires large bandwidths, on the order of 500 MHz or more (unless narrow-linewidth lasers are used).

■ *Injection-locked laser*—In the injection-locked laser (Figure 9.10[b]), the light from the master laser is injected into the cavity of the slave laser. (The optical isolator is a device to prevent reflections from the slave laser from entering the master source and upsetting its stability.) The electric fields of the lasers are coupled together within the slave laser. Okoshi and Kikuchi show (1988) that, if the coupling is strong and the original Δf is within the locking range of the injection-locking process, the slave laser will eventually (dependent on the coupling and the slave laser response time) lock to the frequency of the master laser and, then, to the phase of the master laser. (In fact, the equations of both methods of phase-locking are almost the same (Okoshi and Kikuchi, 1988).)

Both phase-locking methods require that the signal laser and the local oscillator laser have low FM noise characteristics. Implementation of these (and other) techniques is under study in research labs. The simultaneous frequency locking and phase locking of two sources has proved to be a nontrivial task.

9.6 Polarization Manipulation Techniques

The received signal and the local oscillator must have the same state of polarization to produce a maximum strength signal. If the polarizations are orthogonal, then no fringes are formed and the signal strength is zero. For polarizations that are misaligned, the signal strength is less than the maximum that could otherwise be achieved. If the polarization of signal is slowly shifting in time due to environmental effects, the strength of the interference fringes formed will be slowly varying, causing the resulting detected signal strength to be changing. This phenomenon is called *polarization fading*.

A single-mode fiber actually can propagate two orthogonally polarized modes. If the fiber were perfectly round and perfectly straight, then there would be no coupling between the two modes and the initial polarization of a polarized optical signal would be preserved after transmission through the fiber. Nature, of course, does not allow the ideal conditions to be maintained and, as a result, the polarization state of the light in the fiber will change as the light propagates, due to the slightly different propagation constants of the two modes. (The spreading of an input pulse due to this difference in propagation constant is called *polarization dispersion*.)

The existence of the differing propagation constants is called *birefringence* and is due to unequal stresses in the fiber. Assuming a linear polarization at the input,

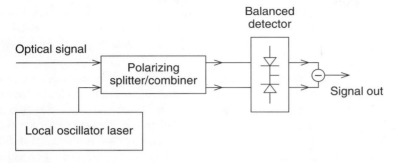

Figure 9.11 Balanced detector used with a polarizing splitter to prevent polarization fading.

an observer would see the polarization change from linear to elliptical and back to linear in an oscillatory fashion as the wave propagates down the fiber.

At any given position, the polarization state of the light can be decomposed into the vector sum of two complex-valued phasor components aligned along orthogonal axes (e.g., along the horizontal and vertical directions). By specifying the amplitude of the two components and the phase difference between the components, we can completely specify the polarization state of the light. The value of these components at the receiver-end of the fiber depends on the initial polarization state and the conditions along the length of the fiber. The measurement of this output polarization is complicated by the fiber conditions varying with time, due to environmental changes along the length of the fiber. (It has been shown experimentally that these changes in the output polarization have time scales on the order of minutes to hours, primarily due to environmental effects such as temperature [Hodgkinson et al., 1982; Imai and Matsumoto, 1988; Okoshi and Kikuchi, 1988; Calvani et al., 1989; Nicholson and Temple, 1989].)

Polarization fading can adversely affect the performance of the system unless measures are taken to remove it.

- One technique is to build a fiber that will maintain the original polarization state; these special *polarization-maintaining fibers* are addressed in a section of Chapter 11.

- A second method uses balanced detector configuration in conjunction with a polarizing splitter (Glance, 1987), as shown in Figure 9.11. Here, a polarizing splitter is used, which divides the signal beam according to its polarization state. That portion of the signal that is polarized in one direction appears at one output; the light polarized along the orthogonal polarization appears at the other output. (The local oscillator beam polarization is adjusted to divide equally between the outputs.) The combined detector outputs will not exhibit any polarization fading effects, since the signal power that does not appear at one output due to its polarization state will appear at the other output.

■ A third technique is the use of polarization-control devices to return the polarization of the signal (or the local oscillator) to its desired state. *Polarization controllers* (Koehler and Bowers, 1985; Okoshi, 1985; Noe et al., 1988; Okoshi and Kikuchi, 1988; Walker and Walker, 1990) are used to adjust the polarization of the fiber's signal to that of the local oscillator. Usually, we want to convert from an elliptical polarization of an unknown orientation to a linear polarization of a known orientation. Without any loss of generality, we will assume in our discussions that the local oscillator polarization is known to be horizontal.

The usual technique is to cascade devices, such as conventional waveplates, that can induce a phase shift between the orthogonal polarizations. (Waveplates are polarization-sensitive devices that perform this phase-shifting function.)

1. *Passive waveplates* that use the naturally occurring birefringence of some materials can be used. (A *birefringent material* is one that has the property, when properly oriented, of propagating two orthogonally polarized waves with slightly different velocities. This induces a phase shift between the polarizations and changes the size and orientation of the elliptical polarization.) One waveplate, called a quarterwave plate, is oriented in such a way as to change the unknown elliptical polarization into linear polarized light, oriented in a nonhorizontal direction. A second waveplate, called a halfwave plate, is oriented to rotate the linear polarization until it is horizontal. Such devices must be mechanically rotated to adjust for a changing polarization and, as a result, are fairly slow in response and subject to mechanical stability problems. The insertion loss of the waveplates is relatively high, compared to some other competing techniques.

2. *Electro-optic polarization-control devices* have an electrically controllable birefringence that allows the polarization to be modified by changing the voltage across the device rather than physically rotating the device. Two devices are usually cascaded with a 45 degree offset in orientation. The first device converts the input elliptical polarization with an unknown orientation into an elliptical polarization with a vertical (or horizontal) orientation; the second device converts the upright elliptical polarization into the desired horizontal linear polarization. These devices are fast responding and can easily be adjusted to handle a

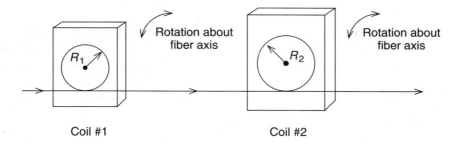

Figure 9.12 Representation of a fiber-coil polarization controller. The coils are rotated about the fiber axis (into or out of the page).

time-varying input polarization. The insertion loss, however, is also relatively high.

3. *Optical fiber coils* (Lefevre, 1980) can also be used to manipulate the polarization of light in a fiber. When coiled into a loop, the bending stress induces a birefringence in the fiber coil. By modifying the diameter of the loop, different amounts of relative phase shift can be designed (Lefevre, 1980). The loops are usually experimentally designed for operation at a specific phase shift for the specific fiber at the desired operating wavelength. Usually the coils are designed for quarterwave operation (a 90 degree phase shift) or halfwave operation (a 180 degree phase shift) and are cascaded in pairs. By rotating the plane of the coil about the axis of the fiber (see Figure 9.12), the angle of the polarization can be controlled. The first coil might be designed to be quarterwave coil (induced phase difference of 90 degree) and the second would be a halfwave coil (induced phase difference of 180 degree). The adjustments need to be made mechanically, so this device has a slow response to changing polarization and is susceptible to mechanical problems. Since it is a fiber device, its insertion losses are only the splice losses and are comparatively low.

4. In a *Faraday controller* (or *magneto-optic controller*), the magneto-optic effect is used to induce a stress-related birefringence. The fibers are mounted on ferro-magnetic coils and a current loop is used to produce a magnetic field. The size of the induced phase shift is related to the strength of the current. In the polarization-controller implementation shown in Figure 9.13, the first Faraday controller uses I_1 to control the variable phase difference required to align the input elliptical polar-

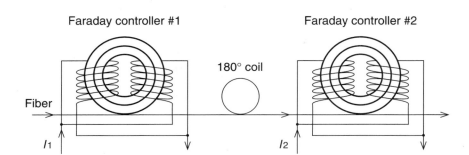

Figure 9.13 Representation of a magneto-optic or Faraday polarization controller.

ization vertically. A fiber coil is then inserted to achieve a 90 degree phase shift, converting the vertical elliptical polarization into a tilted linear polarization. The current of the second Faraday controller, I_2, is set for a 180 degree phase shift, converting the tilted linear polarization into a horizontal linear polarization, as desired. Since this is a fiber device, the insertion loss is low. With electronic control of the Faraday controllers, the response time is fast.

Each of these three polarization manipulation techniques has its proponents and each has advantages and disadvantages, so a clear-cut choice has not emerged. Research continues to explore the techniques.

9.6.1 Optical Amplifiers

Optical amplifiers are desired to regenerate the signal and to overcome losses that are introduced into the system without any deleterious effects on the signal modulation. Technologies for these optical amplifiers were discussed in Section 7.8. It is ironic that the ability to amplify the signal optically at the receiver removes much of the impetus to use coherent systems to achieve the increased receiver sensitivity. The advantages of frequency selectivity of coherent systems is still realized, however.

9.7 Summary

We have seen that coherent detection holds the promise of increased sensitivity, an expansion of options to include various phase and frequency modulations, and greater tunability of the receiver, allowing increased channel density in the fiber medium.

Homodyne detection offers the highest sensitivity, but is also the most difficult to implement. Heterodyne detection offers, for now, a realizable combination of improved system performance and physical achievability.

The advantages offered by coherent detection do not come without the penalty of increased requirements on the components used. The laser transmitters must have decreased linewidth (i.e., increased coherence) and increased stability. The local oscillator lasers must also have the same properties and must be tunable. Research is proceeding on designing lasers or arrays of lasers with these properties. Modulators must be devised to angle-modulate the lasers without disturbing the low linewidth and high stability properties. Receivers must find a way to frequency-lock the local oscillator to the incoming signal and to handle the polarization fading effects.

At present, demonstration systems have been implemented using coherent detection techniques (Brain et al., 1990; Gnauck et al., 1990; Park et al., 1990; Toba et al., 1990). The desired goal is primarily one of increasing the channel capacity of the fiber using frequency-division multiplexing techniques. The goal of increased receiver sensitivity is currently fading away, as fiber amplifiers hold the promise of allowing us to boost the signal power optically whenever desired. It should be noted that many sensing applications of fiber optics use coherent detection. These techniques are described in more detail in Chapter 11.

9.8 Problems

1. The BER of an amplitude-shift–keying (ASK) signal with heterodyne detection followed by asynchronous processing is given as (Barry and Lee, 1990)

$$\text{BER} = \frac{1}{2}e^{-M/4},$$

where M is the average number of photons required to be received by the detector in the bit period when a logical **1** is received.

(a) Find the value of M that is required to achieve a BER of 10^{-9}.

(b) The value of M relates to the peak optical power P_s in the pulse as

$$M = \left(\frac{\eta P_s \lambda}{hc}\right)T_b,$$

where η is the quantum efficiency of the detector and T_b is the bit period of the data. Calculate the required P_s for the value of M found in part (a) if $\eta = 1$, $\lambda = 1300$ nm, and the signal data rate is 100 Mb $\cdot$ s^{-1}.

2. The BER of a differential-phase-shift–keying (DPSK) signal with heterodyne detection is given as (Barry and Lee, 1990)

$$BER = \frac{1}{2}e^{-M},$$

where M is the average number of photons required to be received by the detector in the bit period when a logical **1** is received.

(a) Find the value of M that is required to achieve a BER of 10^{-9}.

(b) Using the relation in the prior problem, calculate the required P_s for the value of M found in part (a) if $\eta = 1$, $\lambda = 1300$ nm, and the signal data rate is 100 Mb $\cdot$ s^{-1}.

(c) Calculate the sensitivity increase (in dB) for the DPSK signal over the ASK signal in the previous problem.

3. Some of the receiver sensitivity expressions found in (Barry and Lee, 1990) are expressed in terms of the Gaussian Q-function $Q(\rho)$, where

$$Q(\rho) = \frac{1}{\sqrt{2\pi}}\int_{\rho}^{\infty} e^{-\frac{x^2}{2}}\, dx .$$

(a) Express the Q-function in terms of the error function erf(ρ) where

$$\text{erf}(z) = \frac{2}{\sqrt{\pi}}\int_{0}^{z} e^{-x^2}\, dx .$$

(b) The BER of an ideal photodetector that uses intensity modulation (i.e., on/off keying) with direct detection is given by (Barry and Lee, 1990)

$$BER = Q(\sqrt{M}) ,$$

where M is the average number of photons required to be received by the detector in the bit period when a logical **1** is received. Using a computer program that can calculate the error function (such as *MATLAB*), find the value of M and P_s (see Problem 1) to achieve a BER of 10^{-9}.

4. (a) Calculate the spectral width of a 1300 nm source having a frequency linewidth of 5 THz.

    (b) How many microwave channels can the optical carrier have, if each channel is 5 GHz wide and a 4 GHz separation is used between channels?

5. Consider a source operating at 1300 nm to be operated with heterodyne detection of a 1 Gb/s signal.

    (a) Calculate the required spectral linewidth $\Delta\lambda$ of the laser if FSK modulation is used with asynchronous postdetection processing. Find the fractional linewidth $\Delta\lambda/\lambda$.

    (b) Calculate the required spectral linewidth $\Delta\lambda$ of the laser if PSK modulation is used with synchronous postdetection processing. Find the fractional linewidth $\Delta\lambda/\lambda$.

    (c) Calculate the required spectral linewidth $\Delta\lambda$ of the laser if PSK modulation is used with homodyne detection. Find the fractional linewidth $\Delta\lambda/\lambda$.

6. Consider a laser with a linewidth-power product of 100 MHz · mW. If this laser produces 5 mW, calculate its linewidth and indicate the modulation, detection, and postprocessing combinations that it can accommodate. (Be sure to specify the range of data rates.)

References

Barry, J. R. and E. A. Lee, "Performance of coherent optical receivers," *Proc. IEEE*, vol. 78, no. 8, pp. 1369–1394, 1990.

Brain, M., M. Creaner, R. Steele, N. G. Walker, J. Mellis, S. Al-Chalabi, J. Davidson, M. Rutherford, and I. Sturgess, "Progress towards field deployment of coherent optical fiber systems," *J. Lightwave Technology*, vol. 8, no. 3, pp. 123–437, 1990.

Calvani, R., R. Caponi, and F. Cisternino, "Polarization measurements on single-mode fibers," *J. Lightwave Technology*, vol. 7, no. 8, pp. 1187–1196, 1989.

Chung, Y., "Frequency-locked 1.3- and 1.5- μm semiconductor lasers for lightwave systems applications," *J. Lightwave Technology*, vol. 8, no. 6, pp. 869–876, 1991.

Glance, B., "Polarization independent coherent optical receiver," *J. Lightwave Technology*, vol. LT-5, no. 2, pp. 274–276, 1987.

Gnauck, A., K. Reichmann, J. Kahn, S. Korotky, J. Veselka, and T. Koch, "4-Gb/s heterodyne transmission experiments using ASK, FSK, and

396

DPSK modulation," *IEEE Photonics Technology Letters*, vol. 2, no. 12, pp. 908–910, 1990.

Hodgkinson, T., R. Wyatt, D. Maylon, B. Nayar, R. Harmon, and D. Smith, "Experimental 1.5 μm coherent optical fiber communication system," in *Tech. Digest of ECOC '82—Eighth European Conference on Optical Communications*, (Cannes, France), pp. 414–418, 1982.

Imai, T. and T. Matsumoto, "Polarization fluctuations in a single-mode optical fiber," *J. Lightwave Technology*, vol. 6, no. 9, pp. 1366–1375, 1988.

Kazovsky, L. G., "Coherent optical receivers: performance analysis and laser linewidth requirements," *Optical Engineering*, vol. 25, no. 4, pp. 575–579, 1986.

Kazovsky, L. G. and D. A. Atlas, "A 1320-nm experimental optical phase-locked loop: performance investigation and PSK homodyne experiments at 140 Mb/s and 2 Gb/s," *J. Lightwave Technology*, vol. 8, no. 9, pp. 1414–1425, 1990.

Kimura, T., "Coherent optical fiber transmission," *J. Lightwave Technology*, vol. LT-5, no. 4, pp. 414–428, 1987.

Koch, T. L. and U. Koren, "Semiconductor lasers for coherent optical fiber communications," *J. Lightwave Technology*, vol. 8, no. 3, pp. 274–293, 1990.

Koehler, B. G. and J. E. Bowers, "In-line single-mode fiber polarization controllers at 1.55, 1.30, and 0.63 μm," *Applied Optics*, vol. 24, no. 3, pp. 349–353, 1985.

Lee, T., "Recent advances in long-wavelength semiconductor lasers for optical communication," *Proc. IEEE*, vol. 79, no. 3, pp. 253–276, 1991.

Lefevre, H., "Single-mode fractional wave devices and polarization controllers," *Electronics Letters*, vol. 16, no. 20, pp. 778–790, 1980.

Nicholson, G. and D. J. Temple, "Polarization fluctuation measurements on installed single-mode optical fiber cables," *J. Lightwave Technology*, vol. 7, no. 8, pp. 1197–1200, 1989.

Noe, R., H. Heidrich, and D. Hoffman, "Endless polarization control systems for coherent optics," *J. Lightwave Technology*, vol. 6, no. 7, pp. 1199–1208, 1988.

Nosu, K. and K. Iwashita, "A consideration of factors affecting future coherent lightwave communication systems," *J. Lightwave Technology*, vol. 6, no. 5, pp. 686–694, 1988.

Nosu, K., "Advanced coherent lightwave technologies," *IEEE Communications Magazine*, vol. 26, no. 2, pp. 15–21, 1989.

Okai, M., T. Tsuchiya, K. Uomi, N. Chinone, and T. Harada, "Corrugation-pitch-modulated MQW-DFB laser with narrow spectral width (170 kHz)," *IEEE Photonics Technology Letters*, vol. 2, p. 529, 1990.

Okoshi, T., "Polarization-state control schemes for heterodyne or homodyne optical fiber communications," *J. Lightwave Technology*, vol. LT-3, no. 6, pp. 1232–1237, 1985.

Okoshi, T., "Ultimate performance of heterodyne/coherent optical fiber communications," *J. Lightwave Technology*, vol. LT-4, no. 10, pp. 1556–1562, 1986.

Okoshi, T., "Recent advances in coherent optical fiber communication systems," *J. Lightwave Technology*, vol. LT-5, no. 1, pp. 45–1255, 1987.

Okoshi, T. and K. Kikuchi, *Coherent Optical Fiber Communications*. Tokyo: KTK Scientific Publishers, 1988.

Olsson, N. and J. Van der Ziel, "Performance characteristics of 1.5-μm external cavity semiconductor lasers for coherent optical communications," *J. Lightwave Technology*, vol. LT-5, no. 4, pp. 510–515, 1987.

Park, Y., J. P. Delavaux, R. Tench, and T. Cline, "1.7 Gb/s-419 km transmission experiment using a shelf-mounted FSK coherent system and packaged fiber amplifier modules," *IEEE Photonics Technology Letters*, vol. 2, no. 12, pp. 917–919, 1991.

Smith, D. W., "Techniques for multigigabit coherent optical transmission," *J. Lightwave Technology*, vol. LT-5, no. 10, pp. 1466–1478, 1987.

Stanley, I., G. Gill, and D. Smith, "The application of coherent optical techniques to wideband networks," *J. Lightwave Technology*, vol. LT-5, no. 4, pp. 439–451, 1987.

Têtu, M., B. Villeneuve, N. Cyr, P. Tremblay, S. Thériault, and M. Breton, "Multiwavelength sources using laser diodes frequency-locked to atomic resonances," *J. Lightwave Technology*, vol. 7, no. 10, pp. 1540–1548, 1989.

398

Toba, H., K. Oda, K. Nakanishi, N. Shibata, K. Nosu, N. Takato, and M. Fukuda, "A 100-channel optical FDM transmission/distribution at 622 Mb/s over 50 km," *J. Lightwave Technology*, vol. 8, no. 9, pp. 1396–1401, 1990.

Wagner, R., N. Cheung, and P. Kaiser, "Coherent lightwave systems for interoffice and loop-feeder applications," *J. Lightwave Technology*, vol. LT-5, no. 4, pp. 429–438, 1987.

Walker, N. G. and G. R. Walker, "Polarization control for coherent communications," *J. Lightwave Technology*, vol. 8, no. 3, pp. 438–458, 1990.

Chapter 10

Wavelength-Division Multiplexing

10.1 Introduction

Despite the high capacity of an optical channel, most systems currently use only one wavelength for carrying information. We have seen, as illustrated in Figure 10.1, that optical fibers have relatively broad regions of low loss that could support the operation of the link with more than one source. Since we have the ability to control the operating wavelength of the semiconductor emitters through the alloy composition or through the grating spacing in the distributed-feedback (DFB) lasers or distributed Bragg reflector (DBR) lasers, it is possible to hypothesize a link that would carry several wavelengths, as illustrated in Figure 10.2. At the receiving end, filters, or other wavelength-sensitive elements such as gratings, would separate the wavelengths and each carrier would fall on a separate receiver for detection. Since the wavelengths are used to share (or multiplex) the channel, the technique is called *wavelength-division multiplexing (WDM)* (Wagner and Kobrinski, 1989; Hill, 1990). The key elements in the link are the devices that combine the separate source emissions (the *couplers* or *multiplexers*) at the transmitter end and the elements that separate the channels (the *demultiplexers*) at the receivers. We also need techniques to set and maintain the wavelength of the sources, so that they are fairly close together.

Bidirectional links are also possible in WDM systems by intermixing sources and receivers at each end, as shown in Figure 10.3.

The optical fiber passband extends roughly from 800 nm to 1600 nm. (There are significant losses associated with the OH$^-$ ions at 1400 nm and 1250 nm.

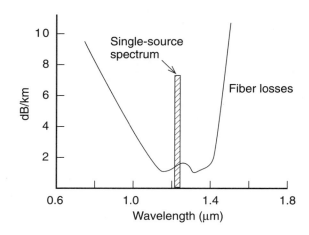

Figure 10.1 Underutilized optical bandwidth in a single-source multimode system.

These losses are not important for short-distance links [up to 10 km or so] but become increasingly important as one tries to go maximum distances.) Three fiber-optic transmission windows are identified near 850, 1300, and 1550 nm.

- Multimode systems operating at 850 nm have bandwidth-distance products of 1 GHz · km (attained with graded-index fibers).

- As seen in Figure 10.4, there are about 14,000 GHz of low-loss bandwidth in the 1300 nm region of a typical single-mode fiber. Single-mode fibers operating at 1300 nm have losses of about 0.4 dB · km^{-1}. The dispersion of a single-mode fiber with a 10 μm core and a Δ of about 0.5% is zero at about 1300 nm, with a representative dispersion in the 1260 nm to 1360 nm

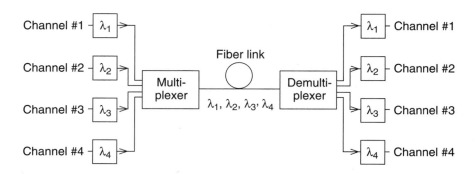

Figure 10.2 System using wavelength-division multiplexing.

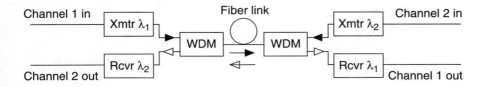

Figure 10.3 Bidirectional WDM system.

window of about 4 ps $\cdot$ km^{-1} $\cdot$ nm^{-1}. (Recall that the effects of dispersion become apparent when the dispersion exceeds about 20% of the bit period.)

- As also seen in Figure 10.4, there are about 15,000 GHz of low-loss bandwidth in the 1550 nm region of a typical single-mode fiber. Single-mode fibers operating at 1550 nm have losses of about 0.2 dB $\cdot$ km^{-1} with a typical dispersion of 18 ps $\cdot$ km^{-1} $\cdot$ nm^{-1}. This dispersion can be reduced if dispersion-shifting techniques are applied to move the minimum-dispersion wavelength up to the 1500 nm window or by decreasing the source linewidth.

For long-distance, high–data-rate links, then, the usable windows are from 1250 to 1350 nm and from 1450 to 1600 nm, allowing a combined optical window of 250 nm.

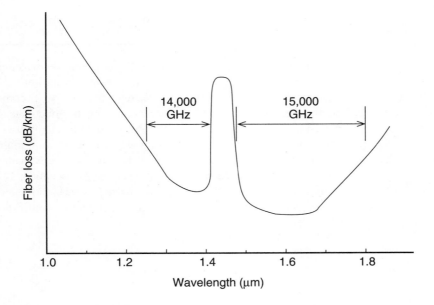

Figure 10.4 Spectral windows in single-mode fibers.

Example:

(a) Calculate the width of the 1300 nm window in nm.

Solution: We know that frequency and wavelength are interrelated by

$$\nu = \frac{c}{\lambda}.$$

(10.1)

So, for the 14,000 GHz window at 1300 nm, we find that

$$\frac{d\nu}{d\lambda} = -\frac{c}{\lambda^2}$$

(10.2)

$$|\Delta\nu| = \frac{c}{\lambda^2}\Delta\lambda$$

(10.3)

$$|\Delta\lambda| = \frac{\lambda^2}{c}|\Delta\nu|$$

(10.4)

$$= \frac{(1300 \times 10^{-9})^2(14,000 \times 10^9)}{3.0 \times 10^8}$$

(10.5)

$$= 7.89 \times 10^{-8} \text{ m}$$

(10.6)

$$= 78.9 \text{ nm}.$$

(10.7)

The 15,000 GHz window at 1550 nm is 120.1 nm wide.

Other factors can limit the use of single-mode fibers in multi-wavelength applications. The fiber ceases to be single-mode if the wavelength of the source is below the cutoff wavelength of the fiber. At longer wavelengths, a fiber becomes more susceptible to microbending losses. Additionally, commercial semiconductor laser sources do not exist over much of the band but probably could be engineered.

The transmission window can be exploited by wavelength-division multiplexing. In these systems, sources operate at various wavelengths within the window are combined for transmission over the channel and are separated for reception, as in Figure 10.2. The prime advantage is that synchronous, asynchronous, and analog signals can all be carried simultaneously over the fiber channel with full utilization of the bandwidth offered by the fiber.

10.2 Wavelength-Selective WDM vs. Broadband WDM

Wavelength-division multiplexing can be divided into wavelength-selective techniques and broadband techniques.

10.2.1 Wavelength-Selective WDM

Wavelength-selective techniques are illustrated in Figure 10.5(a). Each source operates at a separate wavelength. The power from all sources is combined (ideally) without loss. The demultiplexer at the receiving end is wavelength-sensitive; it separates each wavelength into a different channel, unique to that wavelength. (The separation is, again, ideally lossless.) The multiplexing and demultiplexing can be achieved with low insertion loss, as all of the power at a given wavelength is (in theory) directed along only one path; the other wavelengths go along separate paths. Neglecting insertion loss of the couplers, then,

$$P_{\text{out}}(\lambda_j) = P_{\text{in}}(\lambda_j) \ . \tag{10.8}$$

The amount of light that leaks from one channel into another is called the *crosstalk*. The crosstalk CT from channel j into channel i is expressed (in dB) as

$$CT_{ij} = 10 \log \left(\frac{P_{ij}}{P_{ii}} \right), \tag{10.9}$$

where P_{ij} is the power measured in channel i when only channel j is active and P_{ii} is the power in channel i when it is the only channel that is active. The crosstalk can be expressed either pair-wise, as just described, or in total as

$$CT_{\text{total}} = 10 \log \left(\frac{\sum_j P_{ij}}{P_{ii}} \right), \tag{10.10}$$

where the numerator is the sum of the power from all of the other channels as measured in channel i.

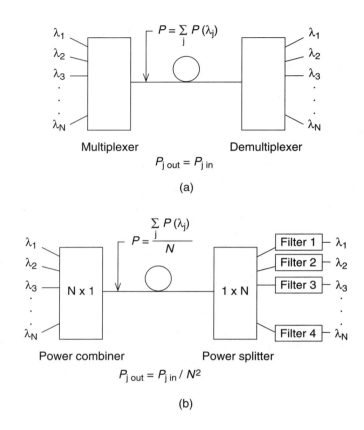

Figure 10.5 (a) Wavelength-selective multiplexing. (b) Broadband multiplexing.

10.2.2 Broadband-Multiplexing Techniques

The *broadband-multiplexing technique* (see Figure 10.5[b]) combines the powers of the sources at the transmitting end and then divides the total signal power at the receiver end (the source end and receiver end may, in fact, be co-located in a star configuration). Simple splitters and combiners are used, resulting in a loss of $1/N$ for each device. The overall loss, then, is $1/N^2$, as illustrated. Hence, we have a significant power loss in each channel, expressed by (neglecting insertion losses)

$$P_{out}(\lambda_j) = \frac{P_{in}(\lambda_j)}{N^2}. \tag{10.11}$$

All wavelengths are broadcast to all receivers. Each receiver requires its own filter to separate the channels. The divided power reaching each receiver is filtered for the desired wavelength, which is then passed to the detector.

The losses of the splitter are $1/N$ or $10 \log N$ dB (excluding any excess losses). While this loss penalty can be severe without the use of compensating optical amplifiers, the broadband technique has some advantages. The sources need only to be tuned within the passband of the spectral filter, a fairly generous spectral width. The minimum channel spacing depends on the steepness of the passband filter characteristics and the amount of crosstalk that can be tolerated. Suitable receiver filtering elements include Fabry-Perot etalons, tuned semiconductor laser amplifiers, integrated filters, and heterodyne receivers, as discussed later. In addition, the broadband system is an agile system for military applications, since the source wavelengths and receiver filters can be easily switched without disturbing the distribution system.

We now want to consider some of the optical components used in WDM systems.

10.3 Multiplexers

Broadband WDM systems use broadband versions of the power splitters and combiners that were described in Chapter 6. The bandwidth of a wavelength-flattened fused-fiber combiner or splitter can be as much as 400 nm (Hill, 1990), which is more than adequate for the postulated window extending from 1250 nm to 1600 nm.

Wavelength-selective WDM systems require multiplexing and demultiplexing that can be carried out by a variety of techniques including diffraction gratings, spectral filters, or directional couplers. The elements that separate the wavelengths are of two types, angularly dispersive devices and filter devices.

10.3.1 Angularly Dispersive Devices

Angularly dispersive devices (e.g., prisms and gratings) are optical devices that transmit or reflect light at an angle that depends on the wavelength of the incident light (assuming a collimated input light beam) (Keiser, 1991). The main parameters that describe such a device are the excess loss and the angular dispersion of the device. The *excess loss* is the loss (in dB) incurred by passing through the device in excess of the expected $1/N$ splitting loss. Since these devices are, by their nature, wideband devices, the excess loss should be specified as a function of the wavelength (Ishio et al., 1984).

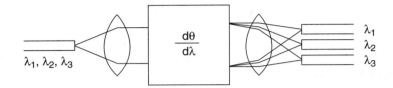

Figure 10.6 Wavelength separation system using an angular disperser and lenses.

The *angular dispersion* of the device is the measure of the angular spread $d\theta$ between two beams that are spatially coincident at the input with wavelengths separated by $d\lambda$. The angular dispersion in given by $d\theta/d\lambda$. Angular dispersion can be converted into lateral displacement by the addition of a lens, as shown in Figure 10.6. Here the dispersive device is placed in the front focal plane of the rear lens and the detector inputs are located in the back focal plane.

The dispersive element requires a collimated beam of light incident on it. Figure 10.6 shows a collimating lens intercepting the beam from the input fiber. The end of the input fiber is in the front focal plane of the lens; the dispersive device is in the back focal plane of the lens.

In most angularly dispersive systems, a grating is the device that is combined with the lenses (Figure 10.7). A grating is a reflecting or transmitting element with a series of close parallel lines engraved or etched into the surface. Excess losses of 1 to 3 dB are typical for these demultiplexing devices; the typical channel separation is 1 to 10 nm.

10.3.2 Filtering Devices

Filtering devices reflect or transmit light depending on the wavelength. Several types of filtering devices can be used to separate the wavelengths (Ishio et

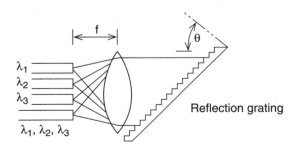

Figure 10.7 Geometry for reflecting grating WDM receiver.

Table 10.1: Typical channel bandwidths for different filtering techniques.

Method	Typical Channel Bandwidth
Interference filters	5 nm
Fabry-Perot filter	0.1–10 nm
Tuned semiconductor laser amplifier	1–10 GHz
Heterodyne receiver	1–10 GHz

al., 1984). Table 10.1 shows representative bandwidths that can be achieved with each type of filter.

1. *Interference filters*—Most optical filters used in this application are multi-layer dielectric stacks (rather than the absorbing filters used in photographic applications) to provide a sharp change in reflectivity at the desired wavelength and to minimize the insertion loss. The filters typically pass one wavelength (with a typical resonance response) and reflect all other wavelengths that are more than 10% (typically) away from the center wavelength of the filter. (With more complex designs, the spacing can be made to be 1% or even 0.1% of the center wavelength.) For operation with more than two channels, a series arrangement of several filters becomes necessary (McCartney et al., 1987), with increasingly large insertion losses for the combination of filters. Only cascades of two or three filters are practical for use, due to their relatively high insertion loss, limiting broadband systems to five or six separate wavelength channels (Ishio et al., 1984).

2. *Fabry-Perot resonator filters*—The Fabry-Perot resonator is a bandpass optical filter that transmits a narrow band of wavelengths and reflects others (Yariv, 1991). The transmission frequencies of the filter occur every $c/2L$, where L is the spacing between the mirrors of the resonator. This period is called the *free spectral range,* FSR, of the interferometer. The 3 dB frequency widths of the filter passband (i.e., the full-width at the half-maximum points) is Δf and is related to the FSR by the *finesse F* of the filter by

$$\Delta f = \frac{\text{FSR}}{F}, \tag{10.12}$$

where

$$F = \frac{\pi\sqrt{R}}{1 - R} \tag{10.13}$$

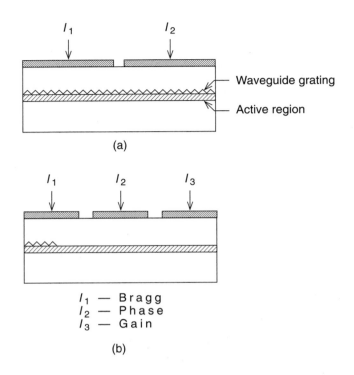

$I_1 \quad\;\; I_2$

Waveguide grating

Active region

(a)

$I_1 \quad\quad I_2 \quad\quad I_3$

I_1 — Bragg
I_2 — Phase
I_3 — Gain

(b)

Figure 10.8 (a) A distributed-feedback (DFB) amplifier (shares its feedback region with the current pump). (b) A distributed–reflector (DFR) amplifier with multiple control electrodes (separates the reflector region from the current-pumped region).

and R is the reflectivity of the resonator mirrors (assumed the same for both mirrors of the resonator filter). The minimum transmission between the peaks is $(\pi/2F)^2$ (Kaminow, 1989).

3. *Diode laser amplifier filters*—A diode laser that is pumped below threshold (i.e., it is not pumped sufficiently for lasing) is a tuned bandpass amplifier with a narrow passband (Suzuki et al., 1990). The center frequency is determined by the composition of the material. Using a distributed-feedback (DFB) laser or a distributed-feedback–reflector (DFR) laser with multiple electrodes (as shown in Figure 10.8) allows one to adjust the amplified wavelength. In the distributed-feedback amplifier (Figure 10.8[a]), one current input can be used for the pump while the other is used to control the nominal center wavelength of the amplifier. In the distributed–feedback-reflector amplifier, the current I_1 varies the optical periodicity of the reflec-

tor (and, hence, the center wavelength of the amplifier), the current I_2 controls the phase of the amplifier, and the current I_3 controls the pump power or gain of the amplifier.

4. *Coherent detection with electronic filters*—Once a source with a linewidth smaller than the signal bandwidth is possible, *coherent detection* techniques become possible. These coherent systems have been discussed in the preceding chapter as frequency-division multiplexing (or FDM). The channel spacing of these systems is determined by the bandwidth of the modulated signal (dependent on the type of modulation used and the data rate) and the receiver arrangement used. This detection method allows the use of narrow channels that are separated in the electronic processing subsystem of the receiver where precision electronic filters can be used, rather than filtering in the optical regime where separation filters are still fairly crude and primitive in terms of the minimum channel width.

10.4 Sources

To make WDM systems possible, we would ideally have a powerful source with sufficient stability and tunability to allow rapid selection of the desired wavelength. Table 10.2 illustrates the spectral width of some candidate sources that we want to discuss.

10.4.1 LEDs

The wide bandwidth of an LED is usually considered a disadvantage due to the resulting fiber dispersion. It is possible, however, to pass the output through a spectral filter to narrow the spectral width of this source; this technique is called *spectral slicing* (Hill, 1990). This method has the advantage of allowing use of sev-

Table 10.2: Typical spectral widths for different transmitters.

Type	Typical Spectral Width	Typical Frequency Width
LED	50–100 nm (or more)	5,000–10,000 GHz
Fabry-Perot laser	3–6 nm	300–600 GHz
Distributed feedback laser	<0.01 nm	10–100 MHz
External cavity laser		<1 MHz

eral of the same relatively inexpensive LEDs as the sources and, using different filters, selecting the operative wavelengths. This selective filtering further reduces the already meager power level available from the LED source, however. The reduction in power is proportional to the bandwidth reduction, making this technique practical only for short-distance, low–data-rate links.

10.4.2 Fabry-Perot Diode Lasers

The *Fabry-Perot (FP) laser* will typically have 6 to 8 modes oscillating with a spacing of $c/2L$ Hz between them. For these lasers the center wavelength has a typical tolerance of ± 3 nm, and a typical linewidth of 6 nm. A representative tolerance on the multiplexer/demultiplexer is ± 1 nm (Hill, 1990), so the smallest available channel spacing is about 14 nm. (Note that the laser needs to be temperature-controlled, since the center wavelength has a temperature coefficient of about 0.4 nm $\cdot$ $°C^{-1}$.) Allowing a guard band between channels that is equal to the channel width itself, we find that we need to allocate about 28 nm per channel and that our 100 nm and 150 nm wide windows can accommodate 3 and 5 channels, respectively, for a total of 8 channels.

10.4.3 DFB Diode Lasers

The *distributed-feedback (DFB) laser* (Lee, 1991; Yariv, 1991) allows single-mode operation of the diode laser and offers a lower temperature sensitivity of about 0.08 nm $\cdot$ $°C^{-1}$, as well. The typical spectral width of the output is 10 to 100 MHz. This is smaller than the signal spectral width (for high data rates) or the wavelength "chirp" that accompanies the output of these lasers when pulsed. (The chirp can be as much as several 10s GHz [or several tenths of a nm]. This chirp can be removed by using a cw laser as a source and using an external modulator to modulate the data onto the light.) The channel width can be estimated from a source center wavelength tolerance of ± 0.5 nm (assuming a temperature-controlled laser), a chirp tolerance of ± 0.2 nm, and a filter tolerance of ± 1 nm (Hill, 1990). The actual channel width is about 3.5 nm and, adding a guard band equal to the channel width, the allocated channel width would be about 7 nm. Therefore, we could have 14 and 21 channels in our respective fiber transmission windows.

10.4.4 Frequency-Locked Lasers

A laser can be phase-locked to an atomic resonance and, with temperature and current feedback controls, can achieve a linewidth as small as 5 MHz. This small linewidth allows the channel bandwidth to be determined only by the signal bandwidth, independent of the source linewidth. (More information on frequency-locking can be found in Chapter 9.) In this case, the demultiplexing cannot be done by fixed filters but must be done by a tunable, narrow-bandwidth element to ensure matching the passband of the filter element precisely to the wavelength of the source. Once frequency-locked, the laser cannot be easily tuned. Other techniques can be used to offset a second laser from the reference-locked laser by an adjustable (or preset) amount.

10.5 Multipoint, Multiwavelength Networks

We now want to discuss some samples of network implementations that use broadband transmission and wavelength-selective techniques.

10.5.1 Broadband Transmission Systems

In the broadband network (see Figure 10.5[b]), any signal transmitted is broadcast to *all* receivers (allowing the maximum in network flexibility). If N is the number of stations in the array, then the power is split into $1/2^N$ and the overall splitting loss is $10 \log 2^N$ dB. The network can be configured as either a linear tree or as a star configuration (as seen in Figure 10.9). In a nonswitched network (i.e., in a network that does not incorporate electronic or optical switching devices), the transmitters and receivers can be assigned unique wavelengths.

10.5.2 Wavelength-Selective Transmission Networks

These networks combine wavelength-division multiplexed systems in various ways. The power transmitted at a particular wavelength is guided to particular receiver through the network. Both the wavelength spectrum and the optical power can be utilized more efficiently, but at a cost of decreased network flexibility.

Examples of these networks include the star network, the chain network, and the ring network.

WDM Star Network

In the wavelength-selective WDM *star network* (Figure 10.10), the receivers have a set wavelength and the sources are tunable. (A configuration with set source-wavelengths and tunable receivers is also possible.) Any source can talk to

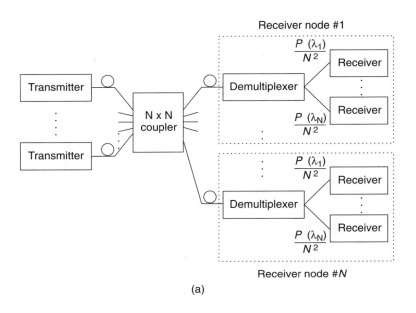

(a)

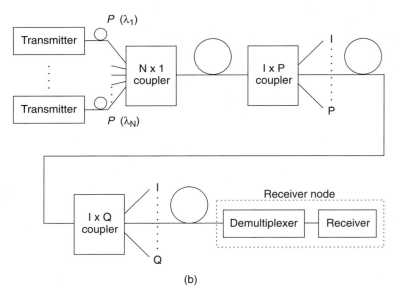

(b)

Figure 10.9 (a) A broadband-transmission star network. (b) A broadband-transmission tree network. (After Hill, G. R., *Proc. IEEE*, vol. 78, no. 1, pp. 121–132, 1990. © IEEE, 1990.)

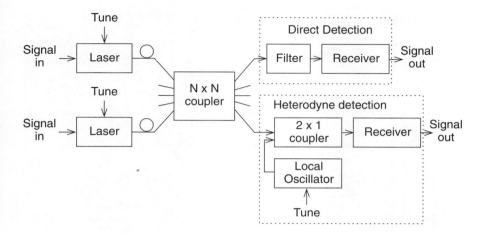

Figure 10.10 Example of a WDM network arranged as a star network. Receivers can
be either direct-detection receivers with spectral filters or heterodyne detection FDM
receivers, as shown. (After Hill, G. R., "Wavelength domain optical network techniques,"
Proc. IEEE, vol. 78, no. 1, pp. 121–132, 1990. © IEEE, 1990.)

any receiver by tuning to the receiver's preassigned wavelength. Two possible
receiver configurations are shown, an optical WDM direct-detection receiver that
uses optical filters to pass the proper wavelength, and a coherent FDM receiver that
uses heterodyne detection to tune to the proper frequency of the receiver. The N×N
coupler in this arrangement is a wideband device that combines the source power
and splits it equally among the receivers.

WDM Chain Network

In the wavelength-selective WDM *chain network* (Figure 10.11), each node
has a splitter that separates the wavelengths. The assigned wavelengths are sensed
and new information is added to the network medium at these wavelengths. Additional wavelengths can also be added to the network (provided that the output coupler of the node has inputs for them). Data can be removed and added to each wavelength channel as the information traverses the network.

WDM Ring Network

In the wavelength-selective WDM *ring network* (Figure 10.12), the configuration is similar to the chain network with its ends tied together. The number of

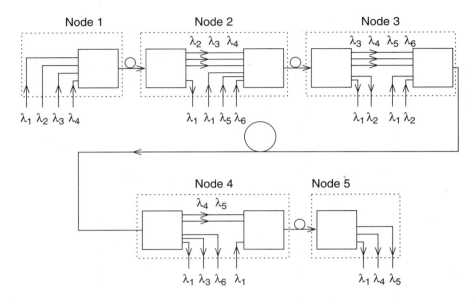

Figure 10.11 A wavelength-division multiplexed chain network.
(After Hill, G. R., "Wavelength domain optical network techniques,"
Proc. IEEE, vol. 78, no. 1, pp. 121–132, 1990. © IEEE, 1990.)

wavelengths added at a node must equal the number received, and all of the couplers and splitters are the same.

In these configurations, the basic idea is that a set of identical WDM elements provides a fixed optical path between transmitter and receiver. The wavelengths can be reused in another section of the network, thereby decreasing the number of wavelengths required to implement the network. It can be shown that a set of N nodes will require only $N(N-1)$ optical channels, rather than the N^2 wavelengths that might be expected to fully interconnect the nodes (i.e., so that any node can talk to any other node). Figure 10.13 illustrates a WDM star network that can interconnect four nodes with eight identical WDM units but requires only three wavelengths. The wavelength assignments are also shown. (Note that it is assumed that transmitter #1 does not want to connect to receiver #1, etc.) Table 10.3 shows the minimum number of wavelengths required (Hill, 1990) to implement a WDM network between N nodes; Figure 10.14 plots the relations shown in the table.

When multiplexing devices are cascaded, it should be noted that the bandwidth of the combination is reduced from the bandwidth of a single element. For example, if the individual bandwidth of an element were Gaussian-shaped with a bandwidth value of BW, then a cascade of N units would be Gaussian-shaped with a

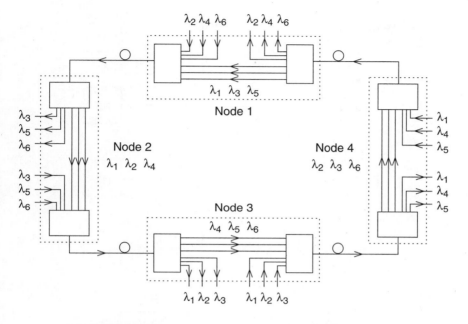

Figure 10.12 A wavelength-division multiplexed ring network.
(After Hill, G. R., "Wavelength domain optical network techniques,"
Proc. IEEE, vol. 78, no. 1, pp. 121–132, 1990. © IEEE, 1990.)

bandwidth value of $BW/\sqrt{N}$, thereby reducing the bandwidth of the combination and placing more emphasis on maintaining the source wavelength within a tighter tolerance.

10.5.3 Switched Networks

Switching in a WDM network can be accomplished by changing the source or the receiver to use a different transmitter-receiver wavelength and, hence, a different path through the network. (We note that a broadband optical path is needed, unless we want to subdivide the network into smaller channels.)

Figure 10.15 shows a switched optical network. The hub is totally passive, consisting only of 1×N and N×1 couplers. The receiver of each node is assumed to be co-located with the transmitter of that node, although shown separately on the figure for clarity. The switching is done by choosing the wavelength of a transmitter to match the appropriate receiver and then routing the traffic to that transmitter of the node. For example, the traffic at node #1 that is to be sent to node #2 is routed into

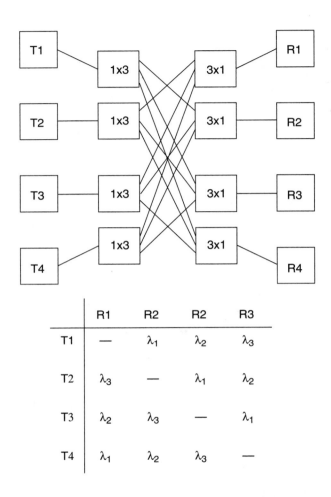

	R1	R2	R2	R3
T1	—	λ_1	λ_2	λ_3
T2	λ_3	—	λ_1	λ_2
T3	λ_2	λ_3	—	λ_1
T4	λ_1	λ_2	λ_3	—

Figure 10.13 A wavelength-division multiplexed star network with four nodes and three wavelengths. (After Hill, G. R., "Wavelength domain optical network techniques," *Proc. IEEE*, vol. 78, no. 1, pp. 121–132, 1990. © IEEE, 1990.)

the laser with wavelength λ_{12}. The WDM coupler at node #1 combines this wavelength with all of the other wavelengths (i.e., all of the traffic for the other nodes) and sends it to the hub. At the hub, the $1\times N$ wavelength-sensitive coupler divides the wavelengths and sends the traffic at λ_{12} to the $N\times1$ coupler, which combines the power at all wavelengths that are to be sent to node #2. At node #2 the channels are split by the wavelength-sensitive $N\times1$ coupler and received simultaneously. The individual messages are time-demultiplexed and routed to the appropriate user via

Table 10.3: Minimum number of wavelengths required to connect N channels with a WDM network.

Method	Minimum Number of Wavelengths
Star	$N - 1$
Chain	$\left(\dfrac{N}{2}\right)^2$ (if N is even) $\dfrac{(N-1)\,(N+1)}{4}$ (if N is odd)
Ring	$\dfrac{N(N-1)}{2}$

the switch at node #2. It can be shown (Wagner and Kobrinski, 1989) that only N wavelengths are required to completely crossconnect N nodes with this topology (rather than the N^2 wavelengths that might seem to be required).

We note from this arrangement that, while the hub has been simplified, the nodes require several sources and receivers. The number of these sources and receivers can be reduced to one if we can tune the source and the receiver filter and limit ourselves to transmitting and receiving only one message at a time. A tradeoff is

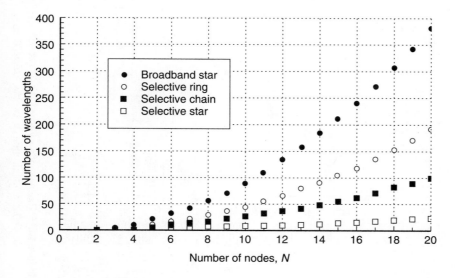

Figure 10.14 Minimum number of wavelengths required to connect N nodes of a network.

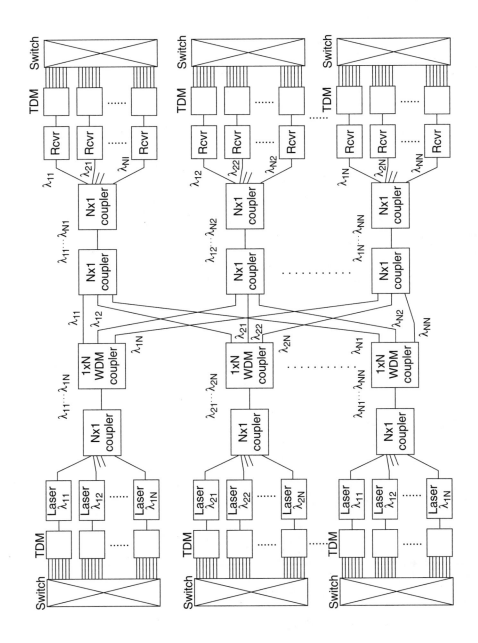

Figure 10.15 Example of a switched WDM network. (After Wagner, S. S. and Kobrinski, H., "WDM applications in broadband telecommunications networks," *IEEE Communications Magazine,* pp. 22–30, March 1989. © IEEE, 1989.)

made between the number of components and the number of simultaneous channels desired.

Various methods have been proposed to tune the sources or the receivers (or both) in a switched network.

10.5.4 Source Tuning

Some mechanisms for rapidly tuning a source have been identified.

- *Temperature/current tuning*—The source wavelength can be changed by a modest amount by changing the temperature of the source (a resulting change of about 0.08 nm · °C^{-1} for a distributed-feedback laser, making a 1 nm change feasible) or by changing the laser current (a resulting change of about 200 MHz · mA^{-1} for switching rates higher than 10 MHz, making major tuning infeasible) (Hill, 1990).

- *DFB and DFR tuning*—Broader tuning ranges can be achieved (Hill, 1990) by combining the DFB laser with a phase-control section and an active section or by fabricating an integrated structure that controllably combines the power of different laser structures oscillating at various wavelengths.

- *External element tuning*—Additionally, it is possible to use a Fabry-Perot laser with an external cavity that has a movable grating controlled by a piezoelectric pusher. Tuning with this technique has been demonstrated over 55 nm (Wyatt and Devlin, 1983), although rapid, accurate tuning of this device still needs to be demonstrated.

10.5.5 Receiver Tuning

To tune the receivers, tunable filters can be based on technology using multilayer filters, Fabry-Perot interferometers, or integrated optic devices.

- A tunable *multilayer filter* device can be made whose transmission properties vary spatially across the face of the filter (e.g., by varying layer thickness). Tuning is accomplished by translating the filter. A filter that uses this technique has reported coverage of 270 nm with a passband of 7 nm (McCartney et al., 1987).

- *Fabry-Perot interferometer filters* are tuned by changing the length of the cavity using a piezoelectric pusher. Devices with 30-channel operation have been demonstrated (Hill, 1990). Tuning time can be a few microseconds.

- *Integrated optic tunable filters* that use TE-to-TM mode convertors (i.e., that connect transverse electric modes to transverse magnetic modes), have been demonstrated in the laboratory (Hill, 1990).

- Distributed feedback lasers operated below threshold (i.e., operated as amplifiers rather than oscillators) present narrow passbands. The devices can be tuned by incorporating a phase-control section. The response times are fast (on the order of nanosecond switching times) and tuning ranges of 71 GHz (for 10-channel operation) have been demonstrated (Hill, 1990).

10.6 Nonlinear Effects on WDM Links

Nonlinear effects, as discussed in Chapter 3, can be deleterious on multisignal data links (Chraplyvy, 1990; Waarts et al., 1990). Chraplyvy (1990) analyzed the effects of nonlinearities on both single-signal and multisignal fibers.

- For multisignal fiber links, he found that *stimulated Raman scattering* would have a negligible effect.

- The effects of *carrier-induced phase noise* (i.e., changes in the optical phase of one signal due to changes in the power in itself or in other light waves in the link) will present problems for phase-modulated signals in a WDM link. Broadband multiplexed links will need to keep their power levels below 20 mW to avoid this effect; selective-wavelength links will need to keep their power below a few milliwatts to successfully operate over 10 or so channels.

- The effects of *Brillouin scattering* depend only on the power level and are independent of the number of channels in the link. Power limits of 10 mW are suggested (Chraplyvy, 1990) for maximum-performance links. (Generally, broadband systems require that each source's power level be degraded by a factor of $1/N$ from the maximum dictated by the nonlinearity [where N is the number of channels in the link]; wavelength-selective links allow *each* source to achieve the power limit dictated by the nonlinearity.)

- *Four-photon mixing* occurs when two propagating light waves mix nonlinearly as they travel through the fiber and produce frequency components at $2\nu_1 - \nu_2$ and $2\nu_2 - \nu_1$. These sidebands grow and, through further nonlinear interaction, produce nine new optical frequencies at $\nu_{ijk} = \nu_i + \nu_j - \nu_k$ with $i, j, k = 1,2,3$. The appearance of these new signals and the power reduction of the original signals degrade the performance of the link. The strength of the effect depends on the channel separation and the fiber dispersion. To minimize the effect, Chraplyvy (1990) calculates that the channel spacing will have to be larger than 50 GHz and that the region of zero-dispersion should be avoided (a surprising result that reduces the data rate–distance product of the link). Source power levels of less than a few milliwatts are recommended. Waarts et al. (1990) perform a similar analysis of the effects of nonlinearities on *coherent* frequency-division multiplexed links where the channels are desired to be very close to each other.

10.7 Summary

In this chapter, we have seen that current fiber systems utilize only a fraction of the spectral width available in both multimode and single-mode fibers. Wavelength-division multiplexing (WDM) offers a method to fill up the fiber's passband with several sources of different wavelengths. WDM techniques hold much promise for allowing vast amounts of data to be transferred over fiber links.

Wavelength-division multiplexers and demultiplexers can be either wavelength-selective or broadband. The wavelength-selective demultiplexers typically use an angularly dispersive element, such as a diffraction grating, to separate the wavelengths. Filters of various types have better control of the wavelength passband but have an insertion loss that limits the number of filters that can be cascaded to just a few filters.

WDM techniques require sources with a controllable operating wavelength and, having been set at a specific wavelength, require long-term stability to maintain that wavelength. Various techniques have been described including filtered LEDs, Fabry-Perot resonator lasers, DFB diode lasers, and frequency-locked diode lasers.

Various sample architectures have been described which allow tunable point-to-point communications or broadcast communications. The architecture for an optical-switched network has also been described.

10.8 Problems

1. Consider the windows in a single-mode fiber.
    (a) Calculate the width in Hz of the optical window extending from 1250 to 1350 nm.
    (b) Calculate the width in Hz of the optical window extending from 1450 to 1600 nm.
    (c) Find the total bandwidth in Hz in both windows combined.

2. Consider the geometric arrangement of Figure 10.6.

    (a) If two wavelengths are separated by $\Delta\lambda$, show that the focussed spots at the receiving fiber locations are separated by $w = f\,\Delta\lambda\,d\theta/d\,\lambda$.

    (b) From the result of the previous part, calculate an expression for the maximum outside diameter of the receiving fibers.

3. (a) The typical channel separation of an interference filter is 5 nm. Calculate the equivalent frequency linewidth $\Delta\nu$.

    (b) Calculate the ratio of $\Delta\nu$ for the filter of part (a) to the 10 GHz frequency linewidth achievable with a heterodyne receiver.

4. Calculate the value of reflectivity required for a Fabry-Perot filter to have a ΔF of 0.1 nm if the mirror separation is 1 cm and the operating wavelength is 1.5 μm.

5. Consider a 600 MB $\cdot$ s^{-1} signal with a 5 nm linewidth source.

    (a) Find the maximum dispersion-limited transmission distance.

    (b) Repeat the calculation if the data rate is 150 Mb $\cdot$ s^{-1}.

6. Make a wavelength assignment table similar to the one shown in Figure 10.13 for the five-node chain network shown in Figure 10.11. (Note that each transmitter can talk only to its downstream neighbors (e.g., transmitter 2 can talk to receivers 3, 4, and 5, but it cannot not talk to 1.)

7. Make a wavelength assignment table similar to the one shown in Figure 10.13 for the four-node ring network shown in Figure 10.12.

8. Table 10.3 gives the minimum number of wavelengths required to connect N channels in a WDM star network. Sketch the star network configuration for $N = 3$.

References

Chraplyvy, A. R., "Limitations on lightwave communications imposed by optical-fiber nonlinearities," *J. Lightwave Technology*, vol. 8, no. 10, pp. 1548–1557, 1990.

Hill, G. R., "Wavelength domain optical network techniques," *Proc. IEEE*, vol. 77, no. 1, pp. 121–132, 1990.

Ishio, H., J. Minowa, and K. Nosu, "Review and status of wavelength-division-multiplexing technology and its application," *J. Lightwave Technology*, vol. LT-2, no. 4, pp. 448–463, 1984.

Kaminow, I. P., "Non-coherent photonic frequency-multiplexed access networks," *IEEE Network*, pp. 4–12, 1989.

Keiser, G., *Optical Fiber Communications,* 2nd Edition. New York: McGraw-Hill, 1991.

Lee, T., "Recent advances in long-wavelength semiconductor lasers for optical communication," *Proc. IEEE*, vol. 79, no. 3, pp. 253–276, 1991.

McCartney, J., P. Lissberger, and A. Roy, "Recent developments in the production of narrowband position tuned interference filters for wavelength multiplexed optical fiber systems," *Optics Communications*, vol. 64, pp. 338–342, 1987.

Suzuki, S., M. Nishio, T. Numai, M. Fujiwara, M. Itoh, S. Murata, and N. Shimosaka, "A photonic wavelength-division switching system using tunable laser diode filters," *J. Lightwave Technology*, vol. 8, no. 5, pp. 660–666, 1990.

Waarts, R. G., A. Friesem, E. Lichtman, H. H. Yaffe, and R. Braun, "Non-linear effects in coherent multichannel transmission through optical fibers," *Proc. IEEE*, vol. 78, no. 8, pp. 1344–1368, 1990.

Wagner, S. S. and H. Kobrinski, "WDM applications in broadband telecommunication networks," *IEEE Communications Magazine*, pp. 22–30, March 1989.

Wyatt, R. and J. Devlin, "10 kHz linewidth 1500 nm InGaAsP external cavity laser with 55 nm tuning range," *Electronics Letters*, vol. 19, pp. 110–112, 1983.

Yariv, A., *Optical Electronics,* 4th Edition. New York: Holt Rinehart and Winston, 1991.

Chapter 11

Fiber-Optic Sensors

11.1 Introduction

The advantages of fiber optics that were listed in Chapter 1 make them ideal candidates for sensor applications in environments where EMP and EMI are harsh, where low-volume, light-weight sensors are required (such as aircraft and space vehicles), and in systems requiring wide bandwidths (Giallorenzi et al., 1982; Davis, 1985; Dandridge and Cogwell, 1991). In this chapter we will find that fiber sensors exist to measure many physical parameters including liquid levels, gas concentrations, pressure (Hocker, 1979; Fields et al., 1980), acoustic fields (Bucaro et al., 1977a and 1977b; Cole et al., 1977; Bucaro and Carome, 1978; Spillman and Gravel, 1980; Rines, 1981; Gardner et al., 1987; Danielson and Garrett, 1989; Lagakos et al., 1990), electric fields, magnetic fields, rotation rates, and temperature (Hocker, 1979; Lagakos et al., 1981a). The effects used for the measurements are widely varied and can use modifications of the light amplitude, phase, or polarization. Amplitude sensors are the easiest to implement; phase sensors show much promise in the laboratory but are more demanding in the optical components used. Polarization sensors, in particular, need to use polarization-maintaining fiber.

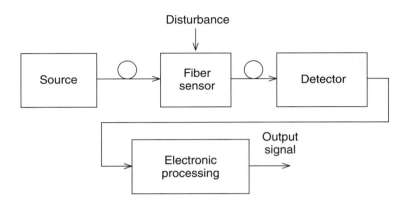

Figure 11.1 Components of a fiber-optic sensor system.

11.2 Sensor Systems

Figure 11.1 shows the components of a fiber-sensor system. All components are the same as a fiber-communications system except for the presence of an optical-modulation element in the fiber that changes an optical property of the transmitted light in some fashion. Various mechanisms and effects have been used in fiber sensors; the trick is to achieve the desired change due the field to be measured with adequate sensitivity, yet to isolate the optical field from other environmental effects.

Figure 11.2 illustrates three different types of sensors. In Figure 11.2(a), the *transmission-gap sensor*, the light enters a space between the fibers. Within this gap the light is perturbed in some fashion by the disturbance being measured. Usually the amplitude of the transmitted light is affected. In Figure 11.2(b), the *evanescent-wave sensor*, a portion of the fiber has its outside jacket and cladding stripped away from the core in the sensing region of the fiber. The presence of the sensed field within the region alters the guiding properties of the fiber (by affecting the propagation properties of the evanescent waves in the cladding) and couples some of the light out of the core, causing a corresponding change in the power at the detector. Figure 11.2(c) shows an *internal-sensing fiber,* where the propagation properties of the light in the core are changed by the sensed quantity (e.g., the phase of the light is altered). The sensing fiber itself is not modified by the user, as is the case in the prior two fiber sensors.

The sensors can be subdivided into amplitude sensors and phase sensors. They are each discussed below.

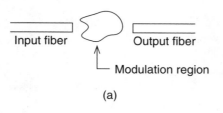

Input fiber

Output fiber

Modulation region

(a)

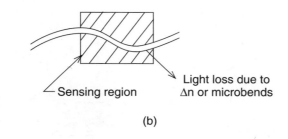

Sensing region

Light loss due to
Δn or microbends

(b)

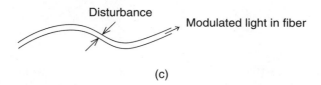

Disturbance

Modulated light in fiber

(c)

Figure 11.2 Three types of fiber sensors. (a) Transmission-gap sensor.
(b) Evanescent-wave sensor. (c) Internal-sensing sensor.

11.3 Amplitude-Modulation Sensors

In an *amplitude-modulation* sensor, the field to be measured causes a change in the amplitude or power of the light in the core. Ideally, this change is linearly proportional to the strength of the disturbing field. These sensors are fairly simple to implement, are low-cost (since they can use multimode components), and have adequate sensitivity for most applications (but less sensitivity than most phase-modulation sensors). There are generally two types of amplitude-modulation sensors, those that use transmission across a gap in the fiber and those that use microbend losses in the fiber.

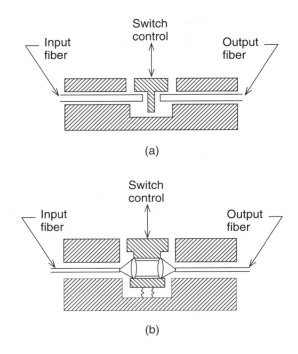

Figure 11.3 Fiber-optic implementations of a microswitch.
(a) With a shutter device. (b) With a set of expanded-beam lenses.

11.3.1 Transmission-Gap Sensors

In a transmission-gap sensor, light is sent down the core of one or more fibers, transmitted across a gap (that might contain a modulating element), and recollected in either another fiber or in the same fiber (for reflective techniques). The collected light varies in amplitude (or power) in proportion to the disturbance.

The simplest transmission-gap sensor is a microswitch sensor, illustrated in Figure 11.3. In Figure 11.3(a), a shutter is placed across the optical path or removed from the path (by a spring mechanism that is not shown). Figure 11.3(b) shows an expanded-beam version of the same device that relaxes the mechanical tolerances that are required to implement the switch.

A reflection version of these devices can be made by replacing the output side of the device with a reflecting surface and requiring the input fiber to couple the reflected light out of the device to the sensor. The same principle can be used with grating shutters, optical shaft encoders, and vibrating structure detection (e.g., vibrating wires, vibrating quartz structures, or vibrating micromachined silicon

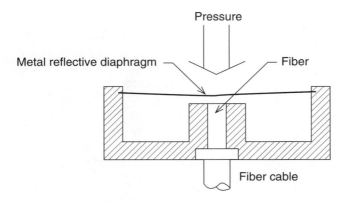

Figure 11.4 A typical fiber-optic pressure sensor, representing a transmission-gap sensor.

structures). To achieve a variable attenuation of the light, the shutter can be replaced by a movable attenuating wedge or attenuating grating.

Reflection methods can include reflection from a movable surface. The amount of light reflected and collected depends on the transmit-receive geometry and the spacing between the fiber(s) and the reflector. To first order, the portion collected is calculated from the ratio of the area of the collecting fiber's core to the size of the expanded beam at the collecting fiber. (Actually, only the light within the collecting fiber's NA will be coupled into the collecting fiber.)

Figure 11.4 shows a typical reflection fiber-optic pressure sensor. The metal diaphragm will flex with increasing pressure. Increased flexure will reflect the light over a wider angle, resulting in less light being picked up by the receiving fiber (the same fiber, in this case). Other means of changing the light amplitude include absorption of the light (either uniformly or at certain wavelengths), excitation of fluorescence at one wavelength by stimulation at another wavelength, movable blocks in the gap, etc.

The effect of the modulating material can be more subtle than described. For example the region between the fibers can contain a gas absorption cell with a sample of gas. While few gas species of interest have absorption lines near the 1.3 µm wavelength used in fiber systems, several species have harmonic absorptions that lie in this region. For example, methane has a third-harmonic absorption at 1.33 µm. To cancel out common path effects, two measurements of the light transmitted through the cell are made, one at the wavelength of the absorption peak and one at a wavelength well away from the peak. (The source diode wavelength needs to be tunable, by changing the diode's temperature, for example.) Taking a ratio of the recorded intensities will reveal the species concentration in the light beam.

The temperature of an environment can be measured by the change of the optical absorption coefficient of some materials (Figure 11.5). The amount of light

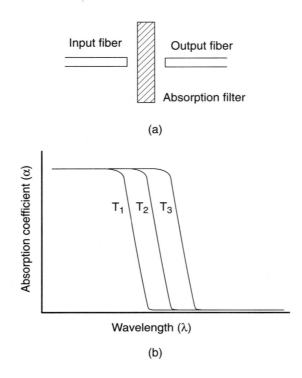

Figure 11.5 (a) A fiber-optic temperature sensor.
(b) Shifts in the absorption coefficient with temperature.

transmitted across the gap is quite sensitive to alterations of the absorption coefficient due to changes in the ambient temperature. The figure shows that, for an operating wavelength in the middle of the T_2 absorption coefficient transition region, the absorption coefficient of the material will be high at T_3, moderate at T_2, and negligible at T_1. These various coefficients convert into changes in transmitted power, and, if we know $\alpha(T)$ (at the wavelength of interest) and the thickness of the material, the measured changes in power can be converted into the corresponding changes in temperature.

11.3.2 Microbend-Loss Sensors

The *microbend-loss sensor* is quite sensitive and was among the first fiber sensors studied (Pitt, 1990). The fiber is mechanically deformed, as shown in Figure 11.6. The deformation is caused by the use of a pair of toothed plates that are

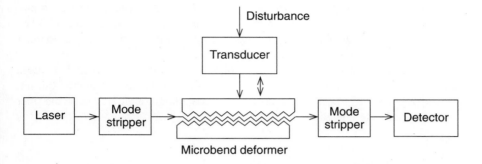

Figure 11.6 Mechanical fiber deformation in a microbend sensor.

clamped onto the fiber. This deformation causes some of the light in the core to leak into the cladding, since the critical angle condition is no longer met. For small deformations (on the order of 1 μm), the amount of light removed from the core is proportional to the deformation, which can be made proportional to the field to be measured. By carefully monitoring the power in the core, the changing light amplitude is measured at the end of the fiber.

In these sensors the inverse of the spacing between the teeth is the *spatial frequency* of the teeth. The *sensitivity of the sensor* is a function of this spatial frequency due to the nature of the coupling of the mechanical motion of the sensor and the modulation of the light in the fiber core. This sensitivity S can be expressed as the ratio of the change in the light power in the core ΔP to the change in the field being measured ΔF, given by

$$S = \frac{\Delta P}{\Delta F}. \tag{11.1}$$

This sensitivity can be expressed (Bucaro et al., 1982) as the product of two terms

$$S = \frac{\Delta P}{\Delta x} \frac{\Delta x}{\Delta F}, \tag{11.2}$$

where $\Delta P/\Delta x$ is the change in power due to the size deformation of the fiber Δx (measured perpendicular to the fiber propagation axis) and $\Delta x/\Delta F$ is the deformation in the fiber for the force F applied to the toothed plate by the field being measured. To maximize $\Delta P/\Delta x$, we use fibers with large NAs, a large number of deformations (i.e., lots of teeth in the device), an optimized spatial frequency of the teeth, a large ratio of fiber core to fiber diameter, and fibers with graded-index construction. To maximize the $\Delta x/\Delta F$ term we want small diameter fiber, fewer deformations (requiring a tradeoff with the prior factor), and a larger spatial frequency. The $\Delta P/\Delta x$ term is well understood; future efforts will concentrate on techniques to maximize the $\Delta x/\Delta F$ term (Bucaro et al., 1982).

Other examples (Pitt, 1990) of amplitude-modulation sensors include those based on fibers that transmit fluorescence intensity, blackbody radiation from hot objects, or blackbody emissions from specially-coated light emitters put on the fiber tips to respond to temperature.

11.4 Evanescent Sensors

In an evanescent-wave sensor, a region of the fiber cable and some of the cladding are stripped away and the bare sides of the fiber are exposed to the surroundings, as illustrated in Figure 11.2(c). If the refractive index of the surrounding material is equal to that of the cladding, then no light will escape from the core. If the index of the surrounding material is different from the cladding index, then some light will escape. The greater the difference in the index of refraction, the more light will escape from the core, causing a reduction in the power transmitted through the fiber. (The reduction in power can also be sensed at the transmitter end by using an optical time-domain reflectometer.)

A classic application of this technique is the liquid-level detector shown in Figure 11.7. Here the tip of the fiber has been cut to ensure total internal reflection of the light when the external medium is air (Figure 11.7a). When the liquid level rises, as in Figure 11.7(b), the index matches that of the core and the total internal reflection is frustrated, thereby decreasing the intensity of the light reflected back to the detector. (The reflection is, in general, now a partial reflection rather than a total reflection.) The loss of power in the core is sensed and converted into a reading of the liquid level. The sensor is in the form of a hook to allow gravity to aid in keeping a droplet from forming on the sensor end, thereby giving an erroneous reading.

Other forms of the evanescent sensor involve the use of coatings on the bare fiber that change optical properties in the presence of whatever is being sensed (e.g., chemical species sensors, color change sensors).

11.5 Phase Sensors

The most sensitive fiber-optic sensors are those that detect changes in the phase of the optical signal due to an external disturbance. These techniques require the use of single-mode fibers and coherent sources.

From source   To detector

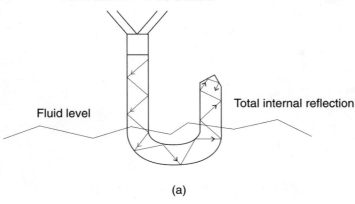

Fluid level

Total internal reflection

(a)

From source   To detector

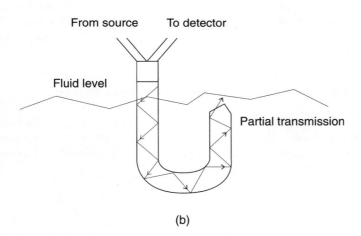

Fluid level

Partial transmission

(b)

Figure 11.7 Liquid-level fiber sensor demonstrating evanescent-wave detection method. (a) During low fluid level the tip of the fiber is in air. (b) When the fluid level rises, the tip is in fluid and the reflection changes.

The phase shift $\Delta\varphi$ encountered by an optical signal in traversing a fiber of length L is

$$\Delta\varphi = knL , \qquad (11.3)$$

where $k = 2\pi/\lambda_0$ and n is the index refraction of the core. Both n and L can be changed by an external disturbance. Various techniques can be used to convert the

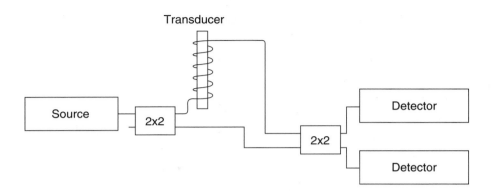

Figure 11.8 Mach-Zender interferometer fiber sensor.

phase modulation to an amplitude modulation, especially the use of interferometers. The four configurations of interferometer are the Mach-Zender interferometer (Figure 11.8), the Michelson interferometer (Figure 11.9), the Fabry-Perot interferometer (Figure 11.10), and the Sagnac interferometer (Figure 11.11).

11.5.1 Mach-Zender Interferometer Fiber Sensor

As an example, consider the use of the Mach-Zender interferometer fiber sensor. At the recombination location, the phase-shifted beam (the *signal beam*) combines with the unshifted beam (the *reference beam*). Using plane-wave representations of the two beams for simplicity, we have

$$E_{\text{total}} = A_1 e^{j(2\pi vt)} + A_2 e^{j(2\pi vt + \Delta\varphi)} \tag{11.4}$$

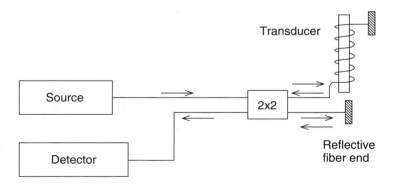

Figure 11.9 Michelson interferometer fiber sensor.

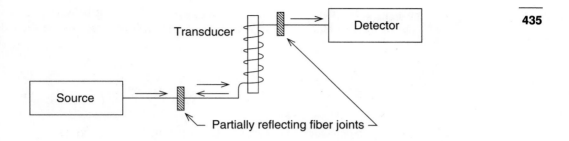

Figure 11.10 Fabry-Perot interferometer fiber sensor.

The output current of either detector is proportional to $|E_{tot}|^2$. So,

$$i_{out\,1} \propto \left\{ A_1 e^{j(2\pi vt)} + A_2 e^{j(2\pi vt + \Delta\varphi)} \right\}$$
$$\cdot \left\{ A_1 e^{j(-2\pi vt)} + A_2 e^{j(-2\pi vt + \Delta\varphi)} \right\} \tag{11.5}$$

$$\propto A_1^2 + A_2^2 + A_1 A_2 e^{j\Delta\varphi} + A_1 A_2 e^{-j\Delta\varphi} \tag{11.6}$$

$$\propto A_1^2 + A_2^2 + 2A_1 A_2 \cos \Delta\varphi . \tag{11.7}$$

If $A_1 = A_2 = A$, then the equation reduces to

$$i_{out\,1} \propto 2A^2(1 + \cos \Delta\varphi) . \tag{11.8}$$

Hence, we find that the output has a term that is proportional to the change in phase. The size of the output will rise and fall, depending on the size of $\Delta\varphi$. (There is no restriction on the size of $\Delta\varphi$; it can range from fractions of a radian to many millions of radians, depending on the effect that is being measured.) Because of the phase

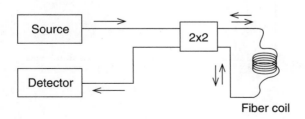

Figure 11.11 Sagnac interferometer fiber sensor.

shift encountered at the final combiner, there is phase shift of π between the two paths out of the combiner (Wilson and Hawkes, 1989). The output of the second detector is

$$i_{\text{out 2}} \propto 2A^2(1 + \cos{(\Delta\varphi + \pi)}) \tag{11.9}$$

$$\propto 2A^2(1 - \cos{\Delta\varphi}) . \tag{11.10}$$

Applying the two detector output currents to a differential amplifier gives

$$v_{\text{out}} \propto (i_{\text{out 1}} - i_{\text{out 2}}) \tag{11.11}$$

$$\propto 4A^2 \cos{\Delta\varphi} . \tag{11.12}$$

This output voltage is directly proportional to the cosine of the phase shift. The application of the differential amplifier has doubled the sensitivity of the system and has advantages when the amplitudes of the two waves are not equal.

The normalized response of the system using the differential amplifier is shown in Figure 11.12. The sensitivity to small changes in phase is largest at the linear portions of the curve, in between the peaks. These are the so-called *quadrature points*. The least sensitivity is obtained at the maximum and minimum peaks of the curve.

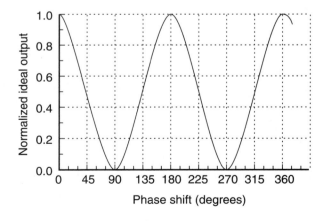

Figure 11.12 Response of a Mach-Zender interferometer fiber sensor.

11.5.2 Enhancing Sensor Response

A major problem in phase sensors is isolating the undesired changes in the phase caused by shifts in the environment (e.g., temperature, pressure) from the desired changes in the signal (e.g., acoustic waves, magnetic fields, temperature changes, pressure changes). Two techniques are used to help isolate the desired response from the other effects, the use of coatings on the sensing fiber and the use of signal processing.

Fiber Coatings

Frequently the fiber's sensing region is coated with special materials to enhance its response to the desired field (Lagakos et al., 1981b). For example, a magnetostrictive material, such as nickel, can be applied to the fiber to increase its response to magnetic fields. The increased stiffening of the fiber will also decrease its response to acoustic fields. Other coatings that can be used are (Pitt, 1990): polymers for enhanced acoustic response, PVDF for electric field measurements, and palladium for gas measurements. Similarly the mounting of the fiber on a support mechanism (e.g., a spool or *mandrel* for a coil of fiber) can enhance the response through the proper selection of the mandrel material, its mechanical design (to take advantage of mechanical resonances), and its orientation with respect to the sensed field (Gardner et al., 1987; Danielson and Garrett, 1989).

Signal Processing

Various signal processing techniques have also been implemented to accomplish this isolation of the sensed field. One frequently used sensor is shown in Figure 11.13. This detector is called an *active phase-tracking homodyne-detection sensor* (Wilson and Hawkes, 1989). The device in the reference arm is an annulus made of piezoelectric material. When voltage is applied in the radial direction across the annulus, the diameter expands, causing an increase in the length of the fiber wrapped around the device. (The induced phase shift in the reference arm can be increased by wrapping more turns around the fiber.) The idea of the sensor is to use feedback to induce a change in the reference arm phase that will just match the change in phase in the sensor arm. The output voltage of the differential amplifier will remain locked to the quadrature point if the device works as proposed; that is, we want to keep the total phase at

$$\Delta\varphi + \Delta\varphi_R = (2m + 1)\frac{\pi}{2}, \qquad (11.13)$$

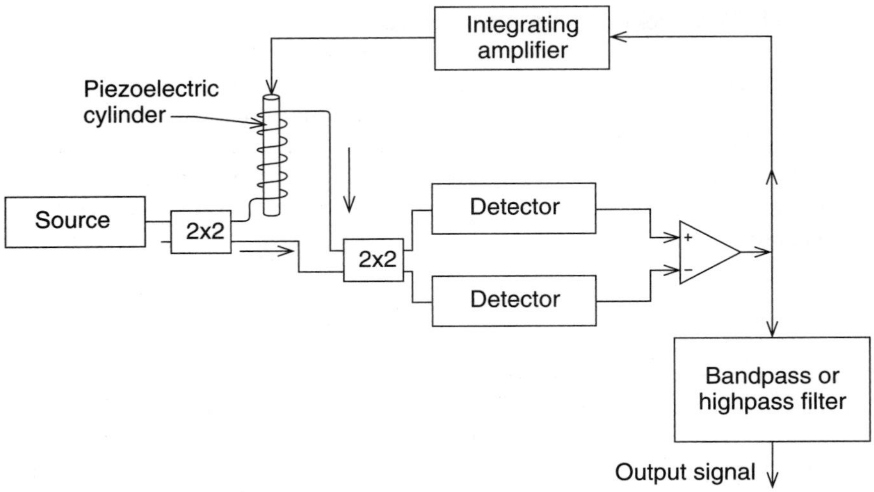

Figure 11.13 An example of an active phase-tracking homodyne-detection sensor.

where $\Delta\varphi_R$ is the change in phase of the reference arm and m is an integer. For small signals ($<< \pi$), the output of the differential amplifier contains the information about the change in phase in the signal arm. (For large changes in phase, the output is nonlinear with respect to the change in phase and an ambiguity can arise because of the periodic nature of the detection process.)

While this technique can be used to cancel out the drift in the phase of the sensor arm due to environmental changes, it also will tend to cancel out the desired information about the quantity to be sensed. Mathematically, the signal term can be broken up into a phase shift due to the desired phenomenon $\Delta\varphi_s$ and the unwanted environmental effects $\Delta\varphi_e$. To measure the signal, we need to resort to some signal processing based on *a priori* information. One assumption that might be used in the sensing of acoustic waves created by a ship, for example, is that the frequency of the ship's sound will probably be at a higher frequency than the slowly changing environmental parameters (e.g., ocean temperature). We can use this information to separate the environmental effects from the signal by integrating the differential amplifier output for a time that is long for the signal but short compared to the change in the environmental parameters. In this way, the signal will integrate out of the feedback signal but the slower environmental signal will passed to the piezoelectric element for cancellation.

The signal can be separated from the output of the differential amplifier by passing through a bandpass filter or a highpass filter to remove the environmental component.

The feedback loop is keeping

$$\Delta\varphi_R + \Delta\varphi_s + \Delta\varphi_e = (2m+1)\frac{\pi}{2}. \tag{11.14}$$

We can rewrite this as

$$\Delta\varphi_R = (2m+1)\frac{\pi}{2} - \Delta\varphi_s - \Delta\varphi_e. \tag{11.15}$$

Since $v(t)$ (the voltage applied to the piezoelectric device) is proportional to $\Delta\varphi_R$, it contains information about $\Delta\varphi_s$ (and $\Delta\varphi_e$). This linearity in the feedback voltage is good for many cycles of 2π, up to the expansion limits of the piezoelectric device. The separation of $\Delta\varphi_s$ from $\Delta\varphi_e$ is usually done with electric filters, as described before.

11.5.3 Induced Phase Shift

We now want to consider the sensitivity of the fiber to the disturbance to be measured. The phase change φ for a mode that travels straight down a fiber of length L is

$$\varphi = \frac{2\pi L n_1}{\lambda_0}, \tag{11.16}$$

where n_1 is the index of the core. The external field (i.e., the quantity to be measured) can change both the length of the fiber and the index of refraction of the medium. The resulting change in phase due to a change in a parameter p, then, is

$$\Delta\varphi = \frac{2\pi L}{\lambda_0}\frac{dn_1}{dp}\Delta p + \frac{2\pi n_1}{\lambda_0}\frac{dL}{dp}\Delta p \tag{11.17}$$

(It can be shown (Wilson and Hawkes, 1989) that the effects of a changing p on the core radius a are negligible compared to the changes in L and n_1.)

The effect of an axial force F will predominately change L. The change in L is dependent on the Young's modulus of the material E and is given by

$$\Delta L = \frac{FL}{AE},$$ (11.18)

where A is the cross-section area of the fiber. (The Young's modulus of silica glass is $E = 2 \times 10^{11}$ Pa.) The corresponding change in the phase is

$$\Delta \varphi = \frac{2\pi n_1 LF}{\lambda_0 AE}$$ (11.19)

for an axial force F.

A fiber that is subjected to a uniform radial pressure P will have its radius decreased (a negligible effect) and its length L increased. The increase in length is given by

$$\Delta L = \frac{2\xi PL}{E},$$ (11.20)

where ξ is the material's Poisson ratio ($\xi = 0.2$ for silica glass). The change in phase, then, is given by

$$\Delta \varphi = \frac{4\pi n_1 L\xi P}{\lambda_0 E}.$$ (11.21)

A change in temperature ΔT changes both the length *and* the refractive index. The change in length is

$$\Delta L = L\alpha \, \Delta T,$$ (11.22)

where α is the coefficient of thermal expansion of the material ($\alpha = 5 \times 10^{-6}$ m $\cdot$ K^{-1} for silica glass). The total change in phase is

$$\Delta \varphi = \frac{2\pi L}{\lambda_0} \left(\frac{dn_1}{dT} + n_1 \alpha \Delta T \right).$$ (11.23)

The coefficient of thermal changes in the index of refraction is $dn/dt = 7 \times 10^{-6}$ K^{-1} for silica glass.

Some the problems at the end of the chapter will show that the temperature effects are much larger than the pressure and longitudinal force effects.

If we can directly measure changes in force, pressure, and temperature, how are other parameters (e.g., magnetic fields) measured? Usually we incorporate a transducer to convert the desired field into a parameter that we *can* measure. For example, we can attach a fiber to a magnetostrictive element to change the fiber's length in response to the desired magnetic field (although linearity in magnetostrictive materials is a problem). Another technique is to cover the fiber with a coating that incorporates magnetostrictive material to convert the changing magnetic field into a change in axial pressure on the fiber. (The problem, of course, is to maintain temperature stability of the fiber arms so that the change in temperature does not mask the effect of the desired field.)

11.6 Fiber-Optic Gyroscopes

A special case of a fiber-optic sensor is the *fiber-optic gyroscope*, used to detect rotation. The primary principle of the gyroscope uses the *Sagnac effect*. We begin by considering a circular coil of fiber around an axis of rotation, as shown in Figure 11.14. The coupler at the input separates the transmitted beam into two equal-amplitude, clockwise and counter-clockwise rotating beams. If the plane containing the fiber is not rotating, then the transit time τ of the beams is given by (Bergh et al., 1984; Wilson and Hawkes, 1989)

$$\tau = \frac{2\pi R}{c} \,, \tag{11.24}$$

where R is the radius of the loop. (For a noncircular loop, we would replace $2\pi R$ by the perimeter length of the fiber; for N loops we would use $2\pi NR$.) If the plane of the fiber were rotating in a clockwise direction at a rate of Ω radians per second, then the clockwise beam will have a slightly longer path to cover before the beams meet and the counter-clockwise beam will have a shorter path. The length of the longer path L' is approximately

$$L' = 2\pi R + R\Omega\tau \tag{11.25}$$

and the length of the shorter path L'' is

$$L'' = 2\pi R - R\Omega\tau \,. \tag{11.26}$$

The difference between the propagation times of the counterpropagating beams is (Bergh et al., 1984; Wilson and Hawkes, 1989)

$$\Delta t = \frac{R(2\pi + \Omega\tau)}{c} + \frac{R(2\pi - \Omega\tau)}{c} \tag{11.27}$$

$$= \frac{4\pi R^2 \Omega}{c^2} \,. \tag{11.28}$$

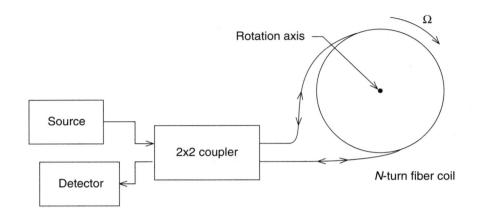

Figure 11.14 Simplified Sagnac interferometer with coil of fiber around a rotation axis.

For continuous waves of frequency ν, the corresponding phase difference $\Delta\varphi$ between the waves is

$$\Delta\varphi = 2\pi\nu\Delta t \tag{11.29}$$

$$= \frac{8\pi^2 R^2 \nu}{c^2}\Omega \tag{11.30}$$

$$= \frac{8\pi A\Omega}{c^2}, \tag{11.31}$$

where A is the area of the loop. (For N loops, we would use NA for the area.) For a fiber of length L wrapped in a circle of diameter D, the phase shift relation reduces to

$$\Delta\varphi = 2\pi\Omega\left(\frac{LD}{\lambda c}\right). \tag{11.32}$$

(Note that the phase shift does *not* depend on the index of refraction of the fiber. This is because the Fresnel-Fizeau drag effect due to the medium's movement cancels the effects of the changed path length [Bergh et al., 1984].)

It is interesting to note that too much coherence in the source can be an undesirable property for this application. As the beams propagate, there is a small amount of backscattered light. If the light is coherent, this backscattered light will interfere with the desired beams, creating a spurious response that can mask the desired interference of the counterpropagating beams. By reducing the coherence of the source (e.g., by using a *superluminescent LED*), only beams that have nearly the same pathlength (i.e., the counter-rotating beams) will be able to interfere. Hence, the backscattered light will have a diminished effect.

While ring-laser gyroscopes (another type of optical gyroscope) have been implemented commercially, fiber-optic gyroscopes are still in the development stage.

11.7 Polarimetric Sensors

Polarimetric sensors are used to detect changes in the polarization state of the light caused by an external disturbance. Figure 11.5 shows a transmission-gap sensor that can be used to measure pressure. Light is introduced into the gap from the collimator on the input fiber and passes through a linear polarizer (that converts the unpolarized light into linear-polarized light) and through a quarterwave plate (that converts the linear-polarized light into circular-polarized light). The circular-polarized light passes through a photo-elastic material (e.g., polystyrene), where the mechanical stress induced in the sheet by the pressure-sensing diaphragm modifies the polarization. The light is then converted back into linear-polarized light by the second quarterwave plate and is analyzed by the second linear polarizer with a known orientation. After being focussed by the output lens system (or by graded-index optics), the light power is measured at the receiver. The amount of received power depends, then, on the pressure applied to the diaphragm.

Recent detection systems that use coherent detection to achieve more sensitivity or to detect some physical field, such as a magnetic field or an acoustic field, require that the polarization of the output light be maintained so that a polarized

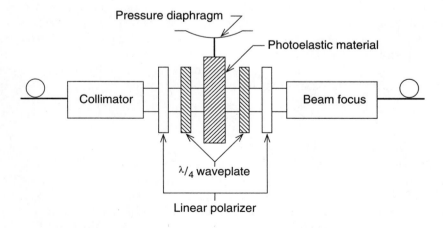

Figure 11.15 Transmission-gap polarization fiber sensor for measuring pressure.

local oscillator can be mixed with the output. have been invented for such uses. *Polarization-maintaining fibers* (Payne et al., 1982; Noda et al, 1986; Sasaki, 1987; Okoshi and Kikuchi, 1988; Sears, 1991) have been invented for such uses.

11.7.1 Polarization-Maintaining Fibers

In single-mode fibers, the single mode that can propagate can have two polarization modes (i.e., it is actually two modes with orthogonal polarizations). At any given location, the single mode that propagates is doubly degenerated with two orthogonal polarizations, the LP_{11}^x mode and the LP_{11}^y mode. (The orthogonal modes could be represented by two counter-rotating circular polarizations, as well.) In an ordinary single-mode fiber, these modes have almost identical propagation velocities. For most communication applications this small difference in velocity is ignored; in sensor applications the difference in velocity can be useful and can even be exaggerated to allow for more sensitivity.

Optical fibers generally do not retain the polarization of the input wave for more than a few meters if illuminated with a linearly polarized light. Various perturbations along most fibers cause the energy to couple between the modes, scrambling the polarization. Removal of these inhomogeneities can be used to make a polarization-maintaining fiber. A second technique for making a polarization maintaining fiber is to use a phenomenon called *modal birefringence.*

Single-Polarization Fibers

If the fiber is perfectly circular and perfectly straight with no imperfections at the core-cladding interface, the two orthogonal modes will remain perfectly separated and will propagate without any cross-coupling. One approach to maintaining the polarization, then, is to make the perfect fiber.

Each mode will have a slightly different propagation velocity as it proceeds through the fiber. The difference in the propagation constant

$$\Delta\beta = |\beta_x - \beta_y|$$

(11.33)

can be used (Okoshi and Kikuchi, 1988) as a figure of merit in evaluating the polarization maintenance of a fiber (with a perfect fiber having $\Delta\beta = 0$). Okoshi and Kikuchi (1988) give a typical value of $\Delta\beta = 10$ degrees/m for a fiber with excellent circularity. Some of this difference in propagation constants is due to stress that is induced between the core and cladding material due to differing thermal expansion coefficients of these fiber regions. (This stress reduces the circularity of the fiber.) In fibers where this stress has been reduced, Okoshi and Kikuchi (1988) indicate that a $\Delta\beta = 2.6$ degrees/m has been achieved. Fibers that have been made from rotating preforms have achieved a low $\Delta\beta$ of 0.4 to 0.6 degrees/m (Okoshi and Kikuchi, 1988).

An alternative approach is to design the fiber so that the transmission losses for the orthogonal modes are widely different. In the ideal case, one of the modes will be below cutoff, while the other mode will be above its cutoff. (Detailed analysis and measurement of the degenerate modes in a "single-mode" fiber reveals that the two modes will have slightly different cutoff wavelengths.) Operation at the proper wavelength will allow only one of the modes to propagate. (This type of fiber has been called "absolutely single-polarization fiber" [Okoshi and Kikuchi, 1988].) Following Okoshi and Kikuchi (1988), we will call this type of fiber "single polarization fiber" and the other type of fibers "polarization-maintaining fibers."

High-Birefringence Fibers

An alternative approach to maintaining polarization in the fiber is to deliberately maximize the difference in the propagation constants of the polarizations. For fibers with a fairly large difference in propagation constants, the modes will remain uncoupled and the launched polarization will remain the same. Silica glass itself has slightly different velocities of propagation for each polarization. (This dual velocity is called *birefringence*.) This birefringence can be due to the properties of the glass (*natural birefringence*) or due to external influences, such as the shape of the core and cladding (*geometry-induced birefringence*) and unsymmetrical stresses introduced into the fiber (*stress-induced birefringence*).

In a fiber exhibiting birefringence, the orthogonal modes that start out in phase will gradually get out of phase as the waves propagate due to the different velocities of the waves. We will let the x axis be aligned along the polarization axis that has the slower velocity and the y axis be aligned along the polarization axis with the faster velocity. The difference in the index of refraction will be defined as the *modal birefringence B*, or

$$B = n_x - n_y . \qquad (11.34)$$

(Fibers with a birefringence of $B > 10^{-5}$ are generally considered to be *high-birefringence fibers*; *low-birefringence fibers* are at values well below this, with representative values of 10^{-9} to as low as 10^{-11}.)

The phase shift φ between the modes due to the birefringence after the modes traverse a length L will be

$$\varphi = (n_x - n_y)\,\frac{2\pi L}{\lambda_0} \tag{11.35}$$

$$= \frac{2\pi B L}{\lambda_0}\,. \tag{11.36}$$

The distance that the waves must travel to produce a phase difference of 2π is used to characterize the amount of birefringence in the fiber. This distance is called the *beat length L_p* of the fiber and is given by (Poole et al., 1989)

$$L_p = \frac{\lambda_0}{n_x - n_y} \tag{11.37}$$

$$= \frac{\lambda_0}{B}\,. \tag{11.38}$$

A smaller beat length indicates a higher degree of birefringence. Typical values of beat length in fibers can range from 100 mm to 5 m or so. (We should note that the beat length of a fiber depends linearly on the wavelength, so the beat length specification should also include the wavelength of measurement to allow fair comparison. The modal birefringence is a better measure of fiber birefringence as it is independent of wavelength.) We will let the propagation constants be $\beta_x = 2\pi n_x/\lambda$ and $\beta_y = 2\pi n_y/\lambda$. The *modal birefringence B_f* is defined by

$$B_f = \frac{\beta_x - \beta_y}{2\pi/\lambda}\,. \tag{11.39}$$

Light that is polarized along one or the other orthogonal axis will retain its linear polarization. Light that is linearly polarized at the input but not along the x or y axis will have elliptical polarization of varying amounts. We show this as follows: The input light can be decomposed into orthogonal components at the fiber input. By definition of linear polarization, the components are in phase. After propagation of a distance L, there will be an induced phase shift between the two components as a result of the slightly different velocities. This phase shift $\varphi(L)$ is given by

$$\varphi(L) = (\beta_x - \beta_y)L\,. \tag{11.40}$$

We require the light to retain phase coherence during this propagation. This will be true if the delay between the two transit times is less than the *coherence time* of the source. This coherence time of the source t_c can be shown to be

$$t_c = \frac{1}{\Delta \nu}, \qquad (11.41)$$

where $\Delta \nu$ is the frequency linewidth of the source. The birefringent coherence will be maintained over a fiber that is of length L_b or less where

$$L_b \approx \frac{c\, t_c}{B_f} \qquad (11.42)$$

$$\approx \frac{c}{B_f\, \Delta \nu} \qquad (11.43)$$

$$\approx \frac{\lambda^2}{B_f\, \Delta \lambda} \qquad (11.44)$$

and $\Delta \lambda$ is the spectral linewidth of the source.

As the wave propagates through the fiber, the phase difference between the components changes from 0 degrees toward 90 degrees onward to 180 degrees and so forth. A phase difference of 0 degrees or 180 degrees is linearly polarized light, a phase difference of 90 degrees or 270 degrees (with equal amplitudes) is circular polarization, and a phase other than these values is elliptical polarization. The periodicity of this behavior, L_B, is the same as the *beat length* of the polarization-maintaining fiber.

The beat length of a bare fiber can be observed externally, since the amount of Rayleigh scattering depends on the polarization state. The observer will see alternating bands of bright and dark along the length of the (jacketless) fiber if a visible source such as a HeNe laser is used to inject light into the fiber. The single polarization technique that eliminates coupling effects (or, in practice, minimizes these effects) requires that the observed beat length be shortened to values of 1 mm or less. The high-birefringence technique that *maximizes* the birefringence to maintain the polarization requires a beat length of 50 m or more.

The two modes in a high-birefringence fiber should (ideally) propagate independently. Imperfections in the fiber will cause these modes to be coupled together in varying degrees. A pure polarization state that is launched at the fiber input will eventually have some of its power coupled from the excited mode into the other mode. In fibers with a large amount of birefringence, only defects that have a short periodicity will couple between modes. Since there are relatively few sources of this type of coupling, these fibers can "hold" the input polarization over a longer distance than an equal length of fiber with less birefringence. The *polarization holding parameter h* is used as a measure of this ability to maintain a pure polarization state. It is defined mathematically as

$$h = \frac{\dfrac{dP_x(z)}{dz}}{P_y(z)} \tag{11.45}$$

and measures the amount of coupling between the polarization modes per unit length. It can be shown (Noda et al., 1986) that the coupling between modes produces a power relation of

$$\frac{P_x}{P_y} = \tanh\,(hL)\,. \tag{11.46}$$

The value of h in a high-birefringence fiber usually ranges from 1×10^{-7} to 1×10^{-5} m^{-1}, so we can frequently use the approximation (valid for small hL) that

$$P_x(L) = hLP_y(L)\,. \tag{11.47}$$

Example: Consider a fiber with $P_x(0) = 0$ W and $P_y(0) = 1$ µW. If $h = 5 \times 10^{-5}$ m^{-1}, calculate the power in the x-oriented polarization after traversing 1 km of fiber.

Solution: We estimate the power as

$$P_x(L) = hLP_y(L)\,. \tag{11.48}$$

We need to find $P_y(L)$ from $P_y(0)$ and to determine the fiber losses before we can use this formula. The input power is 0 dBµ, and we can estimate the fiber losses at, say, 1 dB, since no information has been given about the loss. So $P_y(L) = -1$ dBµ or 0.794 µW. The power in the x-oriented direction is

$$P_x(L) = (5 \times 10^{-5})\,(1 \times 10^{3})\,(0.794) \quad \mu\text{W} \tag{11.49}$$

$$= 3.97 \times 10^{-2} \quad \mu\text{W}\,. \tag{11.50}$$

We can define the crosstalk CT between the modes (in dB) as

$$CT = 10\,\log\left(\frac{P_y}{P_x}\right) \tag{11.51}$$

$$= 10\,\log\left(\tanh\,(hL)\right) \tag{11.52}$$

$$\approx hL\,. \tag{11.53}$$

The main intrinsic imperfections that disturb the polarization-maintaining properties of a fiber are structural imperfections, wavelength effects, and nonlinear effects. The primary extrinsic effects are those due to temperature, mechanical perturbations (such as transverse pressure, bending, twisting, and tension), and electromagnetic effects (such as the Faraday effect and the Kerr effect). These effects are discussed in more detail in Noda et al., (1986). By isolating the fiber from all of the extrinsic effects except the one to be measured, a fiber can be used to sense a change in the desired parameter.

Geometry-Induced Birefringence

One way to induce birefringence by geometry is to form an elliptical core in the preform as shown in Figure 11.16. (Not all of the induced birefringence is due to the geometrical effects alone; a portion is due to the stresses induced in the fiber. The stress-induced birefringence is not as strong in the elliptical-core fiber as in the other designs.)

The "side-pit" structure is illustrated in Figure 11.17. Two regions are doped differently and placed either overlapping the core (as shown in the figure) or adjacent to the core. (The same effect can also be introduced with cavities in the preform in a technique known as the "side-tunnel" structure.) The side-pit structure can also be rectangular if desired. Like the elliptical-core fiber, part of the birefringence is due to the asymmetric core propagation properties and part is due to the induced stress in the fiber.

Extreme care must be taken not to add any additional asymmetric stress to the fiber during jacketing and cabling, as well as during use of the fiber. Sharp turns must be avoided to prevent such strains.

The losses associated with the elliptical-core fabrication technique are fairly large, so this technique is limited to making fibers of 1 m length or less. The other techniques are better for making longer lengths of fiber.

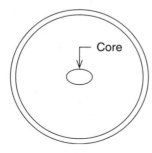

Figure 11.16 Structure of an "elliptical-core" polarization-maintaining fiber.

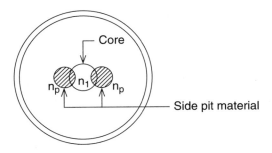

Figure 11.17 Structure of a "side-pit" polarization-maintaining fiber.

Stress-Induced Birefringence

The birefringence in a fiber can be accentuated by inducing a mechanical stress in the fiber that is stronger in the x (or y) direction than in the orthogonal direction. The stress-induced birefringence is maximized by including a stress-inducing element into the fiber design. Figure 11.18 illustrates the *panda fiber* (Sasaki, 1987), where two materials of different composition from the host material are incorporated in the preform and, hence, in the fiber. The localized stress between these regions establishes a birefringence axis between their center with the other axis being perpendicular to this first axis. A related approach is the *bow-tie fiber* shown in Figure 11.19. Here a "bow-tie" shaped region was introduced into the preform. The birefringence was maximized for this approach over the panda fiber.

A different approach is to induce stress through the use of an elliptical cladding around a circular core, as illustrated in Figure 11.20.

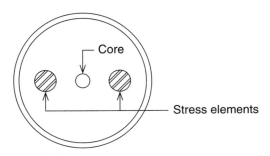

Figure 11.18 Structure of a "panda" polarization-maintaining fiber.

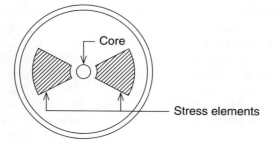

Figure 11.19 Structure of a "bow-tie" polarization-maintaining fiber.

11.7.2 Sensing Applications

In the polarization-maintaining birefringent fibers, there are two orthogonal polarization axes in the fiber along which the light can lie, a fast axis and a slow axis. If each is equally excited at the input end of the fiber, the modes will propagate with two slightly different velocities, building up a phase difference between them. (This same effect is widely used in laser optics to control and manipulate the polarization state of laser beams.) The polarimetric sensor relies on the external field to change the velocity properties of the fiber which results in a phase shift that is to be detected. Before we can detect the phase shift, we have to rotate one of the polarization states by 90 degrees, so that the interference can occur. (One of the conditions for interference is that the interfering beams must be of the same polarization, or, more accurately, that the co-aligned components of the beams will interfere.) One technique for doing this (Wilson and Hawkes, 1989) is to use a combination of a Soleil-Babinet compensator and a Wollaston prism. (A *Soleil-Babinet compensator*

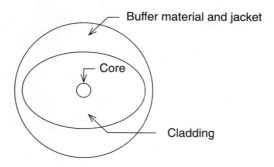

Figure 11.20 Structure of an elliptical cladding polarization-maintaining fiber.

is a phase-shifting device that can transform any state of polarization into any other state. A *Wollaston prism* is an optical prism device that splits the beam into two separated output beams with orthogonal polarization.)

We will let the input beams be represented by the phasors $A_1 e^{j2\pi vt}$ and $A_2 e^{j2\pi vt}$. For equal amplitude components, $A_1 = A_2 = A$. After the Babinet-Soleil compensator the waves are $Ae^{j2\pi vt}$ and $Ae^{j(2\pi vt + (\pi/2) + \varphi)}$, where φ is the relative phase shift between the modes. The Wollaston prism is set up so that it will separate polarization beams that are at angles of +45 degrees and –45 degrees to the fast (or slow) axis of the fiber. We therefore need to compute the components of each fiber polarization state that will be aligned along the axes of the prism. Some thought will show that these components are

$$\frac{Ae^{j2\pi vt} + Ae^{j(2\pi vt + (\pi/2) + \Delta\varphi)}}{\sqrt{2}} \qquad (11.54)$$

for one axis and

$$\frac{Ae^{j2\pi vt} - Ae^{j(2\pi vt + (\pi/2) + \Delta\varphi)}}{\sqrt{2}} \qquad (11.55)$$

for the other axis.

One of these components falls on one detector of a balanced pair; the other component falls on the other detector. The currents out of the detectors are proportional to the magnitude squared of the incident fields.

$$I_1 \propto A^2(1 + \sin \Delta\varphi) \qquad (11.56)$$

$$I_2 \propto A^2(1 - \sin \Delta\varphi) \ . \qquad (11.57)$$

We then form the following relation from the currents,

$$\frac{I_1 - I_2}{2} \propto \sin \Delta\varphi \ . \qquad (11.58)$$

If we let B be the difference between the indexes of refraction of the slow axis and the fast axis, then the phase shift will be

$$\Delta\varphi = \frac{2\pi BL}{\lambda_0} \ . \qquad (11.59)$$

The change in phase due to an external parameter can be written as

$$\frac{d\Delta\varphi}{dP} = \frac{2\pi}{\lambda_0}\left(B\frac{dL}{dP} + L\frac{dB}{dP}\right)$$

(11.60)

For a typical fiber sensor, the first term, which represents the portion of the phase change due to a change in fiber length, is small compared to the second term, which is the change in birefringence (because a typical value of B is on the order of 5×10^{-4}).

11.8 Summary

In this chapter we have introduced the concept of using optical fibers for sensors. The amplitude-changing sensors (i.e., the transmission-gap and the micro-bend sensors) have the advantage of being optically simple and fairly easy to implement. The phase-changing sensors require polarization stability as well as coherent detection and, hence, are fairly difficult to implement. In all cases, fiber-optic sensors are currently more expensive than conventional sensors, so the advantages of fiber optics that were enumerated in the first chapter must be of paramount importance to the user to justify the extra expense. Commercial amplitude-changing sensors are already available. The phase-changing sensors continue to be studied in the laboratory and are reaching prototype development.

11.9 Problems

1. Consider a 125 μm fiber. Find . . .

   (a) . . . the change in length per Newton of increased applied force per meter of fiber length, and

   (b) . . . the change in phase per Newton of increased applied force per meter of fiber length.

2. Consider a 125 μm fiber. Find . . .

   (a) . . . the change in length per Pascal of increased applied pressure per meter of fiber length, and

   (b) . . . the change in phase per Pascal of increased applied pressure per meter of fiber length.

3. Consider a 125 μm fiber. Find . . .

   (a) . . . the change in length per degree change in temperature per meter of fiber length, and

(b) . . . the change in phase per meter of fiber length for a 1 degree temperature increase.

4. Consider a fiber gyroscope with 100 turns of fiber wrapped around a 0.5 m radius loop. Calculate the phase shift that results from a rotation rate that equals the rotation rate of the earth (a frequently used benchmark rotation rate) at a 900 nm wavelength.

5. A gyroscope must be capable of measuring rotation rates of 1×10^{-2} degrees per hour to be considered for use in inertial navigation systems. Calculate . . .

    (a) . . . the phase shift expected from a gyroscope made of 1 km of fiber wrapped into coils with a 10 cm diameter. Assume a wavelength of 1300 nm.

    (b) . . . the equivalent change of fiber length ΔL that would give the same phase shift calculated in the previous part.

6. Consider the fiber in the example on page 443. Find the value of the cross-talk in the fiber.

References

Bergh, R. A., H. Lefevre, and H. J. Shaw, "An overview of fiber-optic gyroscopes," *J. Lightwave Technology*, vol. LT-2, no. 2, pp. 91–107, 1984.

Bucaro, J., H. Dardy, and E. Carome, "Fiber-optic hydrophone," *J. Acoust. Soc. Am.*, vol. 62, no. 5, pp. 1302–1304, 1977a.

Bucaro, J., H. Dardy, and E. Carome, "Optical fiber acoustic sensor," *Applied Optics*, vol. 16, no. 7, pp. 1761–1762, 1977b.

Bucaro, J. and E. Carome, "Single fiber interferometric acoustic sensor," *Applied Optics*, vol. 17, no. 3, pp. 330–331, 1978.

Bucaro, J., N. Lagakos, J. Cole, and T. Giallorenzi, "Fiber optic acoustic transduction," in *Physical Acoustics, Vol. XVI*. New York: Academic Press, 1982.

Cole, J., R. Johnson, and P. Bhuta, "Fiber-optic detection of sound," *J. Acoust. Soc. Am.*, vol. 62, no. 5, pp. 1136–1138, 1977.

Danielson, D. A. and S. L. Garrett, "Fiber-optic ellipsoidal flextensional hydrophones," *J. Lightwave Technology*, vol. 7, no. 12, pp. 1995–2002, 1989.

Davis, C. M., "Fiber optic sensors: an overview," *Optical Engineering*, vol. 24, no. 2, pp. 347–351, 1985.

Dandridge, A. and G. B. Cogwell, "Fiber optic sensors for Navy applications," *IEEE Lightwave Communications Systems (LCS)*, pp. 79–89, 1991.

Fields, J., C. Asawa, O. Ramer, and M. Barnoski, "Fiber optic pressure sensor," *J. Acoust. Soc. Am.*, vol. 67, no. 3, pp. 816–818, 1980.

Gardner, D., T. Hofler, S. Baker, R. Yarber, and S. Garrett, "A fiber-optic interferometric seismometer," *J. Lightwave Technology*, vol. LT-5, no. 7, pp. 953–960, 1987.

Giallorenzi, T. G., J. A. Bucaro, A. Dandridge, J. H. Cole, S. C. Rashleigh, and R. G. Priest, "Optical fiber sensor technology," *IEEE J. Quantum Electronics*, vol. QE-18, no. 4, pp. 626–665, 1982.

Hocker, G., "Fiber-optic sensing of pressure and temperature," *Applied Optics*, vol. 18, no. 9, pp. 1445–1448, 1979.

Lagakos, N., J. Bucaro, and J. Jarzynski, "Temperature-induced optical phase shifts in fibers," *Applied Optics*, vol. 20, no. 13, pp. 2305–2308, 1981a.

Lagakos, N., T. Hickman, J. Cole, and J. Bucaro, "Optical fibers with reduced pressure sensitivity," *Optics letters*, vol. 6, no. 9, pp. 443–445, 1981b.

Lagakos, N., T. Hickman, P. Ehrenfeuchter, J. Bucaro, and A. Dandridge, "Planar flexible fiber-optic acoustic sensors," *J. Lightwave Technology*, vol. 8, no. 9, pp. 1298–1303, 1990.

Noda, J., K. Okamoto, and Y. Sasaki, "Polarization-maintaining fibers and their applications," *J. Lightwave Technology*, vol. LT-4, no. 8, pp. 1071–1089, 1986.

Okoshi, T. and K. Kikuchi, *Coherent Optical Fiber Communications*. Tokyo: KTK Scientific Publishers, 1988.

Payne, D., A. Barlow, and J. R. Hansen, "Development of low and high birefringence optical fibres," *IEEE J. on Quantum Electronics*, vol. QE-18, no. 4, pp. 477–487, 1982.

Pitt, G., "Optical-fiber sensors," in *Fiber Optics Handbook for Engineers and Scientists*, (F. C. Allard, ed.), pp. 8.1–8.78. New York: McGraw-Hill, 1990.

456

Poole, S., J. Townsend, D. Payne, M. Fermann, G. Cowle, R. Laming, and P. Morkel, "Characterization of special fibers and fiber devices," *J. Lightwave Technology*, vol. 7, no. 8, pp. 1242–1255, 1989.

Rines, G. A., "Fiber-optic accelerometer with hydrophone applications," *Applied Optics*, vol. 20, no. 19, pp. 3453–3459, 1981.

Sasaki, Y., "Long-length low-loss polarization-maintaining fibers," *J. Lightwave Technology*, vol. LT-5, no. 9, pp. 1139–1146, 1987.

Sears, F. M., "Polarization-maintenance limits in polarization-maintaining fibers and measurements," *J. Lightwave Technology*, vol. 8, no. 5, pp. 684–690, 1991.

Spillmann, W.B, Jr. and R. Gravel, "Moving fiber-optic hydrophone," *Optics Letters*, vol. 5, no. 1, pp. 30–31, 1980.

Wilson, J. and J. Hawkes, *Optoelectronics: An Introduction.* New York: Prentice Hall, 1989.

Part Five

Fiber Fabrication and Measurement

Chapter 12

Fiber Fabrication
and Characterization

12.1 Introduction

The fabrication of an optical-fiber cable consists of the manufacturing of the glass master (called a "preform"); the drawing of the fiber itself; and the bundling of the fiber, strength members, filler yarns, and other components into a cable. We will consider first the fabrication and testing of the optical fibers and then the concerns in assembling an optical cable. The fabrication of the fiber can be done with either of two classes of techniques, those using high–melting-temperature glasses involving vapor deposition of the glass material and those using low–melting-temperature glasses that make the fiber directly from the melted glass. The fiber is drawn from the stock material by a pulling machine under constant tension and reeled on a spool. Following testing and characterization of the fiber, it is incorporated into an optical cable having the desired environmental protective features.

12.2 Glass

Glass is an amorphous crystalline material that has a wide range of viscosity values as a function of temperature. Unlike ordinary crystals that have a regular periodic lattice of atoms and a well defined melting temperature, glass is an irregular suspension of molecules and has no well defined melting temperature. Instead, as the

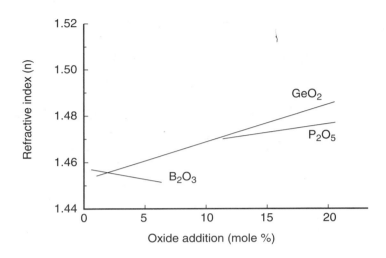

Figure 12.1 Refractive index of silica-based glass vs. concentration of dopant.

temperature increases, the viscosity decreases. Since there is not an abrupt change in the phase of the material, the *melting temperature* becomes a matter of definition rather than a physical observable. The *softening temperature* occurs at a viscosity value of 3.16×10^7 poise and the *working temperature* at a viscosity value of 1.0×10^4 poise.

     For fiber-optics purposes, there are two general kinds of glass. The majority of fibers are made of *fused silica* (SiO_2) to which dopant materials are added to change the index of refraction. Some fibers are made of *multicomponent glasses* (Beales, et al., 1980), i.e., they are made up of a variety of components. Examples of the latter include sodium borosilicate ($Na_2O-B_2O_3-SiO_2$) and soda lime silicate ($Na_2O-CaO-SiO_2$). The working temperature for the silica-based glasses is in the vicinity of 1800 C, while the working temperature of the compound glasses is 900 C. The role of dopants in silica glass is shown in Figure 12.1, which plots the index of refraction of the glass vs. the concentration of dopant material. Here we see that the addition of GeO_2 or P_2O_5 raises the index, while the addition of B_2O_3 lowers the index. Hence, the addition of the proper element in the right proportion can raise or lower the refractive index to the desired value. Note from the figure that we can raise the index by a Δ of 1.4% and lower it by only 0.2%. Because of the wide difference in working temperatures for the two types of glass, there have evolved two funda-mentally different techniques in making fibers, one involving the use of *preforms* and another growing the fibers directly from melted glass.

12.3 Preform Fabrication

Fabrication of fibers using high–melting-temperature silica glass utilizes *preforms* (French et al., 1979; Schultz, 1979), which are cylinders of glass, tens of millimeters in diameter, with an index of refraction profile that is an enlarged replica of the desired profile (i.e., the core area and cladding area are in the same proportion that they will be in the fiber). Both graded-index and step-index fibers can be made by this technique. The preform serves as the stock material for the fiber drawing process and should be as long as possible to maximize the amount of fiber that can be drawn from a single preform. The preforms for the silica-based glass are usually made by a *vapor phase process*.

This process starts with a high-purity silica-base stock. Impurities and glass are then deposited to change the index of refraction in a prescribed fashion. Typically the impurities and glass are introduced as a vapor of chloride combinations that are oxidized in a flame to produce the desired dopant products that combine with silica vapor for the deposition process. Typical reactions include

$$SiCl_4 + O_2 \rightarrow SiO_2 + 2Cl_2$$

$$GeCl_4 + O_2 \rightarrow GeO_2 + 2Cl_2$$

$$2POCl_3 + \frac{3}{2} O_2 \rightarrow P_2O_5 + 3Cl_2$$

$$2BCl_3 + \frac{3}{2} O_2 \rightarrow B_2O_3 + 3Cl_2 .$$

The first products are the desired glass and dopants; the chlorine by-products are extracted in the exhaust. Hydrogen-based flames can also be used for hydrolysis, but the resulting water product must subsequently be removed by careful drying to avoid excess losses in the glass due to the hydroxyl ions.

The glassy product of the vapors is called *soot* and is deposited on the surface of a glass or graphite rod (called the *bait rod*) in the form of a porous material. This porous soot is reduced to smooth clear glass by heating it to approximately 1500 C in a process called *sintering*. The radial dimension is built up layer by layer in this fashion with over 100 layers typically being deposited. For a step-index fiber, the concentration of impurity is maintained constant in all layers. For graded-index fibers, the concentration is gradually changed from layer to layer to build up a step-wise approximation to the desired index profile in the core.

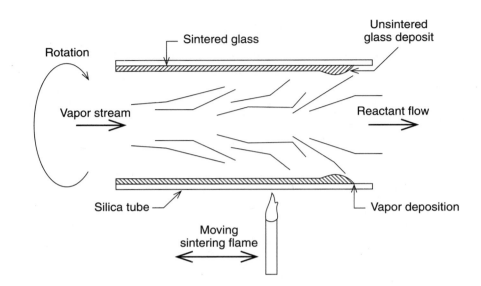

Figure 12.2 Diagram of the MCVD process.

12.3.1 MCVD process

One common method of fabricating the preform uses the *modified chemical vapor deposition (MCVD) process* (Schultz, 1979; MacChesney, 1980; Nagel, 1982; Jablonowoski, 1986; Nagel, 1988), developed by Bell Laboratories. This process is called an *inside process*, since the doped glass is deposited on the inside surface of a hollow rod. (This class of inside processes is called the *inside vapor phase oxidation [IVPO] process.*) The IVPO process has the advantage of confining the interaction region, thereby lowering the chances of contamination by outside impurities. The process is illustrated in Figure 12.2.

The gaseous vapors of the glass and impurities enter the hollow cylinder as hot gases and are attracted to the cool walls of the tube. The tube rotates to preserve the cylindrical symmetry of the deposition. A torch traverses the length of the tube to heat the deposition and sinter the material into smooth glass. Since only the region near the torch is heated, this process is called *zone sintering*.

A variation of this technique, called *modified plasma chemical vapor deposition (MPCVD)* (Lydtin, 1986), uses an RF-generated plasma inside the tube to increase the soot deposition rate while retaining the outside torch to sinter the glass.

A variation on this variation uses a plasma for both internal and external heating of the tube and is called the *Philips process*, after the company developing the technique.

After the deposition process is completed, the entire tube is heated to high temperatures, thereby collapsing the tube and eliminating the central hole. This solid rod is the preform that is then used in the drawing machine to produce the fiber. Primary disadvantages center around the high purity tube stock required for the process. Various techniques are under investigation to provide a low cost, reproducible means of producing the tube stock. Another minor disadvantage is the relatively slow rate of soot deposition in the tube. The development of the faster depositing plasma process was driven by this problem.

12.3.2 Outside Vapor Deposition Process

The *outside vapor phase oxidation (OVPO)* deposition process (Schultz, 1979; Blankenship and Denka, 1982; Vandewoestine et al., 1986) occurs on the outside surface of the host material. The process is subdivided into two types, depending on whether the new material is deposited radially (laterally or longitudinally) on the end of the host. Again, the deposition process is based on the deposition of soot from hot gases containing the glass material and the dopants.

OVPO-Lateral Deposition

The lateral deposition process (Figure 12.3), developed by Corning Glass Works, deposits the soot on a removable host material called a *bait rod*. This rod is typically made of graphite, fused silica, or ceramic. The rod again rotates to assure symmetry. The layers are built up in unsintered form until typically 200 layers have been deposited. After the soot deposition process, the inner bait rod is removed and the deposited material is zone sintered to reduce its radius and to form the smooth glass. The large number of layers in the OVPO process allows more precise control over the refractive index profile. High bandwidth-distance fibers result from this control. The resulting preforms are also longer, allowing more fiber to be drawn (up to 10 km of 125 μm diameter fiber). One difficulty (Schultz, 1979) has been breakage of the glass blanks due to localized stresses where the bait rod has been removed. Another problem has been the presence of high concentrations of hydroxyl

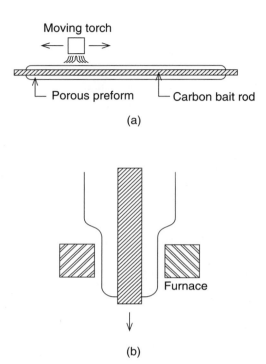

Figure 12.3 OVPO process. (a) Lateral deposition. (b) Sintering process.

ions as a result of the flame burning. This can be removed by careful drying with gaseous chlorine during the sintering process. As shown in Figure 12.4, the hole in the preform is removed during the melting associated with the drawing process (although a dip in the index of refraction along the central axis due to the outward diffusion of the impurities in the heating process is left as an artifact).

Vapor Axial Deposition Process

The *vapor axial deposition (VAD)* process (Schultz, 1979; Izawa and Inagaki, 1980; Inada, 1982; Murata, 1986), developed by Nippon Telephone and Telegraph, simultaneously deposits the core and cladding glass from flames on the end of a rotating fused-silica bait rod. Shown in Figure 12.5, the material is sintered as the rotating rod is drawn through a furnace. Since the growth is continuous, very long preforms can be made from this technique; in fact, preforms yielding in excess of 25 km of fiber have been made. The method also lacks a center hole and, through

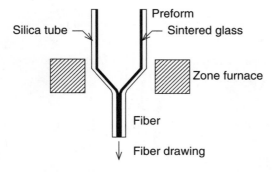

Figure 12.4 Preform hole removal during drawing of fiber.

the combination of the deposition and sintering process, can minimize the contamination problem. This technique has also been used to grow core material that is then placed inside tubes of cladding material (called the *rod-in-the-tube technique*) or that is plastic-clad to produce the preform for fiber drawing. One major problem with this technique is the precision of control of the refractive index variation. Also, the deposition rate of the material (typically 0.5 g/min) is low compared with other techniques.

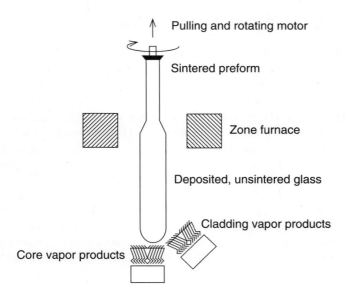

Figure 12.5 Vapor axial deposition.

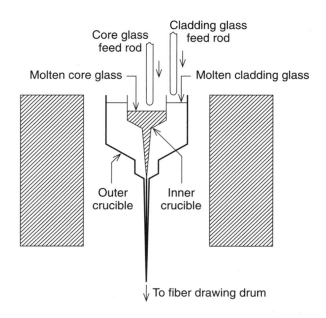

Figure 12.6 Double crucible process.

12.4 Double-Crucible Technique

For low–melting-temperature compound glasses, a preform is not used (Beales et al., 1980; Schultz, 1979). Instead the stock material is purified and grown into rods from seed glass. This seed glass is melted and used in the fiber growing process, as shown in Figure 12.6. Here two concentric platinum heating chambers contain the liquefied glass for the core and cladding. The fiber is drawn from two concentric orifices at the outlet of the crucibles. The feed stock is continuously fed into the crucibles to replenish the supply of glass. Graded-index fibers are made in the double-crucible process by allowing the molten cladding glass to contact the core glass over a relatively long length before the fiber leaves the crucibles. Over this length, diffusion of ions occurs, causing a gradient and, hence, a variation in the index of refraction. The advantages (Schultz, 1979) of the double-crucible process include the abilities to draw long fibers, to produce fibers with high numeric apertures, and to achieve low cost. Disadvantages include the requirement for extreme cleanliness in the crucibles and orifices and a relatively low bandwidth-distance

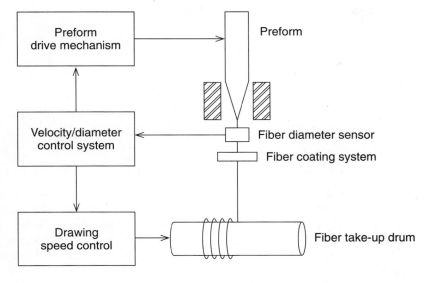

Figure 12.7 Fiber drawing machine.

product (typically <1 GHz · km) (Schultz, 1979). For medium-performance fibers, however, the low production costs ensure a future role for this technique.

12.5 Fiber Drawing

Once a preform has been made by one of the vapor-deposition techniques, the fiber is drawn (Jaeger et al., 1979; Blyler and DiMarco, 1980; Paek, 1986; Hibino and Hanafusa, 1988) in a *drawing engine* or *drawing machine*. A block diagram of such a machine is shown in Figure 12.7. The heat source is typically a graphite or tungsten furnace or a zirconia induction furnace. Uniformity of heating, lack of contaminants, and stable temperatures are the primary requirements of the heating source. When heated, the end of the preform will become molten and neck down into a smaller cross-section, but the glass will preserve the relative core/cladding size relation established in the fabrication of the preform. Since most of the fiber properties (losses, delay, etc.) are sensitive to small changes in the fiber diameter, a sensitive diameter sensor is required. Traditionally a good fiber requires less than ±2% variation in its diameter. Control of the diameter is through an active

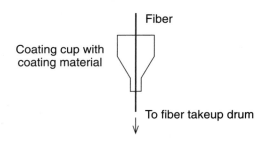

Figure 12.8 Coating process with coating cup.

mechanism that senses the fiber diameter and adjusts the speed of the drawing operation accordingly. (See the problems at the end of the chapter.) Various optical techniques using scanned lasers, scattered light, shadowgrams, and other methods have been developed to achieve the required dimensional sensitivity. In the drawing process, fast drawing velocities cause thin fibers, thereby allowing the feedback control system to maintain the fiber size.

12.6 Fiber Coating

After drawing the fiber, it is important to protect the outside surface of the fiber from the environment immediately (Blyler et al., 1979; Blyler and DiMarco, 1980). This is necessary because any defect or flaws in the outside surface of the cladding will have deleterious effects on the strength of the fiber. The coatings used must be fast drying and provide complete protection from the spooling process. Typically, one of three coatings is used: ultraviolet curable epoxies, lacquers, and thermally cured materials such as silicone. Each coating requires a curing device surrounding the fiber, either an ultraviolet light or a small oven. Figure 12.8 illustrates the application of the liquid coating material through the use of a coating cup.

12.6.1 Hermetic-Coated Fibers

Several application areas for fiber optics require combinations of high strength (Kreidl, 1987; Kurkjian and Inniss, 1991; Bogatyrjov et al., 1991) and long life. Table 12.1 contains a representative list of some of these applications. (Regular fiber strength for ordinary telecommunications applications has a proof strength of

Table 12.1: Typical high-strength fiber applications (After Kreidl, J., *Lightwave*, vol. 4, no. 11, p. 1, November 1987.)

Application	Operating Conditions			Lifetime
	Pressure	**Temperature**	**Strain**	
Oil-well logging	20 kpsi	200C	3% 300 kpsi	4000 hrs
Underseas cable	2–10 kpsi	5C to 25C	1% 100 kpsi	25–30 yrs
Missile tethers	1 atm	–55C to 125C	0.5% to 1% 50–100 kpsi	10 yrs
Chemical sensors	1 atm	Depends on application	0.5% to 1% 50–100 kpsi	10 yrs
Standard telco	1 atm	–25C to 70C	0.5% to 1% 50–100 kpsi	25–30 yrs 10 yrs

about 180 kpsi. The strength drops to 100 kpsi in 10 days, 80 kpsi after 100 days, 65 kpsi after three years, and an estimated 45 kpsi after 30 years.) Fiber coatings also prolong the shelf-life of the fiber by protecting it from the environment. Hermetic coatings may prolong the life from an estimated 15 years to 25–30 years.

Hermetic coatings (Kreidl, 1987; Lu et al., 1988; Nagel, 1988; Kurkjian and Inniss, 1991; Yoshizawa et al., 1991) are a microlayer of inorganic material around the outer circumference of a fiber with a thickness of no more than a few micrometers. The coatings can be applied to high-strength or medium-strength fibers, increasing the lifetime of the coated fiber and improving yields (since fiber that fails to meet any strength tests is discarded). The coating is required to be impermeable, to not allow pinhole defects, and to have thermal expansion properties that match the glass fiber. Chemical vapor deposition techniques are preferred for coating silica fibers. The primary advantage of the hermetic-coated fiber is its higher resistance to stress fatigue, hydrogen permeation, and corrosive chemicals. Additionally, coated fibers using titanium carbide can be soldered or bonded to ensure water-proof operation, allowing the entire fiber system to be hermetically sealed (since sources and receivers are already available in hermetically sealed packages). Efforts are also underway to develop techniques for splicing these fibers while retaining the hermetic seal (Inniss and Krauss, 1991).

Several fiber coatings on silica fibers are under investigation, as seen in Table 12.2. Various coatings are being studied to find an optimum combination of properties and to provide patent protection for the developer.

Table 12.2: Typical hermetic coatings (After Kreidl, J., *Lightwave*, vol. 4, no. 11, p.1, November 1987).

Coating	Thickness (μm)	Median Strength (kpsi)
Aluminum	15–20	529
Titanium oxide	0–5	500
Tin oxide	0.01–0.035	620
Aluminum oxide	0.02–0.04	435
Silicon oxynitride	0.02–0.04	200–300
Silicon carbide	0.02–0.06	500
Titanium carbide	0.025–0.05	400–500
Carbon by CVD	0.02–0.025	600–650
Carbon by plasma	0.020	730
Anhydrous CH	0.025–0.050	500–600

12.7 Polymer-Clad Silica (PCS) Fibers

A third type of fiber is made by heating a rod of purified fused silica and drawing a fiber. This fiber is immediately coated with plastic or silicone and covered by an extruded protective jacket. Here the silica serves as the core material and the coating is the cladding. Such fibers typically have losses on the order of 10 dB/km or more, are very inexpensive to make, and usually have a fairly large core.

12.8 Preform and Fiber-Profile Characterization

Preform characterization (Cohen et al., 1979; Presby and Marcuse, 1979; Saekang et al., 1980; Francois et al., 1982; Stewart, 1982; Morishita, 1986; Raine et al., 1989; Helms et al., 1990; Kapron, 1990) is done to predict the quality of the resulting fiber before drawing begins and to screen against unacceptable materials. The primary parameter to be measured in the preform or fiber is the index of refraction profile, $n(r)$, across the diameter of the glass. For the general case of a graded-index fiber, the preform profile is a scaled replica of the fiber profile and can be given by

$$n(r) = n_1 \sqrt{1 - 2\Delta \, (r/a)^g} \quad \text{(for } r < a\text{)}, \tag{12.1}$$

where r is the radial coordinate, Δ is the relative index difference ($= [n_1 - n_2]/n_1$), and g is the exponent of the power law that determines the profile shaping. Neglecting material dispersion, the optimum value of the profile factor, g_{opt}, that maximizes the fiber bandwidth is given by Equation 3.134, repeated here as

$$g_{\text{opt}} = 2 - \frac{2n_1}{N_{g1}} \frac{\lambda}{\Delta} \frac{d\Delta}{d\lambda}.$$ (12.2)

To achieve the performance theoretically predicted for a fiber, the manufacturer must typically maintain the profile to within 1% of its optimal value (Marcuse and Presby, 1979a). Several effects in the fabrication process counter this attempt:

- As previously described, several fabrication techniques deposit the glass in layers, each with a slightly different index of refraction. Near the core center, the layers have proportionally greater thickness than the outer layers of the preform. In this region near the center, therefore, the index distribution will have ripples, due to the stepwise approximation to the desired profile.

- Another effect causing variation from the ideal index profile is a discontinuity in the index of refraction at the core-cladding interface. This effect is observed in some MCVD preforms.

- When the preform is made by heating and collapsing a tube with a center hole, some of the impurities of the innermost layer evaporate from the glass at the inner surface. This causes a depression in the index of refraction at the center of the preform, as illustrated by the profile of Figure 12.9.

- Occasionally the dopant atoms will clump together due to imperfections in the doping process or inhomogeneities in the soot. Such inhomogeneities tend to cause isolated regions that differ from the desired index profile. The bandwidth of the fiber is more sensitive to this effect when the imperfections are located near the core-cladding interface.

- The refractive-index profile changes over the length of the preform due to variations in the geometry, temperature profile, and doping density.

Deviations from the ideal refraction-index profile, due to these mechanisms, typically cause an order of magnitude reduction in the experimentally observed bandwidth, compared to the predicted bandwidth in modern fibers.

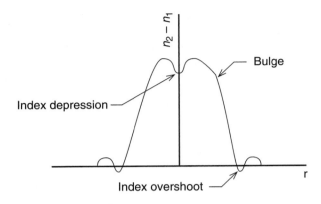

Figure 12.9 Typical refractive index profile.

12.8.1 Reflection Technique

The earliest profile techniques used the reflection coefficient to measure the profile. In this *reflection technique*, a light beam is focused on the end of a cutoff section of a preform (as in Figure 12.10) and the reflected power is measured as the focused beam scans radially. The reflected power is predicted to be

$$\frac{P_r(r)}{P_i} = \left(\frac{n(r) - 1}{n(r) + 1}\right)^2, \tag{12.3}$$

where P_r is the reflected power, P_i is the incident power, and $n(r)$ is the refractive index profile. Measurement of the reflected light power for a constant incident power allows calculation of the $n(r)$ (since $n(r)$ must be positive). The calculation is made easier by continuing the scanning out to the cladding region. Here, the reflectivity is given by

$$\frac{P_{r\,clad}}{P_i} = \left(\frac{n_2 - 1}{n_2 + 1}\right)^2, \tag{12.4}$$

where n_2 is the refractive index of the cladding and $P_{r\,clad}$ is the power reflected from the cladding region of the fiber. Taking the ratio of Equation 12.3 to Equation 12.4 and defining it as $F(r)$, we have

$$F(r) = \frac{P_r(r)}{P_{r\,clad}}. \tag{12.5}$$

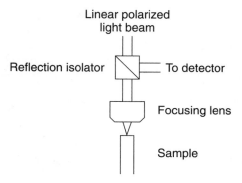

Figure 12.10 Reflection technique setup.

From this definition of $F(r)$, we find that

$$n(r) - n_2 = \frac{(n_2^2 - 1)\left(\sqrt{F(r)} - 1\right)}{(n_2 + 1) - (n_2 - 1)\sqrt{F(r)}}. \tag{12.6}$$

The expected variations in reflectivity in an optical preform are fairly small. The system errors encountered in implementing the reflection technique are sufficiently large to give large errors. Surface contamination effects are also large enough to give errors in the measured reflectivity. In implementing this technique, linearly polarized light is used in combination with a quarter-wave plate (immediately in front of the glass surface) to remove undesirable reflections. Other techniques have generally surpassed the performance of the reflection technique.

12.8.2 Transmitted Near-Field Technique

The next technique, called the *transmitted near-field technique*, uses the properties of the light transmitted from the end of a fiber that has had all modes uniformly excited by the source. If one observes the pattern of light at a location close to the end of the fiber (as in Figure 12.11), it closely resembles the refractive index profile of the fiber. Theory predicts that the relation between the power in the near-field pattern and the refractive index profile is (Cherin, 1983; Senior, 1985)

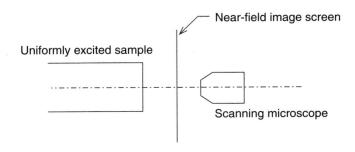

Figure 12.11 Geometry of the transmitted near-field technique.

$$\frac{P(r)}{P(0)} = \frac{n^2(r) - n_2^2}{n^2(0) - n_2^2} \tag{12.7}$$

$$\approx \frac{n(r) - n_2}{n(0) - n_2}, \tag{12.8}$$

where $P(r)$ is the power in the near-field at position r, $P(0)$ is the detected power at the center of the near field distribution, $n(r)$ is the refractive index profile, $n(0)$ is the refractive index at the center, and n_2 is the refractive index of the cladding. Hence,

$$n(r) - n_2 \approx \frac{\left(n(0) - n_2\right) P(r)}{P(0)}. \tag{12.9}$$

The source is usually a large incoherent illuminator that excites all of the fiber modes. A focused microscope is used to scan the pattern close to the end of the fiber (or preform).

12.8.3 Refracted Near-Field Technique

The *refracted near-field technique* (Stewart, 1982) is used to characterize fiber index profiles. This technique is quite accurate and is free of the errors due to leaky waves that affect other transmission techniques. (The *leaky waves* are light waves with exponentially decaying tails that spread into the cladding. While such waves will eventually be attenuated for propagation over a long distance, they can carry energy quite far in some cases, causing errors in the measurement of the refractive index profile by techniques that measure the light transmitted through the fiber.)

The refracted near-field technique removes the effects of these waves by surrounding the fiber with an index-matching solution that matches the cladding index. Additionally, an opaque collar is placed around the fiber (as shown in Figure 12.12) to block a portion of the light that is coupled into the index-matching solution. The light power, coupled into the matching solution, that passes beyond the screen is captured on the detector and is measured. The fiber is excited by a focused spot of light that is scanned to different positions on the fiber end (with the aid of a microscope). If we let $P(r)$ be the total power at the receiver when the *input* spot is at position r, then Snell's law predicts

$$n_2 \sin \theta' = n(r) \sin \theta \qquad (12.10)$$

$$n(r) \cos \theta = n_2 \cos \theta'' . \qquad (12.11)$$

Removing θ from these equations gives

$$n_2 \sin \theta' = \sqrt{n(r)^2 - n_2^2 + n_2^2 \sin^2 \theta''} . \qquad (12.12)$$

The angle θ''_{min} is determined from the size of the collar, as seen in Figure 12.12. The angle θ'_{max} is determined by the microscope objective chosen to focus the illuminating light. It can be shown (Stewart, 1982) that

$$n(r) - n_2 = n_2(\cos \theta''_{min}) \left(\cos \theta''_{min} - \cos \theta'_{max}\right)\left(\frac{P(a) - P(r)}{P(a)}\right), \qquad (12.13)$$

where $P(a)$ is the power measured at the detector when the incident spot is focused on the cladding and $P(r)$ is the power on the detector when the *incident* spot is a distance r from the center.

12.8.4 Interference Techniques

Interference of coherent optical waves offers a way to measure small changes of phase induced by the refractive index profile (Saunders, 1977; Beales et al., 1980). Two interference techniques have evolved.

The first is the *interferometric slab technique,* which uses a piece of excised fiber or preform (cut from the preform tip) that has been polished and prepared with parallel edges. The slab is placed in an interference microscope and coherent light is

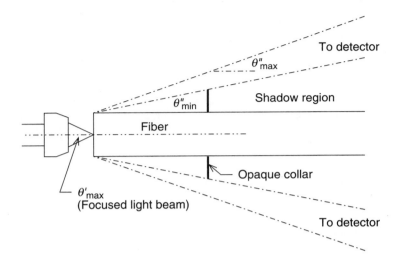

Figure 12.12 Refracted near-field technique.

passed longitudinally through the sample, as in Figure 12.13. When combined with a reference wave, the light forms an interference pattern, as illustrated in Figure 12.14. The interference fringes in the cladding material provide a reference line (and can be used to check the parallelness of the surfaces and the alignment of the instrument). The deflection, $S(r)$, of the interference fringe from the reference line at a radial position along the baseline, r, is measured and recorded. For an illumination wavelength, λ, a slab thickness, d, and a fringe spacing in the cladding of D, interference theory (Beales et al., 1980) predicts a refractive index profile given by

$$n(r) - n_2 = \frac{\lambda S(r)}{d\, D}. \tag{12.14}$$

The primary difficulty with this technique is the time and precision required in the preparation of the sample.

The second interference technique is the *transverse interferometric method* (Iga and Kokobun, 1978), which immerses the fiber in a liquid that matches the cladding index and illuminates the fiber (or preform) from the side with a coherent wave. After combining with a reference wave, the interference pattern is found. Using a computer-controlled pattern scanner, an integral equation can be solved by the computer to yield the refractive index profile. The advantage of this technique is its comparative speed, since minimal sample preparation is required.

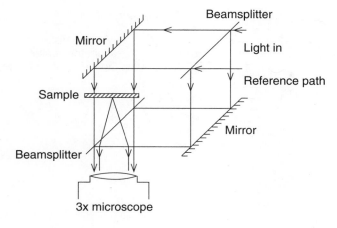

Figure 12.13 Setup for interference microscope.

12.8.5 Ray-Tracing Techniques

Figure 12.15 illustrates the refraction of a light ray incident on an optical fiber or preform from the transverse direction (with the fiber immersed in an index matching solution with index n_2) (Presby and Marcuse, 1980). Obviously, the deflection of the ray is dependent on Snell's Law at the interfaces and on the theory of propagation of rays in media with inhomogeneous velocities. For a circularly symmetric fiber, a ray entering at height, t, (as in Figure12.15) will appear at a screen a distance, L, away from the fiber at a height, $y(t)$, given by

$$y(t) = t + \frac{2Lt}{n_2} \int_t^a \frac{1}{\sqrt{r^2 - t^2}} \frac{\partial n(r)}{\partial r} \, dr \, . \qquad (12.15)$$

With knowledge of $y(t)$, as measured by illuminating the sample with rays (i.e., pencil beams) at various values of t, the refractive index profile can be calculated as

$$n(r) - n_2 = \frac{n_2}{\pi L} \int_r^a \frac{t - y(t)}{\sqrt{t^2 - r^2}} \, dt \, . \qquad (12.16)$$

The primary disadvantage of this method is the need to form the pencil beams and the sequential nature of the measurement. These disadvantages are overcome by the "focusing" method, discussed next.

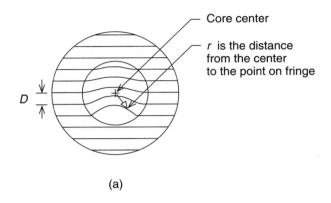

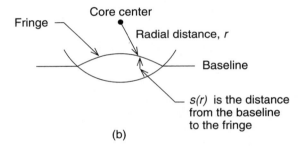

Figure 12.14 Geometry of interference pattern of the interferometric slab technique.

12.8.6 Focusing Method

The *focusing method* (Marcuse and Presby, 1979b; Marcuse, 1979) was developed as an improved version of the ray-tracing technique that would illuminate the preform with a collimated beam, thereby producing all rays from the object simultaneously. The resulting image located a distance, L, is scanned in the y direction producing a power distribution, $P(y)$.

From this power distribution it is possible to deduce the corresponding entry ray height, $t(y)$, with the formula

$$t(y) = \int_0^y p(y')\, dy' . \tag{12.17}$$

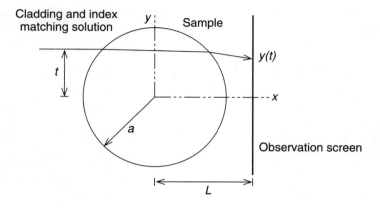

Figure 12.15 Geometry of the ray-tracing technique and the focusing method.

In performing this integral on a computer, the results are pairs of values of y and t. Using these same sets of values with t as the independent variable and y as the dependent variable, the integral of Equation 12.16 can be evaluated to find $n(r)$.

The method is fast and nondestructive. The region in the center is sensitive to error if the there is dip in the index of refraction, since the rays can cross each other before falling on the screen (in violation of the implicit assumptions that are made in using Equations 12.16 and 12.17). The overall shape of the index profile is accurately portrayed, but some errors are made in attempting to portray any ripple features in the profile. Cleanliness is required at the source, at the glass-liquid interface, and at the detector.

12.8.7 Diagnostigram Technique

In this method (Presby and Marcuse, 1980), applied to preforms, a collimated beam of light (either coherent or incoherent) illuminates the sample transversely. Since the preform is a layered structure, there will be successive reflections and transmissions from the different layers. When intercepted on the far side of the sample on a ground glass screen, there is a pattern of (ideally) parallel lines in the preform image (called a *diagnostigram*). Each line in the pattern corresponds to a layer in the preform. As the preform is rotated, the image can be inspected for symmetry and the locations of defects can be noted qualitatively. The technique has been used to measure core size and ellipticity, to measure the core-cladding interface homogeneity, to inspect the uniformity of the layer structures, to detect the presence of an axial depression in the refractive index profile, and to locate imperfections in the core and cladding. The method is rapid, real-time, nondestructive, and non-contacting.

12.9 Cables

The goal in making a cable incorporating an optical fiber is to provide strength and protection while minimizing cable volume and weight and to avoid adding appreciable optical losses (Schwartz et al., 1980; Foord, 1980; Nakahara and Uchida, 1980; Goell, 1981; Kaiser and Anderson, 1986; Gartside et al., 1988; Ramsey, 1990). Additionally, it may be necessary to incorporate power carrying conductors in the cable for some applications, such as underseas cables (Adl et al., 1984; Nagel, 1984; Paul et al., 1984; Runge and Trishitta, 1984; Wagner, 1984; Worthington, 1984; Fukinuki, 1984; Niiro and Yamamoto, 1986; Runge and Bergano, 1988; Trischitta and Chen, 1989). Generally, a list of desirable properties in a cable would include

- minimized stress-produced optical losses,
- high tensile strength,
- immunity to water vapor penetration,
- stability of characteristics over a specified temperature range,
- ease of handling and installation (especially compatibility with current installation equipment), and
- low acquisition, installation, and maintenance costs.

Fiber cables can be as simple or complicated as the application requires. Typically, a cable will consist of some of the following elements:

- *The optical fibers*, which may be single fibers or multiple fibers.
- A *buffering material,* which is a soft substance placed around the fiber to isolate it from radial compressions and other localized stresses applied to the cable. If the fiber is able to move within the buffering material while responding to strain in the cable, then the fiber is said to be *loose-buffered*. If the fiber is rigidly constrained to its position within the cable, then the fiber is said to be *tight-buffered*.
- *Strength members* are high tensile-strength materials that provide the longitudinal strength of the fiber. Separation of the strength member role from the information carrying role is one of the primary advantages of fiber cables, as the selection of each material can be optimized. Typically, high-strength, low-weight materials, such as Kevlar, can be used. The tensile

strength of a fiber cable is the sum of the strengths of the individual parts of the cable, that is,

$$T = S\sum_i E_i A_i \qquad (12.18)$$

where T is the tensile load, S is the maximum allowed strain or elongation (e.g., 1%), E_i is the Young's modulus of the i-th component, and A_i is the cross-sectional area of the i-th component.

If there is a requirement to carry power down the length of the wire, copper conductors or copper-coated high-strength wires could also be utilized. One potential problem is elongation of the cable. A typical fiber will break at an elongation greater than 1%, while a typical stress member can elongate on the order of 20% before breaking. This disparity can result in the requirement to helically coil the fiber within the center region of the cable to allow the cable to elongate without breaking the fiber. A cable-making machine has this capability to wind the fiber, buffer material, and strength members at the proper pitches without producing microbends in the fiber and without inducing a torque in the cable that would hinder the installation process.

■ *Filler yarns*, added to take up space between strength members and also to provide a degree of buffering.

■ A *jacket*, to provide abrasion protection; waterproofing; protection from rodents, fish, and other gnawing animals; resistance to chemical reaction; and other environmental protection. The outer jacket largely determines the installation properties, as it must have the correct friction properties to allow convenient installation.

Typical cable installations involve a wide range of environments including ducts, aerial stringing from posts, burial in trenches, underwater installation, and laying the fiber on the top of the ground.

■ Figure 12.16 shows a representative cable. The outer jacket is a polyurethane plastic with the right properties for fire and smoke retardation, jacket smoothness for fiber drawing, and color for fiber identification. The next layer is a set of Kevlar fibers for load-carrying. These high-strength, low-weight fibers are frequently found in fiber cables. The fiber itself is surrounded by a protective plastic jacket that isolates the fiber from the other parts of the cable design.

■ Figure 12.17 shows a representative cable designed for aerial stringing from post to post. Multiple fibers are present in the cable along with two conducting members. The fibers are in a filling gel.

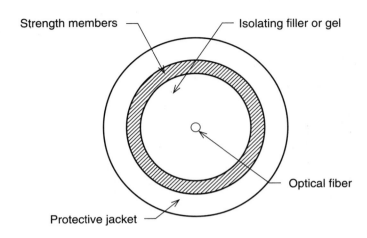

Figure 12.16 Typical fiber-optic cable structure. Representative outside diameters are 2 mm to 3 mm.

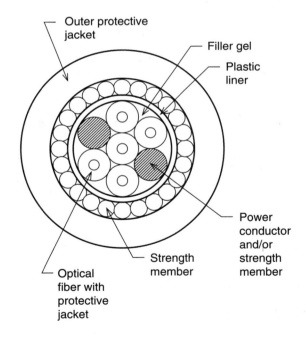

Figure 12.17 Typical aerial fiber-optic cable structure. Representative outside diameters are 6 mm to 7 mm.

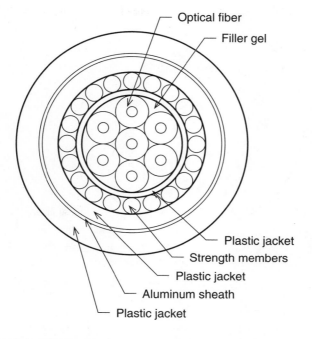

Figure 12.18 Typical fiber-optic cable structure for burial installation. Representative outside diameters are 9 mm to 10 mm.

- Figure 12.18 shows a representative cable designed for burial in trenches. More outer protective layers are evident.

- Figure 12.19 shows a representative cable designed for short-distance undersea transmission. Here, copper-clad steel wires are used both to conduct electrical power through the cable and to provide cable strength for the harsh installation and repair environment. (It should be noted that the presence of electrical power in the cable can provoke defensive behavior from sharks and fish in certain installation areas. The cable may require extra protective layers.)

12.10 Summary

In this chapter we have reviewed the fabrication of optical fibers by both high- and low-temperature techniques, the characterization of both preforms and fibers to measure their refractive index profile, and, finally, the techniques used to measure the operating performance of the optical fibers. The fabrication methods are

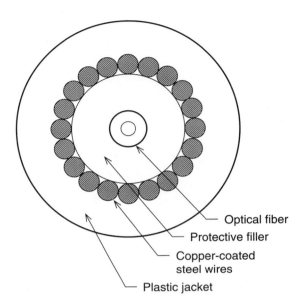

Optical fiber

Protective filler

Copper-coated
steel wires

Plastic jacket

Figure 12.19 Typical short-distance undersea cable structure.
Representative outside diameters are 2 mm to 3 mm.

dominated by the vapor-phase processes that ensure tight control over the purity of
the deposition and the layer thicknesses. To measure the resulting preforms and
fibers, a variety of ingenious techniques have been devised to meet the required
measurement tolerances. Finally, we have briefly described the elements of a fiber
cable and some of the properties that determine the cable's composition.

12.11 Problems

1. Using the principle of conservation of mass, calculate the approximate
   length of fiber that will be obtained from a 15 mm diameter preform that is
   1 m long . . .

   (a) . . . if the outside diameter (O.D.) of the fiber is 125 μm.

   (b) . . . if the O.D. is 200 μm?

2. Assume an ideal fiber drawing process.

   (a) Calculate the diameter of a fiber drawn from a preform with an outside
   diameter of 12 mm moving with a 0.1 mm/s velocity if the fiber take-up
   velocity is 0.75 m/s?

   (b) Find the velocity of the preform feed mechanism required to make a fiber with an outside diameter of 125 μm.

3. A silica tube with an inner diameter of 3 mm and an outer diameter of 4 mm is to be used in a MCVD process to make preform for a fiber with a 50 μm core diameter and an outside diameter of 125 μm.

   (a) Calculate the thickness of the material that should be deposited inside this tube.

   (b) Calculate the diameter of the core material and the diameter of the preform after the center hole has been removed.

   (c) If this preform is to be used in a drawing machine that feeds in the preform at a rate of 1 cm/s, what should the fiber drawing speed be (assuming an ideal process)?

4. The interferometric-slab technique is used to measure the refractive index of a fiber. The fiber is found to have an index profile given by $n(r) = n_1[1 - \Delta(r/a)]$ (i.e., a triangular profile). Sketch the output from the interference microscope for this case.

References

Adl, A., T. Chien, and T. Chu, "Design and testing of the SL cable," *J. Lightwave Technology*, vol. LT-2, no. 6, pp. 824–832, 1984.

Beales, K., C. Day, A. Dunn, and S. Partington, "Multicomponent glass fibers for optical communications," *Proc. IEEE*, vol. 68, no. 10, pp. 1191–1194, 1980.

Blankenship, M. and C. Denka, "The outside vapor deposition method of fabricating optical waveguide fibers," *IEEE J. Quantum Electronics*, vol. QE-18, no. 10, pp. 1418–1423, 1982.

Blyler, L.L., Jr., B. R. Eichenbaum, and H. Schonhorn, "Coatings and jackets," in *Optical Fiber Telecommunications*, (S. E. Miller and A. G. Chynoweth, eds.), pp. 299–341, New York: Academic Press, 1979.

Blyler, L.L., Jr. and F. DiMarco, "Fiber drawing, coating, and jacketing," *Proc. IEEE*, vol. 68, no. 10, pp. 1194–1198, 1980.

Bogatyrjov, V., M. Bubnov, E. Dianov, S. Rumyantzev, and S. Semjonov, "Mechanical reliability of polymer-clad and hermetically coated fibers based on proof testing," *Optical Engineering*, vol. 30, no. 6, pp. 690–699, 1991.

486

Cherin, A., *Introduction to Optical Fibers*. New York: McGraw-Hill, 1983.

Cohen, L. G., P. Kaiser, P. D. Lazay, and H. M. Presby, "Fiber characterization," in *Optical Fiber Telecommunications*, (S. E. Miller and A. G. Chynoweth, eds.), pp. 343–399, New York: Academic Press, 1979.

Foord, S., "Fibre-optic cables," in *Optical Fibre Communication Systems*, (C. Sandbank, ed.), pp. 70–85, New York: Wiley, 1980.

Francois, P., I. Sasaki, and M. Adams, "Practical three-dimensional profiling of optical fibers," *IEEE J. Quantum Electronics*, vol. QE-18, no. 4, pp. 524–535, 1982.

French, W. G., R. E. Jaeger, J. B. MacChesney, S. R. Nagel, K. Nassau, and A. D. Pearson, "Fiber preform preparation," in *Optical Fiber Telecommunications*, (S. E. Miller and A. G. Chynoweth, eds.), pp. 233–261, New York: Academic Press, 1979.

Fukinuki, H., T. Ito, M. Aiki, and Y. Hayashi, "The FS-400M submarine system," *J. Lightwave Technology*, vol. LT-2, no. 6, pp. 754–760, 1984.

Gartside, C. H. III, P. D. Patel, and M. R. Santana, "Optical fiber cables," in *Optical Fiber Telecommunications II*, (S. E. Miller and I. P. Kaminow, eds.), pp. 217–261, New York: Academic Press, 1988.

Goell, J. E., "Optical fiber cable," in *Fundamentals of Optical Fiber Communications*, (M. F. Barnoski, ed.), pp. 109–146, New York: Academic Press, 1981.

Helms, J., J. Schmidtchen, B. Schüppert, and K. Petermann, "Error analysis for refractive-index profile determination from near-field measurements," *J. Lightwave Technology*, vol. 8, no. 5, pp. 625–633, 1990.

Hibino, Y. and H. Hanafusa, "Consolidation-atmosphere influence on drawing-induced defects in pure silica optical fibers," *J. Lightwave Technology*, vol. 6, no. 2, pp. 172–178, 1988.

Iga, K. and Y. Kokobun, "Formulas for calculating the refractive index profile of optical fibers from their transverse interference patterns," *Applied Optics*, vol. 17, no. 12, pp. 1972–1974, 1978.

Inada, K., "Recent progress in fiber fabrication techniques by vapor-phase axial deposition," *IEEE J. Quantum Electronics*, vol. QE-18, no. 10, pp. 1424–1431, 1982.

Inniss, D. and J. Krause, "Hermetic splice overcoating," *Optical Engineering*, vol. 30, no. 5, pp. 776–779, 1991.

Izawa, T. and N. Inagaki, "Materials and processes for fiber preform fabrication—vapor-phase axial deposition," *Proc. IEEE*, vol. 68, no. 10, pp. 1184–1187, 1980.

Jablonowoski, D. P., "Fiber manufacture at AT&T with the MCVD process," *J. Lightwave Technology*, vol. LT-4, no. 8, pp. 1016–1019, 1986.

Jaeger, R. E., A. D. Pearson, J. C. Williams, and H. M. Presby, "Fiber drawing and control," in *Optical Fiber Telecommunications*, (S. E. Miller and A. G. Chynoweth, eds.), pp. 263–298, New York: Academic Press, 1979.

Kaiser, P. and W. T. Anderson, "Fiber cables for public communications: State-of-the-art technologies and the future," *J. Lightwave Technology*, vol. LT-4, no. 8, pp. 1157–1166, 1986.

Kapron, F. P., "Fiber-optic test methods," in *Fiber Optics Handbook for Engineers and Scientists*, (F. C. Allard, ed.), pp. 4.1–4.54, New York: McGraw-Hill, 1990.

Kreidl, J., "A surge in military demand for hermetic fiber," *Lightwave*, vol. 4, no. 11, p. 1, November 1987.

Kurkjian, C. B., J. T. Krause, and M. J. Matthewson, "Strength and fatigue of silica optical fibers," *J. Lightwave Technology*, vol. 7, no. 9, pp. 1360–1370, 1989.

Kurkjian, C. and D. Inniss, "Understanding mechanical properties of lightguides: a commentary," *Optical Engineering*, vol. 30, no. 6, pp. 681–689, 1991.

Lu, K., G. Glasemann, R. Vandewoestine, and G. Kar, "Recent developments in hermetically coated fiber," *J. Lightwave Technology*, vol. 6, no. 2, pp. 240–244, 1988.

Lydtin, H., "PCVD: A technique suitable for large-scale fabrication of optical fibers," *J. Lightwave Technology*, vol. LT-4, no. 8, pp. 1034–1038, 1986.

MacChesney, J., "Materials and processes for preform fabrication—modified vapor deposition and plasma vapor deposition," *Proc. IEEE*, vol. 68, no. 10, pp. 1181–1184, 1980.

Marcuse, D., "Refractive index determination by the focusing method," *Applied Optics*, vol. 18, no. 1, pp. 9–13, 1979.

488

Marcuse, D. and H. Presby, "Effects of profile deformations on fiber bandwidth," *Applied Optics*, vol. 18, pp. 3758–3763, 1979a.

Marcuse, D. and H. Presby, "Focusing method for nondestructive measurement of optical fiber index profiles, " *Applied Optics*, vol. 18, no. 1, pp. 14–22, 1979b.

Morishita, K., "Index profiling of three-dimensional optical waveguides by the propagation-mode near-field technique," *J. Lightwave Technology*, vol. LT-4, no. 8, pp. 1120–1124, 1986.

Murata, H., "Recent developments in vapor phase axial deposition," *J. Lightwave Technology*, vol. LT-4, no. 8, pp. 1026–1033, 1986.

Nagel, S. R., "An overview of the modified chemical vapor deposition (MCVD) process and performance," *IEEE J. Quantum Electronics*, vol. QE-18, no. 4, pp. 459–476, 1982.

Nagel, S. R., "Review of the depressed cladding single-mode fiber design and performance for the SL undersea system application," *J. Lightwave Technology*, vol. LT-2, no. 6, pp. 792–801, 1984.

Nagel, S. R., "Fiber material and fabrication methods," in *Optical Fiber Telecommunications II*, (S. E. Miller and I. P. Kaminow, eds.), pp. 121–215, New York: Academic Press, 1988.

Nakahara, T. and N. Uchida, "Optical cable design and characterization in Japan," *Proc. IEEE*, vol. 68, no. 10, pp. 1220–1226, 1980.

Niiro, Y. and H. Yamamoto, "The international long-haul optical-fiber submarine cable system in Japan," *IEEE Communications Magazine*, vol. 24, no. 5, pp. 24–32, 1986.

Paek, U., "High-speed high-strength fiber drawing," *J. Lightwave Technology*, vol. LT-4, no. 8, pp. 1048–1060, 1986.

Paul, D., K. H. Greene, and G. A. Koepf, "Undersea fiber optic cable communications system of the future: Operational, reliability, and systems considerations," *J. Lightwave Technology*, vol. LT-2, no. 6, pp. 414–425, 1984.

Presby, H. and D. Marcuse, "Optical fiber preform diagnostics," *Applied Optics*, vol. 18, no. 11, pp. 23–30, 1979.

Presby, H. and D. Marcuse, "The index-profile characterization of fiber preforms and drawn fibers," *Proc. IEEE*, vol. 68, no. 10, pp. 1198–1203, 1980.

Raine, K., J. Barnes, and D. Putland, "Refractive index profiling—State of the art," *J. Lightwave Technology*, vol. 7, no. 8, pp. 1162–1169, 1989.

Ramsay, M., "Fiber-optic cables," in *Fiber Optics Handbook for Engineers and Scientists*, (F. C. Allard, ed.), pp. 2.1–2.50, New York: McGraw-Hill, 1990.

Runge, P. K. and P. R. Trischitta, "The SL undersea lightwave system," *J. Lightwave Technology*, vol. LT-2, no. 6, pp. 744–753, 1984.

Runge, P. and N. S. Bergano, "Undersea cable transmission systems," in *Optical Fiber Telecommunications II*, (S. E. Miller and I. P. Kaminow, eds.), pp. 879–909, New York: Academic Press, 1988.

Saekang, C., P. Chu, and T. Whitbread, "Nondestructive measurements of refractive-index profile and cross-sectional geometry of optical fiber preforms," *Applied Optics*, vol. 19, no. 2, pp. 2025–2030, 1980.

Saunders, M., "Nondestructive interferometric measurement of the delta and alpha of optical fibers," *Applied Optics*, vol. 16, no. 9, pp. 2368–2371, 1977.

Schultz, P., "Progress in optical waveguide process and materials," *Applied Optics*, vol. 18, pp. 108–117, 1979.

Schwartz, M., P. Gagen, and M. Santana, "Fiber cable design and characterization," *Proc. IEEE*, vol. 68, no. 10, pp. 1214–1219, 1980.

Senior, J., *Optical Fiber Communications: Principles and Practice*. Englewood Cliffs, NJ: Prentice Hall, 1985.

Stewart, W., "Optical fiber and preform profiling technology," *IEEE J. Quantum Electronics*, vol. QE-18, no. 10, pp. 1451–1466, 1982.

Trischitta, P. R. and D. T. Chen, "Repeaterless undersea lightwave systems," *IEEE Communications*, pp. 16–21, March 1989.

Vandewoestine, R. V. and A. J. Morrow, "Developments in optical waveguide fabrication by the outside vapor deposition process," *J. Lightwave Technology*, vol. LT-4, no. 8, pp. 1020–1025, 1986.

Wagner, R. E., "Future 1.55- μm undersea lightwave systems," *J. Lightwave Technology*, vol. LT-2, no. 6, pp. 1007–1015, 1984.

Worthington, P., "Cable design for optical submarine systems," *J. Lightwave Technology*, vol. LT-2, no. 6, pp. 833–838, 1984.

490

Yoshizawa, N., H. Tada, and Y. Katsuyama, "Strength improvement and fusion splicing for carbon-coated optical fiber," *J. Lightwave Technology*, vol. 9, no. 4, pp. 417–421, 1991.

Chapter 13

Measurement of Fiber Parameters

13.1 Introduction

After fabrication of the fiber, the manufacturer and user are interested in measuring the fiber parameters. In this chapter we will describe the techniques that are used to measure the fiber attenuation, the fiber dispersion, and the parameters of a single-mode fiber, including the cutoff wavelength and mode-field diameter.

13.2 Attenuation Measurements

The typical user is interested primarily in the attenuation per kilometer of a fiber at the required operating wavelength. The following techniques (Barnoski and Personick, 1978; Cannell et al., 1980; Cohen et al., 1980; Kapron, 1990) can be used to measure the fiber attenuation losses.

13.2.1 Insertion Loss Method

The fiber losses can be determined by the *insertion loss method*. Here, the output powers transmitted by two different lengths of fiber are measured, with the input and output coupling losses held constant. The power lost in the longer fiber (in dBm or dBμ) is subtracted from the power lost in the shorter fiber, and this lost power is attributed to the extra length. Assuming that the additional losses are uniformly distributed, dividing this dB figure by the increased length of the longer fiber will give the losses in dB/km. The spectral distribution of the fiber losses is obtained by using a tunable source such as a dye laser or a filtered white light source to vary the exciting wavelength.

When used with multimode fibers, one key implicit assumption of the insertion loss technique (and other loss measurement techniques) is that all modes are equally excited, either by using special purpose devices (called *mode mixers* or *mode scramblers*) or ensuring enough length of fiber to allow sufficient mode mixing to occur. A physical test to examine whether all modes are excited is to look at the far-field radiation pattern of the light exiting at two different lengths of the fiber. (The far-field pattern can be observed conveniently by using a lens positioned one focal length from the end of the fiber to collect the light. The far-field pattern will then be imaged at a distance of one focal length behind the lens.) If the half-cone angles of the observed far-field patterns are equal, then sufficient equality in mode excitation exists. Unequal angles imply insufficient mixing. (If measurement of the loss is made with unequal angles, then the results are valid only for the particular length of fiber between the points that were used and for the particular excitation condition of the test. While this information is of some use for characterizing an existing system that is in place, it is otherwise limited because of its lack of generality.)

13.2.2 The Cutback Method

In the *cutback method*, the power out of a long length of fiber is measured. The fiber is then cut to a short length and the measurement is repeated. The ratio is expressed in dB and divided by the length of the piece that was removed to compute the average loss per length. This method is accurate but cannot be used to characterize the fiber loss in a working link. The following technique does not require any destructive cuts to the fiber.

13.2.3 Optical Time-Domain Reflectometry (OTDR)

The third technique for measuring losses in fibers is *optical time-domain reflectometry (OTDR)* (Danielson, 1985; Tateda and Horiguchi, 1989; Kapron, 1990; So et al., 1990). This technique has the primary advantage of not requiring any cuts

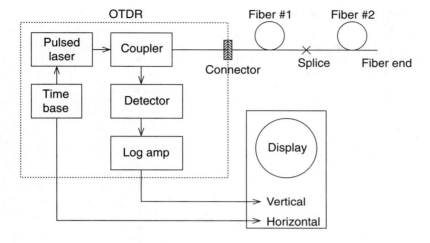

Figure 13.1 Diagram of an optical time-domain reflectometer.

in the fiber, as it works by measuring backscattered light rather than transmitted light. It also requires access at only one end of the fiber.

Figure 13.1 illustrates an OTDR setup. The pulsed excitation from the transmitter is coupled into the fiber (while the detector is blanked to avoid saturation). The pulse is transmitted down the length of the fiber. In an ideal fiber, Rayleigh backscattering will occur uniformly along the length of the fiber. In a fiber without losses, the returned radiation would be constant in time as the excitation pulse traverses the length of the fiber. For fibers with a uniform loss, the returned radiation declines with a constant slope (when the measured power is expressed on a logarithmic or dB axis). The loss per length can be readily measured by determination of the slope.

Figure 13.2 illustrates representative results of the experiment of Figure 13.1. The trace begins with the detection of the light reflected from the connector where the fiber under test connects to the OTDR. (In some systems, the detector is saturated by the partial reflections of the strong outgoing pulse from the coupler. During this saturation time the detector cannot detect the backscatter light or any reflected pulses. The length of fiber corresponding to the time of detector saturation is called the OTDR's *dead zone*. A length of fiber is frequently included within the OTDR between the coupler and the OTDR connector to allow sufficient time for the detector to come out of saturation. However, the losses of this piece of fiber decrease the power available to the test fiber.) The backscattered light is measured from fiber

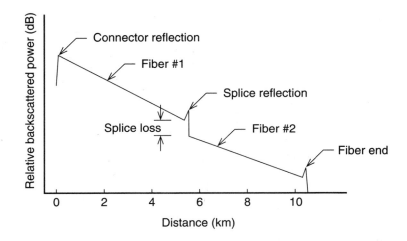

Figure 13.2 Typical result from an OTDR experiment.

#1 and diminishes with time (and distance). The slope of segment #1 is the loss factor for that fiber (in dB/km). At the splice the discontinuity will cause a pulse in reflected light. The difference between the optical power level (in dBm) just before the splice and the power level just after the splice is a measure of the loss introduced by the splice. The light continues to be backscattered in segment #2 of the fiber in a fashion similar to segment #1. A different fiber loss will be reflected as a different slope. The discontinuity at the end of the fiber produces a large reflected pulse. Other fiber segments can be added. The limit on the total losses that can be measured is determined by the dynamic range of the detector (i.e., the difference between the maximum power [in dBm or dBμ] that can be detected reliably without saturation and the minimum power that can be detected as different from the receiver noise). One problem with current OTDR techniques is the limited dynamic range available (approximately 20 dB in commercially available devices), thereby limiting the length of fiber that can be characterized. Advanced techniques for increasing the OTDR sensitivity are discussed in Healey (1985), King et al., (1987), and So et al., (1990).

Note that accurate position information is also available from this trace. The technique also can be used to localize breaks, abrupt dislocations, and other localized sources of loss. Obviously, the position resolution is determined by the pulsewidth of the optical source. Typically, pulsed semiconductor lasers with pulsewidths on the order of 10 ns are used (providing a resolution of 6 m). For increased accuracy, pulse-compression techniques can be applied to increase the resolution of the system.

13.2.4 Mode Losses

The losses in each mode (*modal losses*) are investigated by preferentially exciting a single mode at a time and measuring the losses incurred. In step-index fibers, the excitation of an individual mode is done by irradiating the end of the fiber with a plane wave inclined at a varying angle, since there is a unique correspondence between the angle of incidence and the mode that is excited. For graded-index fibers, the individual modes are excited by focusing a spot of light on a particular position, (r, θ), on the end of the fiber. As r is changed, the various fiber modes will be individually excited.

13.2.5 Mode Stripping

In the loss measurements involving short lengths of fiber (i.e., less than a few hundred meters), removal of the cladding modes is required to obtain an accurate measurement. This is accomplished by immersing a section of the bare fiber in an index matching liquid that matches the cladding index of refraction. Since the cladding modes are no longer reflected at the outer boundary, they are effectively removed. This technique is called *mode stripping*.

13.2.6 Scattering Losses

Techniques exist to measure the absorption and scattering contributions to the loss separately. Scattering loss can be measured by stripping the fiber and collecting the side-scattered radiation with a spherical collector called an *integrating sphere*. Absorption losses are typically computed by carefully measuring the scattering losses and subtracting them from the total losses.

13.3 Dispersion Measurements

Modal delay distortion can be measured either in the time domain (impulse response) or in the frequency domain (transfer function) (Barnoski and Personick, 1978; Cannell et al., 1980; Cohen et al., 1980; Kapron, 1990).

In the *time-domain measurement* of the *impulse response*, a short-duration laser pulse is injected into the end of a fiber sample and the spread pulse is measured at the output end. As would be expected, the selection of a fast source and receiver is of paramount importance, although deconvolution techniques can be used to remove the residual influence of the combined source-detector time response. The laser

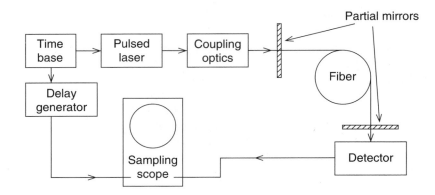

Figure 13.3 Experimental setup for measurement of multiple-reflection impulse response.

sources can be of the large optical cavity (LOC) design, which produce a typical pulse width of 100 ps (Olshansky and Keck, 1976). Typical pin diode detection response times are 500 ps, with faster devices available at increased expense. Again, uniform excitation of the modes is required to obtain an unbiased result. In fibers of low delay, for reasons of little dispersion or short length, it is desirable to simulate longer lengths in order to get a measurable pulse width. This can be achieved by *shuttle-pulse measurements* (as shown in Figure 13.3), where a partially reflective surface is mounted at each end of the test fiber and the pulse makes several round trips, building up dispersion as it traverses the increased effective length of the fiber. In both the single-pass and shuttle-pass methods, we measure the RMS value of the pulses as an average of several fairly reproducible pulses.

In measuring the *transfer function* of a fiber in the frequency domain, one modulates the amplitude of a cw laser source sinusoidally at varying frequencies. As the frequency is increased, the amplitude of the detected modulation becomes smaller, once the frequency limitation of the fiber is approached. (Again, it is assumed that the frequency response of the source and detector will not be the limiting factor.) The frequency of interest is that for which the detected modulation wave amplitude is 0.707 times the input modulation. This is the 3 dB frequency, since the power in the modulation is cut in half. This technique is usually easier to implement in hardware than the impulse-response measurement setup. Spectral effects can also be measured by using a broad-spectrum source in concert with various optical filters.

Once the frequency response or impulse response of a fiber has been measured, one can use the following guidelines in the application of the fiber in a system to ensure that delay distortion will not limit the performance. (These are general-purpose guidelines; specific applications might tolerate more or less delay distortion.)

Either the modulation frequency of an analog signal must not exceed one-half of the upper 3 dB frequency of the fiber's frequency response, or the RMS value of the measured impulse response should be less than one-fourth of the pulse spacing. Mathematically we require

$$f_{max} \leq \frac{1}{2B_{3-dB}} \qquad (13.1)$$

for analog signals, where f_{max} is the maximum modulation frequency and B_{3-dB} is the measured 3 dB frequency of the fiber's response. For digital signals we require

$$DR \leq \frac{1}{4\sigma_{fiber}}, \qquad (13.2)$$

where DR is the data rate of the signal and σ_{fiber} is the measured rms pulse spread for the fiber, as discussed in the dispersion section in Chapter 3. (If the pulse is Gaussian, then the rms pulsewidth is equal to the 1/e points on the voltage of the output.) These criteria determine the maximum data rate or bandwidth without having limitations due to the fiber's delay distortion.

Measurements separating the modal delay component and the material dispersion component can be made when both are significant and nonnegligible. One technique is to inject pulses at two separate wavelengths and measure the resulting delay in each. The difference in the pulse spreading can be attributed to material dispersion; the average spreading can be attributed to modal delay. Alternatively the modal effects can be measured with a narrow-spectrum source (thereby eliminating spectral effects) and then a broadband source can be used to find the additional delay due to the material dispersion.

13.4 Single-Mode Fiber Parameters

13.4.1 Cutoff Wavelength

As defined in Chapter 2, the *cutoff wavelength* is where the fiber ceases to behave as a single-mode fiber. The lowest order electromagnetic mode is called the LP_{01} mode; the next higher mode is the LP_{11}. The cutoff wavelength is where the LP_{11} mode is cut off. For wavelengths *larger* than the cutoff wavelength, the fiber is a single-mode fiber; for wavelengths smaller than the cutoff wavelength, the fiber will support more than one mode. The measurement of the cutoff wavelength is difficult and depends on the technique and environmental conditions, such as the fiber bends, length, and cabling (Shah and Curtis, 1989; Kapron, 1990). Four measurement techniques of cutoff wavelengths have been identified (Coppa et al., 1985; Kapron, 1990):

- The theoretical cutoff wavelength, as calculated from the desired fiber profile,
- a measured value from a short fiber section,
- a measured value from a long fiber section, and
- a measured value from a sample of spliced fiber sections.

Currently the preferred technique is the second method. The present CCITT standard definition of cutoff wavelength is that wavelength where the LP_{11} mode ceases to propagate in a 2 m length of fiber wound in a single loop of 140 mm radius (Kapron, 1990). The power, P_1, transmitted through the loop of fiber is compared with the power, P_0, transmitted through the same fiber bent into a circle with a 12.25 mm radius. For long wavelengths, well beyond cutoff, the ratio of these powers is small and is independent of the wavelength. As the measurement wavelength is reduced, the power ratio rises as the cutoff wavelength is approached and passed. The cutoff wavelength is defined as that wavelength where the power ratio has risen by 0.1 dB above its constant long-wavelength value (Shah and Curtis, 1989).

Cutoff Wavelength and Dispersion

The cutoff wavelength and waveguide dispersion can be also determined by measuring the near-field diameter of a fiber as a function of excitation wavelength (Kapron, 1990). A spectral measurement of $d_n(\lambda)$ shows a dramatic change when the field goes from a single mode to beyond cutoff (i.e., where the LP_{11} begins to propagate). The wavelength where this change occurs is the cutoff wavelength of the single-mode fiber.

A spectral measurement of the far field *MFD* $d_f(\lambda)$ also gives the waveguide dispersion, $\Delta\tau_{Wg}$, since it has been shown (Sansonetti, 1982; Petermann, 1985) that

$$\frac{\Delta\tau_{Wg}}{L\,\Delta\lambda} \approx \frac{2\lambda}{\pi^2 n_1 c}\, \frac{d(\lambda/d_f^2)}{d\lambda}, \tag{13.3}$$

where n_1 is the refractive index of the cladding. The measurement of d_f^2 is made and plotted. The derivative of (λ/d_f^2) is calculated on a computer and scaled by the multiplier to compute this dispersion.

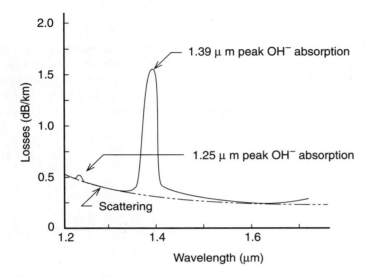

Figure 13.4 Representative spectral losses in a silica single-mode fiber.

13.4.2 Single-Mode Fiber Attenuation

Single-mode fibers are primarily used at 1300 nm (minimum dispersion) or 1550 nm (minimum attenuation). Figure 13.4 illustrates the losses in this long-wavelength region of the spectrum. (An advanced technique to measure and plot the spectral loss of a single-mode fiber is described in Walker, 1986.) The absorption peaks that are shown are due to harmonics of stronger OH⁻ ion absorptions that occur in mid-IR. The strength of these peaks can be reduced by carefully drying the fiber to remove the ions. Fortuitously, the two primary wavelengths of interest lie distant from these absorption peaks. The slight upturn in the loss characteristics above the scattering-limited "floor" is due to the presence of losses caused by macrobends in the fiber. These losses are discussed in the next section.

The optical time-domain reflectometer (OTDR) remains the instrument of choice to characterize the losses in single-mode fibers. Since the core area is so small, however, the backscattered light is reduced and is more difficult to sense. Efforts to improve the performance of OTDRs have focussed on improving the dynamic range of the device by increasing the power of the source (while reducing the pulsewidth to improve resolution) or using pulse compression techniques borrowed from the radar field (Danielson, 1985; Healey, 1985; Tateda and Horiguchi, 1989; Kapron, 1990).

Macrobending Losses

Single-mode fibers are particularly susceptible to *macrobending losses* (i.e., losses to twists and turns in the fiber) because of their small core diameter. Consider a fiber of length L that has T turns in it of radius R. Jeunhomme (1983) provides a formula that gives the dependence of the loss coefficient, α_M (in dB/km), on the fiber parameters, the geometry, and the operating wavelength, as

$$\alpha_M = \frac{A_m}{\lambda'^2} e^{\left(\frac{M_1 \left(2.748 - M_2 \lambda' \right)^3}{\lambda'_c} \right)}, \qquad (13.4)$$

where

$$A_m = 60\pi \left(\frac{T}{L'} \right) (\Delta)^{1/4} \sqrt{R'} \lambda_c'^{3/2} , \qquad (13.5)$$

$$M_1 = -0.705(\Delta)^{3/2} R' , \qquad (13.6)$$

and

$$M_2 = \frac{0.996}{\lambda'_c} . \qquad (13.7)$$

In these equations λ' is the operating wavelength (*in* μm), λ'_c is the cutoff wavelength of the fiber (*in* μm), L' is the fiber length (*in km*), Δ is the fractional difference in the index of refraction of the core and the cladding, and T is the number of turns in the fiber of radius R' (*in* μm). (Here we have used the prime symbol to indicate quantities that are *not* in MKS units.) If the entire length of the fiber is wrapped around a spool or bobbin, then the relationship between L and T is

$$L = 2\pi R T . \qquad (13.8)$$

For this special case, the expression for A_m is

$$A_m = (3 \times 10^{10}) \left(\frac{(\Delta)^{1/4}}{\sqrt{R'}} \right) \lambda_c'^{3/2} \qquad (13.9)$$

When the bends are localized (i.e., $L > 2\pi R T$), the expression gives an equivalent loss coefficient that distributes the same losses over the length of the fiber.

The measurement of the bending losses is described in Kapron (1990). When several regions exist with different bend characteristics, the fiber must be analyzed in pieces and the results superimposed.

Example: Consider a 3 km long, single-mode fiber that is wrapped on a spool with a 12 inch diameter. Calculate the macrobending loss coefficient at 1.3 μm and 1.5 μm if $\Delta = 0.4\%$ and the cutoff wavelength of the fiber is 1.2 μm.

Solution: We begin by converting the 12 inch diameter into a 6 inch radius ($R = 0.1536$ m or $R' = 153{,}600$ μm). The number of turns, T, is

$$T = \frac{L}{2\pi R} \tag{13.10}$$

$$= \frac{3 \times 10^3}{2\pi(0.1536)} \tag{13.11}$$

$$= 3110 \text{ turns .} \tag{13.12}$$

We now want to calculate the parameters A_M, M_1, and M_2. We find A_M as

$$A_M = 60\pi \left(\frac{T}{L'}\right)(\Delta)^{1/4}\sqrt{R'}\lambda_c'^{3/2} \tag{13.13}$$

$$= 60\pi \left(\frac{3110}{3}\right)(0.004)^{1/4} \sqrt{1.536 \times 10^5}(1.2)^{3/2} \tag{13.14}$$

$$= 9.88 \times 10^6 \text{ dB} \cdot \mu\text{m}^2 \cdot \text{km}^{-1} . \tag{13.15}$$

We find the value of M_1 to be

$$M_1 = -0.705(\Delta n)^{3/2}R' \tag{13.16}$$

$$= -0.705(0.004)^{3/2}(1.536 \times 10^5) \tag{13.17}$$

$$= -4.18 \ \mu\text{m} , \tag{13.18}$$

and the value of M_2 to be

$$M_2 = \frac{0.996}{\lambda'_c} \tag{13.19}$$

$$= \frac{0.996}{1.2} \tag{13.20}$$

$$= 0.83 \ \mu m^{-1} . \tag{13.21}$$

Turning now to the expression for the loss coefficient, we have

$$\alpha_M = \frac{A_m}{\lambda'^2} e^{\frac{M_1(2.748 - M_2\lambda')^3}{\lambda'_c}} \tag{13.22}$$

$$= \left(\frac{9.88 \times 10^6}{\lambda'^2}\right) e^{\frac{(-4.18)(2.748 - (0.83)\lambda')^3}{1.2}} . \tag{13.23}$$

For $\lambda' = 1.3$,

$$\alpha_M = \left(\frac{9.88 \times 10^6}{\lambda'^2}\right) e^{\frac{(-4.18)(2.748 - (0.83)\lambda')^3}{1.2}} \tag{13.24}$$

$$= \left(\frac{9.88 \times 10^6}{1.3^2}\right) e^{\frac{(-4.18)(2.748 - (0.83)(1.3))^3}{1.2}} \tag{13.25}$$

$$= 0.542 \ \text{dB/km} , \tag{13.26}$$

and for $\lambda' = 1.5$, we find

$$\alpha_M = \left(\frac{9.88 \times 10^6}{\lambda'^2}\right) e^{\frac{(-4.18)(2.748 - (0.83)\lambda')^3}{1.2}} \tag{13.27}$$

$$= \left(\frac{9.88 \times 10^6}{1.5^2}\right) e^{\frac{(-4.18)(2.748 - (0.83)(1.5))^3}{1.2}} \tag{13.28}$$

$$= 32.1 \ \text{dB/km} . \tag{13.29}$$

We note that the excess losses due to the macrobends rise rapidly as one moves to operating wavelengths that are further away from the cutoff wavelength. This increased susceptibility to macrobends is one of the reasons why we do *not* want the cutoff wavelength to be much smaller than the operating wavelength.

13.4.3 Single-Mode Fiber Dispersion

The total dispersion (i.e., the sum of the waveguide dispersion and the material dispersion) limits the information capacity of the single-mode optical fiber. The dispersion is measured in units of $ps \cdot nm^{-1} \cdot km^{-1}$ at the nominal wavelength of operation. Two measurement techniques have evolved, one in the time domain and one in the frequency domain (Coppa et al., 1985; Kapron, 1990). The time-domain technique (Gloge et al., 1974; Cohen and Lee, 1977; Kapron, 1990) uses an Nd:YAG Raman fiber laser and measurements of the time delay. (A different technique is described in Barlow et al., [1987].) The frequency-domain technique (Costa et al., 1982a; Costa et al., 1982b) measures the phase shift of the pulses and can use simpler lasers or LEDs as sources.

It is important to be able to separate the waveguide dispersion from the material dispersion. The waveguide dispersion can be measured independently from scans of the transmitted near-fields at different wavelengths. Also, spectral measurements of the mode-field diameter can lead to a measured value of the waveguide dispersion.

13.4.4 Refractive-Index Profile of Single-Mode Fibers

The single-mode fiber does not require the refractive-index profile, $n(r)$, to characterize its propagation characteristics, unlike the multimode fiber. The profile does contain useful information on the fabrication characteristics of the fiber, however (Morishita, 1986; Raine et al., 1989; Helms et al., 1990). The usual method of measuring $n(r)$ is the refracted near-field technique (Helms et al., 1990; Kapron, 1990), discussed in the previous chapter. Intercomparison test results described in Raine et al., (1989) show differences in measurement of fiber parameters, such as a depressed-cladding fiber profile and the numerical aperture of various test fibers. While axial interferometric techniques, described in the previous chapter, can achieve high accuracy, the preparation of the samples is extremely laborious and subject to some error (Raine, 1989). If only the core and innermost cladding region are required, it is possible to use the near-field scan technique to deduce the profile (Coppa et al., 1983; Kapron, 1990).

13.4.5 Mode-Field Diameter

Having mathematically defined the mode-field diameter (MFD) in Chapter 2 and pointed out its use in predicting the performance of a fiber link, we consider the measurement of this important parameter (Artiglia et al., 1989; Kapron, 1990).

Since there are several definitions of mode-field diameter, it is no surprise to find that there are several ways to measure it.

To measure the far-field MFD d_f, we usually turn to scans of the near-field pattern or the far-field pattern. *Direct measurement* of the MFD involves optical scanning of the field with point receivers, either point photodetectors or fiber pigtails, or imaging the field pattern with an imaging detector, such as a vidicon or CCD. *Indirect methods* of measuring the MFD use suitable functions that are dependent on the MFD. These indirect methods are described in more detail later.

Direct Method I: Far-Field Scanning

Scanning the far-field pattern is a direct method of measuring the MFD (Artiglia et al., 1989; Kapron, 1990). It is easily implemented, as shown in Figure 13.5. (This setup is the same arrangement used to measure the numerical aperture of a fiber.) The far-field pattern is measured by the small detector as a function of receiver angle. Because of the requirement that the measurement be in the far field of the fiber where the light intensity is decreased by beam spreading, an intense source is required, usually a laser diode. (Usually a spacing of a few centimeters from the end of the fiber is required for a regular photodiode detector; a spacing of a few millimeters is needed for a photodiode with a fiber pigtail.) To obtain accurate measurements, a relatively wide-angle field of ±20 degrees to ±25 degrees is required, especially for fibers with an exceptionally broad field. Dispersion-flattened fibers can have an especially broad angular field extending beyond ±20 degrees. (The tails of the distribution are important due to the ρ^3 weighting factor applied in the integral definition of d_f in Equation 2.57.) To record the field power at these wide angles, the detector is required to have a wide dynamic range. Artiglia et al. (1989) indicate that a dynamic range of 30 dB will give an acceptable result.

The far-field MFD d_f is calculated from Equation 2.57 by inserting the measured far-field data into a numerical integral algorithm. The primary source of error in this measurement technique is alignment between the fiber end and the detector. Usually, the measurement of d_f can be accurate to within ±0.05 μm.

Due to the relation between $E(\rho)$ and the near-field MFD, d_n, as expressed in Equation 2.58, we can also use the far-field measurement to compute d_n. (This calculation requires an increased dynamic range due to the differentiation done in the definition.) Artiglia et al. (1989) cite a 40 dB dynamic range as sufficient for this calculation.

Interlaboratory tests cited in Artiglia et al. (1989) and Drapela et al. (1989) show a reproducibility in measurement of the MFD with the far-field scan technique of ±0.15 μm and a preference for the far-field MFD definition, as opposed to the Gaussian MFD.

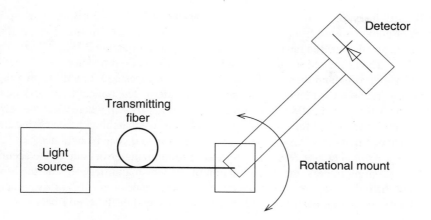

Figure 13.5 Representation of far-field scan measurement.

Direct Method II: Near-Field Scanning

Another way to measure the MFD is the *near-field scanning technique* (Artiglia et al., 1989; Kapron, 1990). In this technique, a lens is used to magnify the fiber end to measure the field with a point detector. Figure 13.6 represents a typical setup that uses a fiber pigtail as the scanning receiver. Careful alignment and precision scanning of the detector are required to maintain accuracy. The data are numerically integrated according to Equation 2.54 to compute the near-field MFD d_n.

A critical component is the magnifying imaging lens. It must have a large numerical aperture (NA > 0.5) to allow measurement of the required tails of the field. Good values of d_n can be obtained with detector dynamic ranges of 30 dB to 40

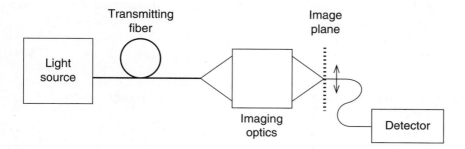

Figure 13.6 Setup for the near-field scanning technique.

dB (Artiglia et al., 1989). With dynamic ranges of 40 dB to 45 dB and the application of numeric noise filtering routines, the near-field measurements of $e(r)$ can also be used to calculate the far-field MFD d_f according to Equation 2.59.

The near-field scanning technique can also be used to determine the refractive index profile and then d_∞. The d_∞ MFD can be measured directly from a high–dynamic-range measurement of the near-field pattern. The optical field, $e(r)$, can be shown (Artiglia et al., 1989) to be proportional to $K_0(4r/d_\infty)$, where K_0 is the zeroth-order modified Bessel function of the second kind. By matching the measured fields with curve-fitted estimates, a best-fit approximation to d_∞ can be obtained.

Artiglia et al. (1989) state that, after minimizing errors due to the accuracy of the scanning motors, the determination of the lens magnification, and the limitation of the finite aperture of the collecting detector, the error in MFD measurement is ± 0.05 µm for step-index fibers and ± 0.2 µm for dispersion-flattened fibers.

Indirect Method I: Transverse Offset

The *transverse offset method* is the most popular method of measuring the MFD (Artiglia et al., 1989; Kapron, 1990). It measures the power transmitted between two overlapped fibers with the same profile. The fraction of the power transmitted through the overlapping fibers is proportional to the overlapped area of the modes. The fraction transmitted, η, is measured as a function of the transverse offset, x, between the centers of the fiber. The fraction of power, $\eta(x)$, transmitted through a fiber offset by a distance x is given by (Artiglia et al., 1989)

$$\eta(x) = \left(\frac{A(x)}{A(0)}\right)^2 ,$$
(13.30)

where $A(x)$ is the overlapped mode area for a center displacement of x. The area overlap is

$$A(x) = \iint_s E(x_0,y_0)E(x_0 - x,y_0) \, dx_0 \, dy_0 .$$
(13.31)

It can be shown (Artiglia et al., 1989) that the far-field MFD is found from

$$d_f = 2\sqrt{2\eta(0)/(- d^2\eta(x)/dx^2)} .$$
(13.32)

This expression has been used to define the far-field MFD, d_f, in terms of the offset by measurement of the normalized power transmitted through an offset fiber. This

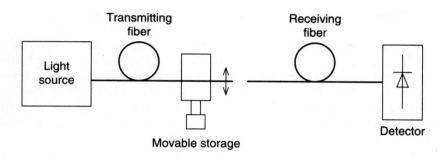

Figure 13.7 Experimental setup for the transverse offset method.

technique requires a careful determination of the center of the fiber, small reproducible offsets, and a careful interpolation through the measured data points (since the second derivative must be calculated from the measured data) (Artiglia et al., 1989).

The experimental setup is illustrated in Figure 13.7. The light source is usually a halogen white-light source. The fiber overlap is controlled with a high-precision micropositioner. The detector has a wide aperture to collect all of the power in the transmission fiber; its dynamic range is not important. If the white light is passed through a monochrometer before entering the fiber, spectral transmission properties can be measured, allowing determination of the cutoff of the LP_{11} mode and the waveguide dispersion. The fiber ends must be carefully prepared and the alignment of the system is critical to accurate measurements.

The small amount of diffraction occurring as the wave transits the gap between the fibers leads to a systematic overestimation of the MFD from this technique. If the wave is approximated as Gaussian, perfect alignment could be achieved, and $d_f(s)$ is the measured value. A correction found in Artiglia (1989) can be applied to obtain a better estimate of d_f as

$$d_f(s) = d_f\sqrt{1 + 4\lambda s/\pi d_f^2} \ . \tag{13.33}$$

To keep $d_f(s)$ and d_f within 1% of each other for typical MFD in the range of 8 μm to 10 μm, the separation, s, must be on the order of 5 μm to 10 μm or smaller (Artiglia et al., 1989). A disadvantage of the transverse offset method is that nonsymmetric fibers require care to ensure that the fiber axes are aligned. Mirror techniques have been devised to help circumvent this problem. Round-robin testing quoted in Artiglia et al., (1989) by different laboratories indicates that this technique suffers from major spreads in the measured values of MFD with a bias toward the measured values being larger than the actual value. This implies that the technique will not prove to be useful in the precision measurement of MFD.

Indirect Methods: Using Spatial Filter Measurements

The MFD is defined in terms of a weighted average of the fiber field. Various techniques have been developed to obtain the MFD by taking a weighted integral of the field. These include the use of an expanding aperture in front of the detector, detecting the field after passing a knife edge, and the use of a weighting mask. In all of these techniques, since the detector spatially integrates the light passing the obstacle, the source is not required to be as powerful as in the cases using a scanned point detector.

Indirect Method II: Variable-Aperture Technique

In this measurement technique, a circular aperture is centered between the fiber and the detector. An aperture of radius, a, located a distance, D, from the fiber end (in the fiber far field), will subtend an angle of $\theta = \tan^{-1}(a/D)$. Letting $v = k \sin \theta$, the power that passes through the aperture and is measured at the detector will be

$$P(v) = 2\pi \int_0^v E^2(\rho) \, \rho \, d\rho , \tag{13.34}$$

where $E(\rho)$ is the far-field pattern. The far-field MFD, d_f, can be calculated from the measured $P(v)$ by using the expression (Artiglia et al., 1989)

$$d_f = 2\sqrt{2} \Big/ \sqrt{1/P(v_{max}) \int_0^{v_{max}} P'(v) \, v^2 \, dv} , \tag{13.35}$$

where $P' = dP/dx$.

An alternative expression for d_f that allows easier numerical computation of the far-field MFD is (Artiglia et al., 1989)

$$d_f = 2 \Big/ \sqrt{\int_0^{v_{max}} \left[1 - P(v)/P(v_{max}) \right] v^2 \, dv} . \tag{13.36}$$

While v_{max} should go to a value of k, the limited aperture sizes restrict the maximum value. The maximum aperture needs to be large enough to include the significant portions of the far-field distribution tails to ensure accurate measurements.

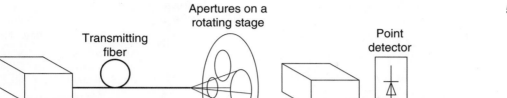

Figure 13.8 Geometry for the variable-aperture technique using a rotating aperture wheel.

The aperture size is variable, either by allowing it to open and close or by mounting a series of fixed apertures of different sizes on a rotating wheel, as in Figure 13.8. Typically, there are on the order of 12 to 20 apertures with NAs from 0.02 to 0.25 and larger (Parton, 1989). A preferred alternative experimental arrangement is to move a fixed aperture along the optical axis toward the detector, thereby decreasing the angular aperture as the distance from the fiber end is increased (see Figure 13.9). The light passed through the aperture is focussed onto a photodetector. A white-light source is used or a filtered source can be used to obtain spectral data, if desired. Detectors having a dynamic range of 40 dB are used (Artiglia et al., 1989). Errors are usually associated with alignment of the aperture and limited aperture sizes, leading to an overstatement of the MFD by about 5% (Artiglia et al., 1989). Since the variable aperture integrates the field, this technique is insensitive to the small discrepancies in the fiber's circular geometry.

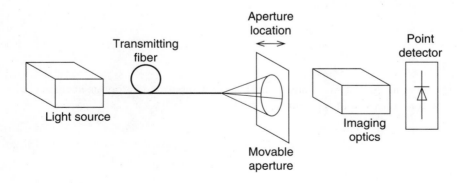

Figure 13.9 Geometry for the variable-aperture technique using a sliding aperture.

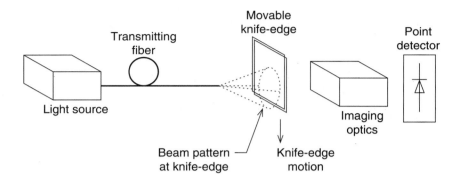

Figure 13.10 Geometry for scanning knife-edge techniques of measuring the mode-field diameter.

Indirect Method III: Knife-Edge Scan Technique

In the knife-edge scan technique, a half-plane light block (called a *knife-edge*) is placed a distance D (about 2 to 3 cm) in front of the detector in the fiber far-field. The knife-edge is scanned transversely across the field with the transmitted power focussed and measured by a photodetector. The MFD is calculated from the measured power as a function of scan position. The geometry is shown in Figure 13.10. A white-light source (e.g., a halogen lamp) is used, which may be filtered if a spectral measurement of MFD is desired.

The knife-edge removes the circular symmetry of the problem. We can erect a Cartesian coordinate system with the orthogonal x–y axes being perpendicular to the propagation direction. With $r^2 = x^2 + y^2$, we let the intensity distribution at the location of the knife-edge be $I(r)$. When the knife-edge is located at position x_0, the unblocked power, $P(x_0)$, is (Artiglia et al., 1989)

$$P(x_0) = 2\int_{x_0}^{\infty}\int_{x}^{\infty} \frac{I(r)\, r}{\sqrt{r^2 - x^2}}\, dr\, dx .$$

(13.37)

The inner integral is Abel's integral. The following inversion for this integral allows us to find $I(r)$ when $P(x)$ is known (Artiglia et al., 1989);

$$I(r) = \frac{1}{\pi}\int_{r}^{\infty} \frac{P''(x)}{\sqrt{x^2 - r^2}}\, dx ,$$

(13.38)

where $P''(x) = d^2P(x)/dx^2$. Further substitution allows us to find the far-field MFD d_f from the measured $P(x)$ as (Artiglia et al., 1989)

$$d_f = D \sqrt{2P(0)/\int_0^\infty P(x)\, x\, dx} \;, \tag{13.39}$$

where D is the distance from the fiber end to the knife-edge plane. This expression allows us to compute the far-field MFD from measurement of the power transmitted past the knife-edge as the edge is scanned transversely to the optical axis.

Typical accuracy for this technique is found to be ± 0.04 to 0.07 μm (Artiglia et al., 1989), with errors due to the light collection system and the presence of noise in the center of the scanned region caused by the weighting factor used in performing the Abel transform.

Indirect Method IV: The Mask Technique

The *mask technique* for measuring MFD is based on the observation that the near-field MFD depends on the power that can be detected after passing the light through a near-field spatial filter that weights the power properly (i.e., with an r^2 dependence). If P_1 is this weighted power and P is the total power emitted by the fiber in the near-field, then the near-field MFD definition (see Equation 2.49) can be written as (Artiglia, 1989)

$$d_n = \frac{2\sqrt{2}\; R_0}{M} \sqrt{P_1/P} \;, \tag{13.40}$$

where R_0 is the filter radius and M is the magnification of the setup.

When the filter is located in the far-field, then the far-field MFD, d_f, can be found as

$$d_f = \frac{\sqrt{2}\; \lambda D}{\pi R_0} \sqrt{P/P_2} \;, \tag{13.42}$$

where D is the distance between the fiber and the filter and P_2 is the power measured through the filter.

Artiglia et al., (1989) show two filters that have been devised for use in this technique. We note that some error will be introduced, since the filters cannot be infinitely wide. This error can be reduced to less than 2% by making $R_0/d_n < 4$. Other errors come from alignment of the mask and diffraction effects as the light propagates from the mask to the detector.

13.5 Summary

In this chapter, we have described methods that are used to measure the fiber attenuation, the fiber dispersion, and the parameters of a single-mode fiber. After a period when manufacturers and users improvised their own techniques for these measurements, the techniques are becoming standardized by organizations such as the CCITT and the Electronics Industry Association (EIA). Fiber attenuation is best measured in the laboratory with the cut-back method; in the field, the OTDR is ubiquitous, since only one end of the link is easily available. In measuring losses in multimode fiber, we noted that care must taken to ensure that all of the modes are excited. Dispersion losses can be measured in the frequency domain or in the time domain. Most techniques are focusing on the time domain, using the propagation of short pulses to characterize the fiber. Measurement of the properties of single-mode fibers is fairly taxing due to the dimensional tolerances allowed. Measurement techniques are being rapidly standardized for these fibers as problems become evident and solutions are made and disseminated.

13.6 Problems

1. Consider an OTDR that injects a laser pulse with a pulse width of τ seconds into the test fiber at time $t = 0$.

   (a) Show that, at time $t = t_1$, the light returning to the input was back-scattered from a segment of the fiber that is Δz long, $\Delta z = (c/n_1)\,(\tau/2)$.

   (b) Show that, at $t = t_1$, the edge of the fiber segment (of size Δz) furthest from the input is a distance of $(c/n_1)\,(t_1/2)$ away from the input.

   (c) The size of Δz is the *resolution* of the OTDR. Calculate the resolution if the pulse width is 50 ns.

   (d) Calculate the pulse width required to have an OTDR resolution of 1 m.

   (e) Calculate the round-trip time for light that is backscattered from a fiber segment located 10 km from the input of the test fiber.

2. Consider a fiber with an attenuation coefficient of α (per unit length) that has a rectangular pulse of power P_i and width τ coupled into it. The scattered light power from a segment of the fiber that is Δz long is given by $P(z)\alpha_s\,\Delta z$, where $P(z)$ is the incident light power at the location of the fiber segment and α_s is the scattering loss coefficient for material ($\alpha_s \approx 0.7\text{km}^{-1}$ for glass fibers). Not all of the scattered light, however, is guided by the fiber back to the input end of the test fiber; only a fraction, S, is. The value of S depends on the fiber properties as (Danielson, 1985)

$$S = \left(\frac{3}{16}\right)\left(\frac{NA^2}{n_1^2}\right)\left(\frac{g}{g+1}\right)$$

for multimode fibers and

$$S = \left(\frac{3}{16}\right)\left(\frac{2\lambda}{\pi n_1\ MFD}\right)$$

for single-mode fibers, where NA is the fiber numerical aperture, g is the fiber profile parameter, and MFD is the mode-field diameter. Hence, the backscattered light that is collected by the fiber is $SP(z)\alpha_s\,\Delta z$.

If Δz is given by the expression given in the previous problem, show that the backscattered power arriving back at the input at time t_1 is given by

$$P(t_1) = P_i S\alpha_s \left(\frac{\tau}{2}\right)\left(\frac{c}{n_1}\right)e^{-\alpha\left(\frac{c}{n_1}\right)t_1}.$$

(Note that there is a tradeoff between the power returned to the input and the resolution of the OTDR. To minimize the resolution, we want to shorten the pulse width, τ; this reduces the backscattered power returned to the OTDR detector.)

3. Consider a step-index multimode fiber with NA $= 0.2$, $n_1 = 1.5$, $\alpha_{fiber} = 0.6$ dB $\cdot$ km^{-1}, and $\tau = 50$ ns.

   (a) Using the results of the previous problem, calculate the ratio of the backscattered power to the input power for $t_1 = 0$ (i.e., for light backscattered from the segment of the fiber located right at the input of the test fiber).

   (b) Calculate the ratio of the backscattered power to the input power for a segment of fiber located 10 km from the input of the fiber.

   (c) Calculate the ratio of the backscattered power from the segment located 10 km from the input to the backscattered power from the segment located right in front of the input.

   (d) The *dynamic range*, DR, (in dB) of an OTDR is defined as

   $$DR = 10\log\left(\frac{P_{s\,max}}{P_{s\,min}}\right),$$

   where $P_{s\,max}$ is the power of the strongest backscattered signal that can be detected at the OTDR receiver and $P_{s\,min}$ is the power of the weakest signal that can be detected. If the dynamic range of the detector is 20 dB, calculate the maximum operating range of the OTDR, assuming that fiber loss is the only loss encountered (i.e., there are no connector or splice losses). You may assume that the coupler that connects the laser to the OTDR fiber pigtail (and the detector to the pigtail) has a 3 dB loss for each pass through it.

4. Consider a single-mode fiber operating at 1300 nm with a mode-field diameter of 9 μm, $n_1 = 1.5$, and $\alpha = 0.5$ dB $\cdot$ km^{-1}. Assume, again, that $\tau = 50$ ns. Repeat the calculations of the previous problem.

5. A break occurs in a fiber with a loss of 3 dB/km. The output power from an OTDR set used to locate the break is 250 mW and the detected echo power is 1 μW. Approximately 10 dB of loss are encountered in coupling the OTDR signal into the fiber and the returned signal encounters 6 dB of loss at the optical splitter (see Figure 13.1). The reflectivity of a perpendicular break in the fiber is approximately 4%; the average reflectivity of a non-perpendicular break is about 0.5%. Using the latter value of reflectivity, calculate the distance to the break in the fiber.

6. A time-domain technique is used to measure the bandwidth of the fiber (i.e., the optical power 3 dB frequency). The measured voltage at the detector load is Gaussian with a 1/e full width of 1.02 ns. Calculate the maximum data rate that can be used without the fiber presenting a limitation.

4. Consider the single-mode fiber described in the example in the macrobend loss section. This fiber has $\Delta = 0.4\%$ and a cutoff wavelength of 1200 nm. The length of 3 km is wrapped around a spool that is 12 inches in diameter.

   (a) Using a computer, plot the value of the loss coefficient due to the macrobending losses vs. wavelength for values of wavelength extending from 1.2 μm to 2.0 μm.

   (b) Using a computer, find the value of the wavelength where the loss coefficient equals 1 dB/km.

   (c) Using a computer, plot the value of the loss coefficient α_M vs. the spool radius R' for an operating wavelength of 1300 nm. Let the vertical axis (α_M) range from 0 to 10,000 and the horizontal axis (R') also range from 0 to 10,000. (Recall that R' is in units of μm.) Using a computer, find the value of spool radius required to keep the value of loss coefficient at or below 1 dB/km for the previous section of this problem.

5. Consider a single-mode fiber with $\Delta = 0.4\%$ and a cutoff wavelength of 1200 nm. The international telecommunications group (CCITT) recommendation for single mode fibers is that their excess loss caused by

macrobending should be below 1 dB when 100 meters of the fiber are wrapped around a 7.5-cm diameter spool.

(a) Calculate the length L of the fiber wrapped around the spool.

(b) Calculate the required loss coefficient α_M to meet the 1 dB loss specification.

(c) Calculate A_M, M_1, and M_2.

(d) Using a computer, find the maximum value of operating wavelength that will allow the 1 dB loss specification to be met for this fiber.

References

Artiglia, M, G. Coppa, P. DiVita, M. Potenza, and A. Sharma, "Mode field diameter measurements in single-mode optical fibers," *J. Lightwave Technology*, vol. 7, no. 8, pp. 1139–1152, 1989.

Barlow, A. J., R. S. Jones, and K. W. Forsyth, "Technique for direct measurement of single-mode fiber chromatic dispersion," *J. Lightwave Technology*, vol. LT-5, no. 9, pp. 1207–1213, 1987.

Barnoski, M. and S. Personick, "Measurements in fiber optics," *Proc. IEEE*, vol. 66, no. 4, pp. 429–441, 1978.

Cannell, C. J., R. Worthington, and K. C. Byron, "Measurement techniques," in *Optical Fibre Communication Systems*, (C. Sandbank, ed.), pp. 106–155, New York: Wiley, 1980.

Cohen, L. and C. Lee, "Pulse delay measurements in the zero material dispersion wavelength region for optical fibers," *Applied Optics*, vol. 16, no. 12, pp. 3136–3139, 1977.

Cohen, L., P. Kaiser, and C. Lin, "Experimental technique for evaluation of fiber transmission loss and dispersion," *Proc. IEEE*, vol. 68, no. 10, pp. 1203–1209, 1980.

Coppa, G., P. Di Vita, and U. Rossi, "Characterization of single-mode fibers by near-field measurements," *Electronics Letters*, vol. 19, pp. 293–294, 1983.

Coppa, G., B. Costa, and P. Di Vita, "Single-mode optical fiber characterization," *Optical Engineering*, vol. 24, no. 4, pp. 676–680, 1985.

516

Costa, B., D. Mazzoni, M. Puleo, and E. Vezzoni, "Phase-shift technique for the measurement of chromatic dispersion in optical fibers using LEDs," *IEEE J. Quantum Electronics*, vol. QE-18, pp. 1509–1515, 1982a.

Costa, B., M. Puleo, and E. Vezzoni, "Phase-shift technique for the measurement of chromatic dispersion in single-mode optical fibers using LEDs," *Electronics Letters*, vol. 19, pp. 1074–1076, 1982b.

Danielson, B. L., "Optical time-domain reflectometer specifications and performance testing," *Applied Optics*, vol. 24, no. 15, pp. 2313–2322, 1985.

Drapela, T., D. Franzen, A. Cherin, and R. Smith, "A comparison of far-field methods for determining mode field diameter of single-mode fibers using both Gaussian and Petermann definitions," *J. Lightwave Technology*, vol. 7, no. 8, pp. 1153–1157, 1989.

Gloge, D., E. Chinnock, and T. Lee, "GaAs twin-laser setup to measure mode and material dispersion in optical fibers," *Applied Optics*, vol. 13, pp. 261–263, 1974.

Healey, P., "Review of long wavelength single-mode optical fiber reflectometry techniques," *J. Lightwave Technology*, vol. LT-3, no. 4, pp. 876–886, 1985.

Helms, J, J. Schmidtchen, B. Schüppert, and K. Petermann, "Error analysis for refractive-index profile determination from near-field measurements," *J. Lightwave Technology*, vol. 8, no. 5, pp. 625–633, 1990.

Jeunhomme, L., *Single-Mode Fiber Optics*. New York: Marcel Dekker, 1983.

Kapron, F. P., "Fiber-optic test methods," in *Fiber Optics Handbook for Engineers and Scientists*, (F. C. Allard, ed.), pp. 4.1–4.54, New York: McGraw-Hill, 1990.

King, J., D. Smith, K. Richards, P. Timson, R. Epworth, and S. Wright, "Development of a coherent OTDR instrument," *J. Lightwave Technology*, vol. LT-5, no. 4, pp. 616–624, 1987.

Morishita, K., "Index profiling of three-dimensional optical waveguides by the propagation-mode near-field technique," *J. Lightwave Technology*, vol. LT-4, no. 8, pp. 1120–1124, 1986.

Olshansky, R. and D. Keck, "Pulse broadening in graded-index optical fibers," *Applied Optics*, vol. 15, no. 12, pp. 483–491, 1976.

Parton, J. R., "Improvements to the variable aperture method for measuring the mode-field diameter of dispersion-shifted fibers," *J. Lightwave Technology*, vol. 7, no. 8, pp. 1158–1161, 1989.

Petermann, K., "Constraints for fundamental-mode spot size for broadband dispersion-compensated single-mode fibers," *Electronics Letters*, vol. 19, no. 18, pp. 712–714, 1983.

Raine, K., J. Barnes, and D. Putland, "Refractive index profiling—State of the art," *J. Lightwave Technology*, vol. 7, no. 8, pp. 1162–1169, 1989.

Sansonetti, P., "Modal dispersion in single-mode fibres: Simple approximation issued from mode spot size spectral behaviour," *Electronics Letters*, vol. 18, pp. 647–648, 1982.

Shah, V. and L. Curtis, "Mode coupling effects of the cutoff wavelength characteristics of dispersion-shifted and dispersion-unshifted single-mode fibers," *J. Lightwave Technology*, vol. 7, no. 8, pp. 1181–1186, 1989.

So, V. C., J. W. Jiang, J. A. Cargil, and P. J. Vella, "Automation of an optical time domain reflectometer to measure loss and return loss," *J. Lightwave Technology*, vol. 8, no. 7, pp. 1078–1083, 1990.

Tateda, M. and T. Horiguchi, "Advances in optical time-domain reflectometry," *J. Lightwave Technology*, vol. 7, no. 8, pp. 1217–1224, 1989.

Walker, S. S., "Rapid modeling and estimation of total spectral loss in optical fibers," *J. Lightwave Technology*, vol. LT-4, no. 8, pp. 1125–1131, 1986.

Index